advanced Geography

2nd edition

AQAB

advanced Geography

2nd edition

David Redfern

Malcolm Skinner

To Judith for all your help, Malcolm.
To my family, David.
Additional thanks to Phil Banks for help with the section on Module 2.

Philip Allan Updates
Market Place
Deddington
Oxfordshire
OX15 0SE

Tel: 01869 338652
Fax: 01869 337590
e-mail: sales@philipallan.co.uk
www.philipallan.co.uk

© Philip Allan Updates 2005

ISBN-13: 978-1-84489-230-3
ISBN-10: 1-84489-230-1 ✓

This textbook has been written specifically to support students studying AQA (B) Advanced geography. The content has been neither approved nor endorsed by AQA and remains the sole responsibility of the authors.

All efforts have been made to trace copyright on items used.

Front cover photographs reproduced by permission of Corel Corporation, Photodisc, Dr David Millar/SPL, Dominic Halleux/Still Pictures.

Design by Juha Sorsa
Printed by Information Press, Eynsham, Oxford

Environmental information
The paper on which this title is printed is sourced from managed, sustainable forests.

Contents

Introduction

This textbook provides a resource specifically for AQA specification B in AS and A2 geography. It covers the subject content of the specification, module by module, as it is laid out in the specification document, and forms a backbone for studies of AQA (B) geography. However, it should be supplemented by reference to topical sources of information including newspapers, television, periodicals specifically aimed at post-16 geography students and the internet.

The following are key features of the content:

- concepts are clearly and concisely explained, and related issues are explored and analysed
- relevant, up-to-date and detailed case studies are provided
- a variety of stimulus material is provided, including full-colour maps, graphs, diagrams and photographs
- sample examination questions are included at the end of each module. For some of the A2 questions, examiner commentary and examples of mark schemes are also provided to aid understanding of the assessment process
- the skills required for the A2 modules concerned with synopticity, coursework and practical abilities are covered in depth

Course coverage and how to use the book

A complete AS/A-level course is provided in these pages, covering in detail the core Modules 1 and 4, the synoptic Module 5, and the optional Modules 6 and 7. Slightly less detailed information is provided for the two optional Modules 2 and 3, as none of this will be relevant to all students, but it is ample for a sound basis of study.

The sample questions included can be used for formal or informal assessment. Mark schemes and notes for answers to the assessment exercises are provided in the *Teacher Answer Guide* available from Philip Allan Updates. Further advice and guidance on assessment strategies can be found in the Unit Guides for AQA (B) Module 1 and Module 4, also published by Philip Allan Updates.

2nd edition

This 2nd edition incorporates the revisions made to the specification for examinations from 2006 onwards.

An overview of AQA (B) geography

This specification provides students with the opportunity to study geographical features and contemporary events they can relate to the world around them. In Modules 1–4 it delivers the content through the separate strands of:

- physical geography
- people and the environment
- human geography

Synoptic assessment occurs in Module 5. The examination involves an issue evaluation exercise in which students are expected to use their geographical experiences and skills to complete a set of tasks. These may lead to the formulation and justification of a decision in a real-world context.

There is a choice between the final two modules, 6 and 7. Module 7 involves completing a geographical investigation and submitting it as a written report. Module 6 is a written examination which is an alternative to coursework, but assesses the same geographical skills.

Assessment Objectives

Like other geography specifications, AQA (B) has four Assessment Objectives. Candidates should:

- show knowledge of the specified content
- show critical understanding of the specified content
- apply knowledge and critical understanding to unfamiliar contexts
- select and use a variety of skills and techniques, including communication skills, appropriate to geographical studies

In addition to these, A2 candidates are required to develop:

- a deeper understanding of the connections between different aspects of geography represented in this specification
- a greater ability to synthesise geographical information in various forms and from various sources

These last two objectives are assessed in the synoptic module, Module 5.

Scheme of assessment

The scheme of assessment is modular. AS and A2 each consist of three modules and make up 50% of the total award. Each of the six modules is assessed through an associated assessment unit. These are described below.

AS

Unit 1 Based on Module 1. Length of exam 1 hour, consisting of four compulsory questions, one of which assesses fieldwork skills.

Unit 2 Based on Module 2. Length of exam 1 hour, consisting of one structured question from a choice of three options:
- *Option P* Glacial environments
- *Option Q* Coastal environments
- *Option R* Urban physical environments

Unit 3 Based on Module 3. Length of exam 1 hour, consisting of one structured question from a choice of two options:
- *Option S* Urban change in the UK and the wider world in the last 30 years
- *Option T* The historical rural and urban landscapes of England and Wales

A2

Unit 4 Based on Module 4. Length of exam 1 hour 30 minutes, consisting of one from two structured questions and one from two essay questions.

Unit 5 The synoptic module. Length of exam 1 hour 30 minutes. All questions are compulsory.

Unit 6 The written alternative. Length of exam 1 hour 30 minutes, consisting of two compulsory questions.

Or **Unit 7** Coursework. An investigation of approximately 3,500–4,000 words.

Command words used in the examinations

One of the major challenges in any examination is interpreting the demands of the questions. Thorough revision is essential, but an awareness of what is expected in the examination itself is also required. Too often candidates attempt to answer the question they think is there rather than the one that is actually set. Answering an examination question is challenging enough, without the self-inflicted handicap of misreading the question.

Correct interpretation of the **command words** of a question is therefore very important. In AQA (B) geography examination papers, a variety of command words are used. Some demand more of the candidate than others; some require a simple task to be performed; others require greater intellectual thought and a longer response.

The notes below offer advice on the main command words that are used in both AS and A2 examinations.

Identify..., What...? Name..., State..., Give...
These words ask for brief answers to a simple task, such as:
- identifying a landform from a photograph
- giving a named example of a feature

Do not answer using a single word. It is always better to write a short sentence.

Define..., Explain the meaning of..., What is meant by...? Outline...
These words require a relatively short answer, usually two or three sentences,

giving the precise meaning of a term. Use of an example is often helpful. The size of the mark allocation indicates the length of answer required.

Describe...

This is one of the most widely used command words. A factual decription is required, with no attempt to explain. Usually the question will give some clue about exactly what is to be described. Some examples are given below.

Describe the characteristics of...

In the case of a landform, for example, the following sub-questions can be useful in writing the answer:
- what does it look like?
- what is it made of?
- how big is it?
- where is it in relation to other features?

Describe the changes in...

This command often relates to a graph or a table. Good use of accurate adverbs is required here — words such as rapidly, steeply, gently, slightly, greatly.

Describe the differences between...

Here only differences between two sets of data will be credited. It is better if these are presented as a series of separate sentences, each identifying one difference. Writing a paragraph on one data set, followed by a paragraph on the other, forces the examiner to complete the task on your behalf.

Describe the relationship between...

Here only the links between two sets of data will be credited. It is important, therefore, that you establish the relationship clearly in verbal form, and that the link is clearly stated. In most cases the relationship will either be positive (direct) or negative (inverse).

Describe the distribution of...

This is usually used in conjunction with a map or set of maps. A description of the location of high concentrations of a variable is required, together with a similar description of those areas with a lower concentration. Better answers will also tend to identify anomalous areas or areas which go against an overall trend in the distribution, for example a spot of high concentration in an area of predominantly low concentration.

Compare...

This requires a point by point account of the similarities and differences between two sets of information or two areas. Two separate accounts do not make up a comparison, and candidates will be penalised if they present two such accounts and expect the examiner to do the comparison on their behalf. A good technique is to use comparative adjectives, for example larger than, smaller than, more steep than, less gentle than. Note that 'compare' refers to similarities and differences, whereas the command word 'contrast' just asks for differences.

Explain..., Suggest reasons for..., How might...? Why...?

These commands ask for a statement about why something occurs. The command word tests your ability to know or understand why or how something happens. Such questions tend to carry a large number of marks, and expect candidates to write a relatively long piece of extended prose. It is important that this presents a logical account which is both relevant and well organised.

Using only an annotated diagram..., With the aid of a diagram....

Here the candidate must draw a diagram, and in the first case provide only a diagram. Annotations are labels which provide additional description or explanation of the main features of the diagram. For example, in the case of a hydrograph, the identification of 'a rising limb' would constitute a label, whereas 'a steep rising limb caused by an impermeable ground surface' would be an annotation.

Analyse...

This requires a candidate to break down the content of a topic into its constituent parts, and to give an in-depth account. As stated above, such questions tend to carry a large number of marks, and candidates will be expected to write a relatively long piece of prose. It is important that candidates present a logical account that is both relevant and well-organised.

Discuss...

This is one of the most common higher-level command words, and is used most often in questions which carry a large number of marks and require a lengthy piece of prose. Candidates are expected to build up an argument about an issue, presenting more than one side of the argument. They should present arguments for and against, making good use of evidence and appropriate examples, and express an opinion about the merits of each side. In other words, they should construct a verbal debate.

In any discussion there are likely to be both positive and negative aspects — some people are likely to benefit (the winners), and others are likely not to benefit (the losers). Candidates are invited to weigh up the evidence from both points of view, and may be asked to indicate where their sympathies lie.

Sometimes, additional help is provided in the wording of the question, as shown below.

Discuss the extent to which...

Here a judgement about the validity of the evidence or the outcome of an issue is clearly requested.

Discuss the varying attitudes to...

Here the question states that a variety of views exists, and candidates are required to debate them. There is often a range of people involved in an issue, including those responsible for the decision to go ahead with an idea or policy (the decision makers), and those who will be affected, directly or indirectly. Each of these will have a different set of priorities, and a different viewpoint on the outcome.

Evaluate... Assess...

These command words require more than the discussion described above. In both cases an indication of the candidate's viewpoint, having considered all the evidence, is required. 'Assess' asks for a statement of the overall quality or value of the feature or issue being considered, and 'evaluate' asks the candidate to give an overall statement of value. The candidate's own judgement is requested, together with a justification for that judgement.

The use of 'critically' often occurs in such questions, for example 'Critically evaluate…'. In this case the candidate is being asked to look at an issue or problem from the point of view of a critic. There may be weaknesses in the argument and the evidence should not be taken at face value. The candidate should question not only the evidence itself but also where it came from, and how it was collected. The answer should comment on the strengths of the evidence as well as its weaknesses.

Justify...

This is one of the most demanding command words. At its most simplistic, a response to this command must include a strong piece of writing in favour of the chosen option(s) in a decision-making exercise, and an explanation of why the other options were rejected.

However, decision making is not straightforward. All the options in a decision-making scenario have positive and negative aspects. The options that are rejected will have some good elements, and equally, the chosen option will not be perfect in all respects. The key to good decision making is to balance the pros and cons of each option and to opt for the most appropriate based on the evidence available.

A good answer to the command 'justify' should therefore provide the following:

- for each of the options that are rejected: an outline of their positive and negative points, but with an overall statement of why the negatives outweigh the positives
- for the chosen option: an outline of the negative and the positive points, but with an overall statement of why the positives outweigh the negatives

Developing extended prose and essay-writing skills

For many students essay writing is one of the most difficult parts of the exam. But it is also an opportunity to demonstrate your strengths.

Before starting to write a piece of extended prose or an essay you must have a plan of what you are going to write, either in your head or on paper. All such pieces of writing must have a beginning (introduction), a middle (argument) and an end (conclusion).

The introduction

This does not have to be too long — a few sentences should suffice. It may define the terms in the question, set the scene for the argument to follow, or provide a brief statement of the idea, concept or viewpoint you are going to develop in the main body of your answer.

The argument

This is the main body of the answer. It should consist of a series of paragraphs, each developing one point only and following on logically from the previous one. Try to avoid paragraphs that list information without any depth, but do not write down all you know about a particular topic without any link to the question set. Make good use of examples, naming real places (which could be local to you). Make your examples count by giving accurate detail specific to those locations.

The conclusion

In an extended prose answer the conclusion should not be too long. Make sure it reiterates the main points stated in the introduction, but now supported by the evidence and facts given in the argument.

Should you produce plans in the examination?

If you produce an essay plan at all, it must be brief, taking only 2 or 3 minutes to write on a piece of scrap paper. The plan must reflect the above formula — make sure you stick to it. Be logical, and only give an outline — retain the examples in your head, and include them at the most appropriate point in your answer.

Other important points

Always keep an eye on the time. Make sure you write clearly and concisely. Do not provide confused answers, endlessly long sentences, or pages of prose with no paragraphs.

Above all: *read the question and answer the question set.*

How are questions marked?

Examination questions for AQA (B) are marked according to levels based on fixed criteria. These criteria assess:

- **knowledge (K)** of concepts, case studies and terminology
- **understanding (U)** of processes, theories, ideas and issues
- **skills (S)** including the use of language (grammar, spelling, punctuation — GSP) and logic, as well as geographical skills such as using maps and diagrams

In most cases the above criteria are assessed at one of three levels which can be summarised as follows:

Level 1

K — generalised material, no or limited case studies or just simple statements of examples; inappropriate or incorrect use of terms; irrelevance

U — weak grasp of concepts and ideas, description only

S — basic, not always logical; clear expression of simple ideas

Level 2

K — some use of case studies, but a lack of depth and detail

U — some understanding of processes by explanation; some evidence of evaluation

S — some illustrative material; logical; GSP is satisfactory, accurate use of English

Level 3

K — good use of case studies, appropriate and detailed; good locational knowledge

U — good understanding of processes and issues; evaluative comment

S — logical and coherent argument; the essay is said to 'flow'; almost faultlessly accurate English

Level 4

In some essay questions Level 4 may exist (see the mark scheme provided for Module 5). It will occur when some overall evaluation of a concept or issue is required, often in response to the higher level commands 'discuss' and 'justify'. Candidates who have the ability to synthesise and draw together their argument(s) will access this level and gain very high marks.

AS

Module 1

The dynamics of change

Physical geography
Shorter term and local change

Atmospheric processes

Depressions

Atmospheric depressions have the following characteristics:

- they are areas of relatively low atmospheric pressure, often below 1,000 mb
- they are represented on a weather map by a system of closed isobars with pressures decreasing towards the centre
- they usually move rapidly from west to east across the British Isles
- isobars are usually close together, producing a steep pressure gradient from the outer edges to the centre
- winds are often strong and generally flow inwards
- in the British Isles the winds flow anticlockwise around the centre of the depression
- a place in the southern part of the British Isles will often experience a change of wind direction from south to southwest to west to northwest — the wind is said to **veer**
- a place in the northern part of the British Isles will often experience a change of wind direction from southeast to east to northeast to north — the wind is said to **back**

Formation of a depression

A depression affecting the British Isles originates in the North Atlantic where two different **air masses** meet along the polar front (see Figure 1.1). An air mass is a large body of air in which there is only a gradual horizontal change in temperature and humidity. The two air masses involved here are:

- polar maritime air from the northwest Atlantic which is cold, dense and moist
- tropical maritime air from the southwest which is warmer, less dense and also moist

As these two bodies of air move towards each other the warmer, less dense air from the south rises above the colder, dense air from the north. The rising air is removed by strong upper atmosphere winds (known as a **jet stream**), but as it rises the Earth's rotational spin causes it to twist. This twisting vortex produces a wave at ground level in the polar front, which increases in size to become a depression.

Two separate parts of the original front have now developed (stage 2 in Figure 1.1):

■ the **warm front** at the leading edge of the depression where warm, less dense air rises over the colder air ahead
■ the **cold front** at the rear of the depression where colder dense air pushes against the warmer air ahead

In between these two fronts lies the **warm sector** — an area of warm and moist air. As the depression moves eastwards, the cold front gradually overtakes the warm front to form an **occlusion** in which the warmer air is lifted off the ground.

Weather conditions

The weather conditions associated with a depression will therefore depend on whether the area in question has polar maritime air or tropical maritime air

Figure 1.1
The stages of
a depression

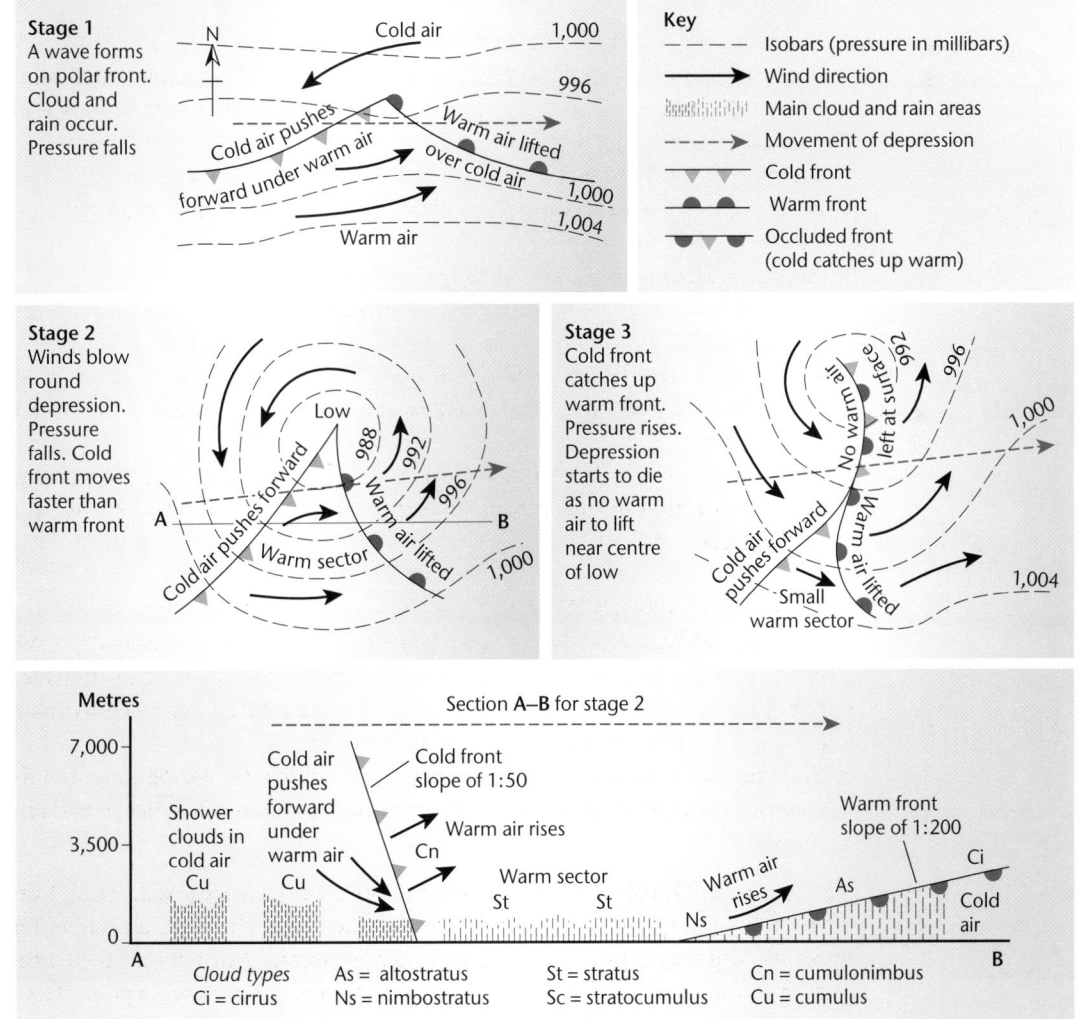

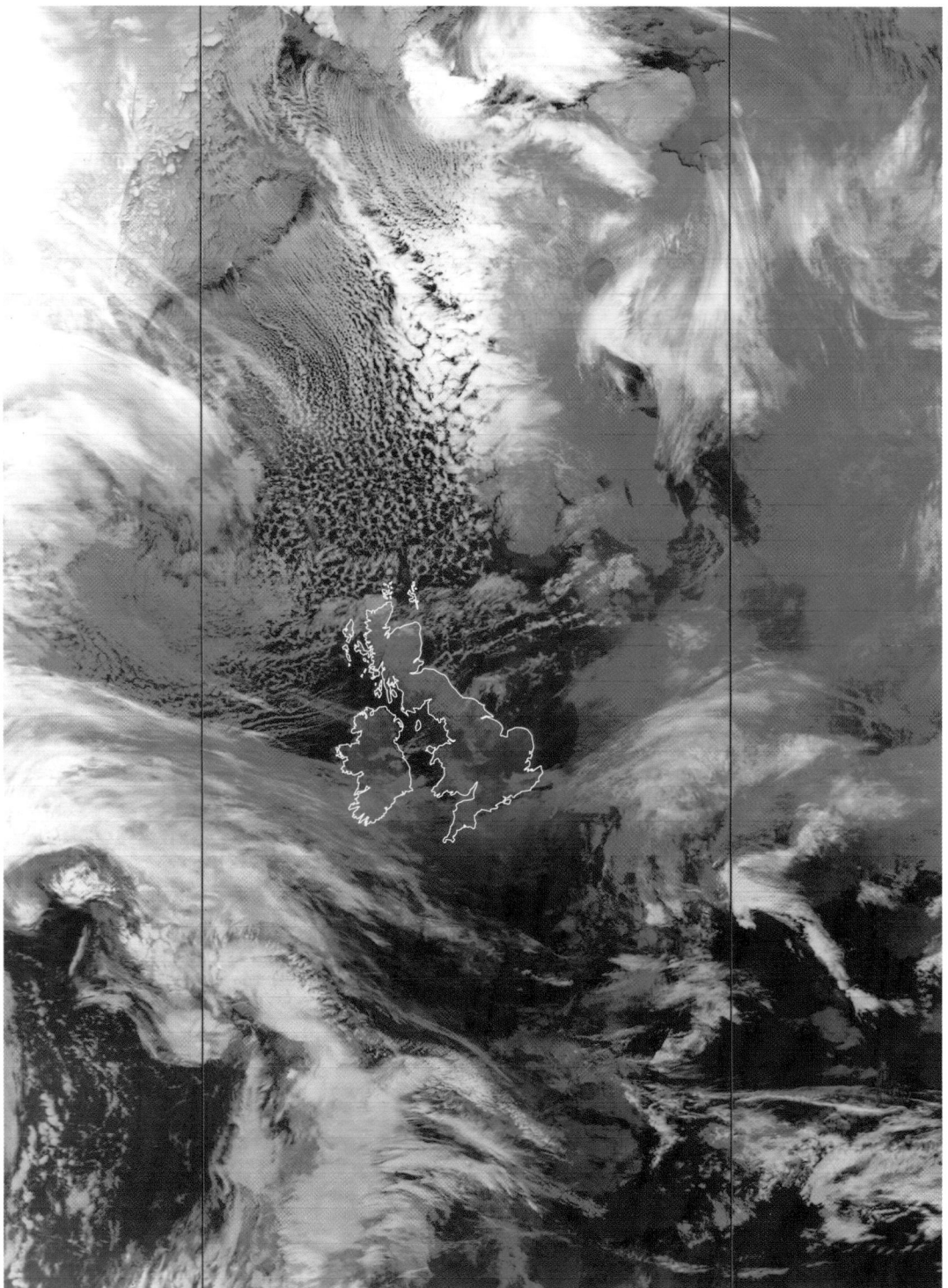

Photograph 1.1 Satellite photograph showing a depression to the west of the British Isles

over it. Polar maritime air brings average temperatures for the season in winter (5–8°C in January) but noticeably cooler temperatures for the season in summer (16–18°C in July). Showers of rain are common in both seasons, with the possibility of sleet in winter.

Tropical maritime air brings humid and mild weather in winter, with temperatures well above the seasonal norm at 12–14°C in January. Low stratus cloud and fog are also common. In summer it may cause advection fog in western coastal areas, but elsewhere temperatures will be warm — 25°C in July. There will be the threat of showers and possibly thunderstorms due to the combination of high humidity levels and low pressure.

Table 1.1 summarises the main weather changes associated with the passage of a depression over an area in the British Isles. It should be used in conjunction with the cross section through a depression shown in Figure 1.1.

Table 1.1
Weather changes
associated with
the passing of a
depression

| Weather element | Cold front | | | Warm front | | |
	In the rear	*At passage*	*Ahead*	*In the rear*	*At passage*	*Ahead*
Pressure	Continuous steady rise	Sudden rise	Steady or slight fall	Steady or slight fall	Fall stops	Continuous fall
Wind	Veering to northwest, decreasing speed	Sudden veer, southwest to west. Increase in speed, with squalls	Southwest, but increasing in speed	Steady southwest, constant	Sudden veer from south to southwest	Slight backing ahead of front. Increase in speed
Temperature	Little change	Significant drop	Slight fall, especially if raining	Little change	Marked rise	Steady, little change
Humidity	Variable in showers, but usually low	Decreases sharply	Steady	Little change	Rapid rise, often to near saturation	Gradual increase
Visibility	Very good	Poor in rain, but quickly improves	Often poor	Little change	Poor, often fog/mist	Good at first but rapidly deteriorating
Clouds	Shower clouds, clear skies and cumulus clouds	Heavy cumulo-nimbus	Low stratus and strato-cumulus	Overcast, stratus and stratocumulus	Low nimbo-stratus	Becoming increasingly overcast, cirrus to altostratus to nimbostratus
Precipitation	Bright intervals and scattered showers	Heavy rain, hail and thunder-storms	Light rain, drizzle	Light rain, drizzle	Rain stops or reverts to drizzle	Light rain, becoming more continuous and heavy

Anticyclone An area of high atmospheric pressure that brings settled weather with dry and calm conditions. Anticyclones tend to linger over an area for several days or even weeks. The amount of cloud depends on the humidity of the air, but in most cases they produce clear skies and long periods of sunshine.

Depression An area of low atmospheric pressure that brings unsettled weather, with rapid changes of cloud, precipitation, wind direction and temperature. Depressions travel across an area relatively quickly and the weather changes they bring are associated with zones called **fronts**. At a front the air rises, causing it to cool. Water vapour in the air condenses, producing cloud and precipitation.

Anticyclones

Anticyclones have the following characteristics:

- they are areas of relatively high atmospheric pressure
- they are represented on a weather map by a system of closed isobars with pressures increasing towards the centre
- anticyclones move slowly and may remain stationary over an area for several days or even weeks
- the air in an anticyclone **subsides** (falls from above), warming as it falls. This produces a decrease in its relative humidity which leads to a lack of cloud development, and dry conditions
- isobars are usually far apart, and therefore there is little pressure difference between the centre and edges of the anticyclone
- winds are weak, and flow gently outwards
- in the UK the winds flow clockwise around the centre of the anticyclone

Weather conditions

In winter, anticyclones result in:

- cold daytime temperatures — from below freezing to a maximum of 5°C
- very cold night-time temperatures — below freezing with frosts
- generally clear skies by day and night. Low-level cloud may linger and **radiation fogs** (caused by rapid heat loss at night) may remain in low-lying areas
- high levels of atmospheric pollution in urban areas, caused by a combination of subsiding air and lack of wind. Pollutants are trapped by a **temperature inversion** (when the air at altitude is marginally warmer than air at lower levels, see Photograph 1.2)

Jane Buekett

Photograph 1.2
A winter anticyclone traps pollution in the air over Burnley. The polluted air can be seen as a brown haze over the town

In summer, anticyclones mean:

- hot daytime temperatures — over 25°C
- warm night-time temperatures — may not fall below 15°C
- generally clear skies by day and night
- hazy sunshine in some areas
- early morning mists which disperse rapidly
- heavy dew on the ground in the morning
- the east coast of Britain may have sea **frets** or **haars** caused by onshore winds
- thunderstorms may occur when the air has high relative humidity

Anticyclones which establish themselves over Britain and northwest Europe and remain stationary for many days are described as **blocking** anticyclones. Depressions which would normally travel across the British Isles on a westerly airstream are steered around the upper edge of the high, and away from the area. Extreme weather conditions are then produced — dry and freezing weather in winter, and heat waves in summer.

Fogs

Fogs are common features of anticyclonic conditions. Fog is cloud at ground level which restricts visibility to less than 1 km. It consists of tiny water droplets suspended in the atmosphere. There are two main types of fog — radiation fog and advection fog.

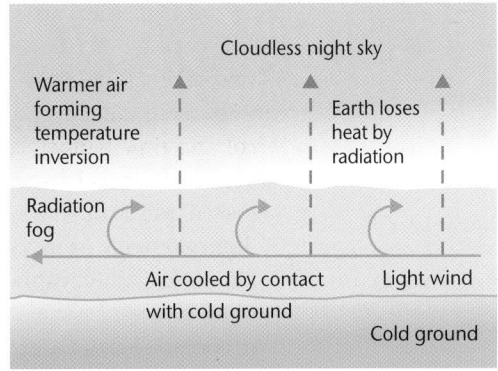

Radiation fog

Radiation fog forms under clear night skies when a moist atmosphere cools through the radiation of heat from the ground surface. The cooling extends some distance above the ground surface and is encouraged by light winds that allow slight mixing of the air. The air is cooled to its **dew point**, at which condensation occurs (Figure 1.2).

Figure 1.2 How radiation fog forms

This type of fog is common in winter, when long hours of darkness allow maximum cooling. In such cases, the fog may persist all day. It disperses either through an increase in wind speed or through a warming of the air (and subsequent evaporation).

Radiation fog is common under temperature inversions, which often occur in valleys (Figure 1.3). In the evening, with clear skies and high humidity, the air on the upper slopes chills quicker than that in the valley bottom. Cooling increases the density of the air, and it begins to move downslope. The cooler air accumulates in the valley bottom, pushing the warmer air upwards. The cold air now in the valley bottom will cool to its dew point, and create a dense fog that can last all day, and cause a severe frost.

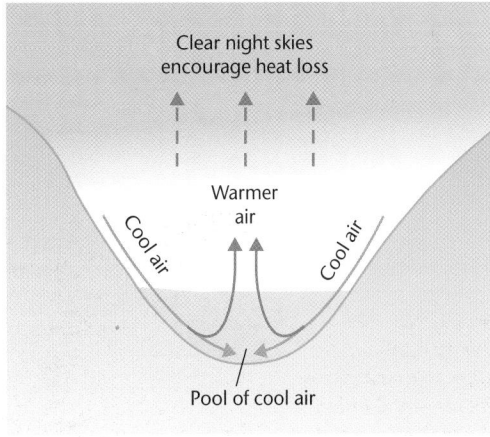

Figure 1.3 Fog formation under a temperature inversion in a valley

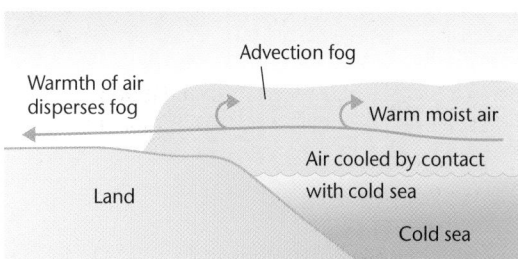

Figure 1.4 Advection fog

Advection fog

Advection fog forms when a mass of relatively warm air moves horizontally across a cooler surface. The air is cooled to its dew point, and condensation occurs. This type of fog is most common around coasts and over the sea in summer. In such areas it is sometimes called a **fret** or a **haar**. As the fog moves inland, it is warmed and evaporates (Figure 1.4).

Hydrology and rivers

Features of a drainage basin system

A drainage basin is the land area or catchment area drained by a single river and its tributaries. An imaginary line called the **watershed** delimits one drainage basin from another. The watershed frequently follows a ridge of high land; any water falling on the other side of the ridge flows into the adjacent drainage basin.

The drainage basin hydrological cycle summarises the characteristics of a single river basin. It is an ideal unit of study because its boundaries are distinctly formed by a watershed and the sea. It is an open system with inputs and outputs, as shown in Figures 1.5 and 1.7.

A drainage basin as a system includes the following:

Figure 1.5
The drainage basin
hydrological cycle

- **inputs**: energy from the sun for evaporation, precipitation (rain, snow)
- **outputs**: evaporation and transpiration from plants (collectively called

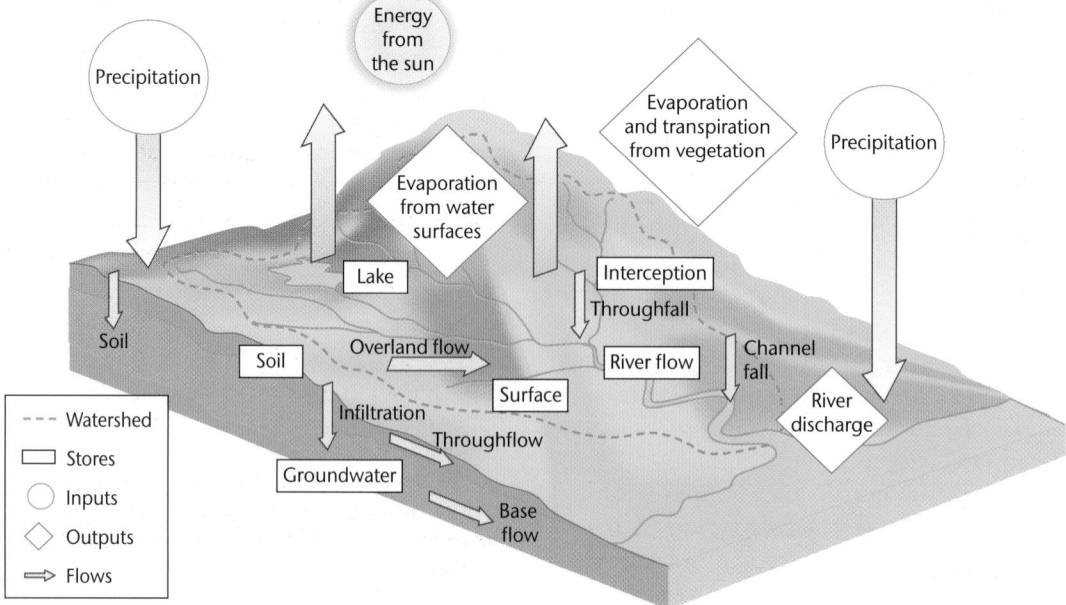

Key terms

Base flow That part of a river's discharge that is produced by groundwater seeping slowly into the bed of the river. It usually increases slightly during a prolonged spell of wet weather.

Infiltration The passage of water into the soil. Water is drawn into the soil by gravity and capillary action. The maximum rate at which this can take place is called the **infiltration capacity**, measured in millimetres per hour. Infiltration takes place at a higher rate at the start of a rainfall event, but as the soil becomes more saturated the rate decreases until it reaches a constant value (Figure 1.6). The rate of infiltration is affected by the nature of the soil and by the amount of rain that has fallen previously (the **antecedent rainfall**). Soils that are rich in clay or silt have lower infiltration rates than sandy soils due to the differences in relative soil pore sizes. Ground that is wet from previous rainfall may become saturated sooner and infiltration rates will be lower. The rate is also affected by land use (Table 1.2). When the infiltration rate is exceeded the surplus water flows over the ground as overland flow.

Interception The process by which raindrops are prevented from falling directly on to the soil surface by the presence of a layer of vegetation. Precipitation can be intercepted by plant leaves, stems and branches and by grass cover.

Land use	Infiltration rate (mm h^{-1})
Permanent pasture (ungrazed)	57
Permanent pasture (heavily grazed)	13
Cereals	9
Bare ground/baked hard	6

Table 1.2 Infiltration rates for different land uses

The amount of water intercepted depends on vegetation type and season. For example, deciduous trees intercept less than coniferous trees, and they intercept less in winter than in summer. Interception rates also decrease with time — a tree will intercept more at the beginning of a rainfall event than at the end because eventually water drips off leaves, or runs down the trunks of trees to the ground.

River discharge The amount of water passing a gauging station in a river at a particular time. It is calculated by the formula:

$$Q = A \times V$$

where
Q = discharge in cubic metres per second (or cumecs)
A = the cross-sectional area of the river at that point (square metres)
V = the velocity of the river (metres per second)

The overall discharge from a drainage basin is the product of the relationship between precipitation, evapotranspiration and storage factors. This relationship can be summarised as follows:

drainage basin discharge = precipitation − evapotranspiration ± changes in storage

Runoff All the water that enters a river and flows out of the drainage basin. It is quantified by measuring the discharge of the river.

Throughflow The movement of water downslope within the soil. It is particularly effective where further downward movement of water is prevented by an impermeable soil layer or by rock. Throughflow may also take place down routes created by tree roots and animal burrows.

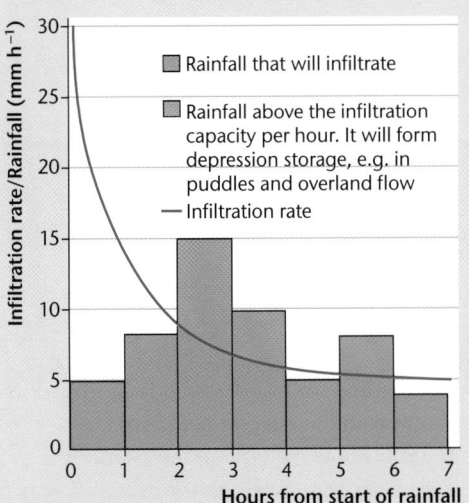

Rainfall that will infiltrate

Rainfall above the infiltration capacity per hour. It will form depression storage, e.g. in puddles and overland flow

— Infiltration rate

Figure 1.6 An infiltration curve

Figure 1.7 A flow diagram of drainage basin hydrology

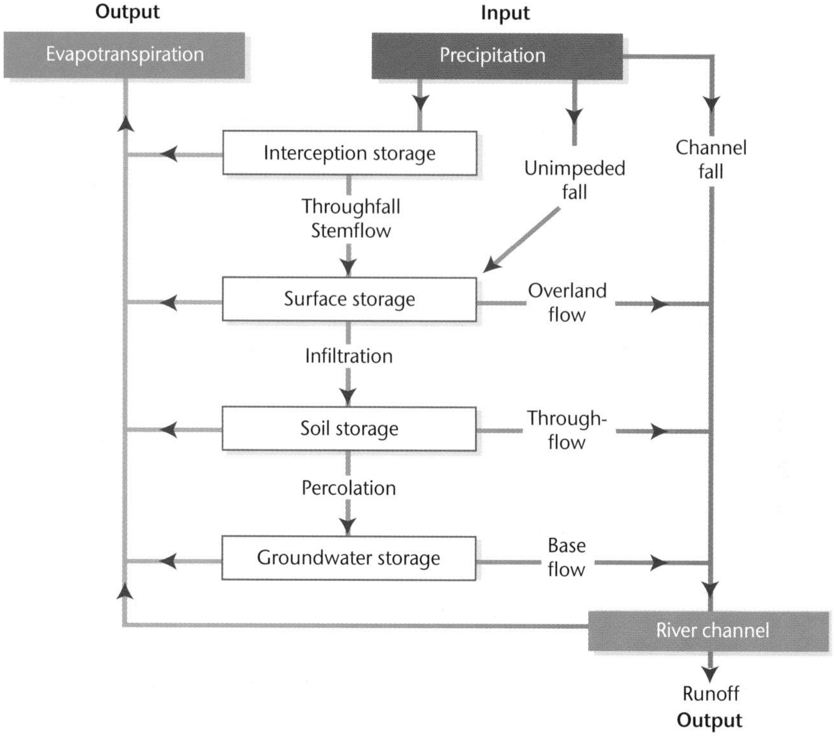

evapotranspiration), runoff into the sea, water percolating deep into underground stores where it is effectively lost from the system

- **storage zones**: surface stores such as puddles, rivers and lakes, soil storage, groundwater stores, storage on vegetation after precipitation

The storm hydrograph

The storm hydrograph (Figure 1.8) shows variations in a river's discharge over a short period of time, usually during a rainstorm. The starting and finishing level show the **base flow** of the river. As storm water enters the drainage basin the discharge rises, shown by the **rising limb**, to reach the **peak discharge** which indicates the highest flow in the channel. The **receding limb** shows the fall in the discharge back to the base level. The time delay between maximum rainfall amount and peak discharge is the **lag time**.

The shape of the hydrograph is influenced by a number of factors:

- The intensity and duration of the storm — if both are high they produce a steep rising limb as the infiltration capacity of the soil is exceeded.
- The antecedent rainfall — heavy rain falling on a soil which is saturated from a previous period of wet weather will produce a steep rising limb.
- Snow — heavy snowfall may not initially show on a hydrograph since the water is being 'stored' in snow on the ground. Indeed, water levels in a river may fall during a prolonged period of snowfall and cold weather. When temperatures rise and melting occurs, massive amounts of water are released, greatly

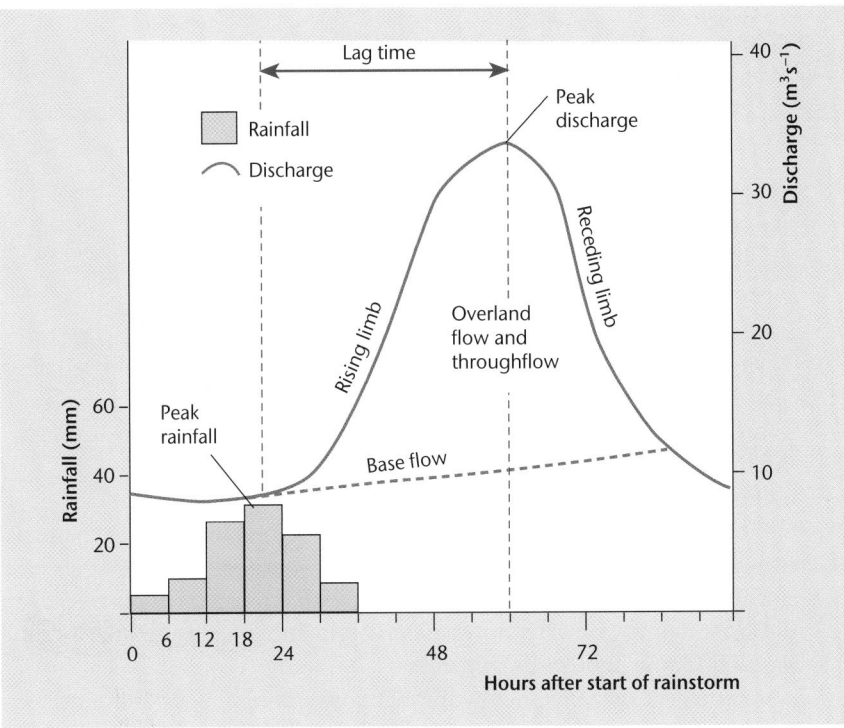

Figure 1.8
A storm hydrograph

increasing discharge. This water may reach the river channel even quicker if the ground remains frozen and restricts infiltration.

- Porous soil types and/or permeable rock types, such as limestone — these produce less steep (or less flashy) hydrographs because water is regulated more slowly through the natural systems.
- Impermeable rock types such as granite and clay — these tend to have higher densities of surface streams (higher drainage densities). The higher the density the faster the water reaches the main river channel, causing rapid increases in discharge.
- Size of drainage basin — a small drainage basin tends to respond more rapidly to a storm than a larger one, so the lag time is shorter.
- Slope angle — in steep-sided upland river basins the water reaches the channels much more quickly than in gently-sloping lowland river basins, producing a steeper rising limb and shorter lag times.
- Temperature — high temperatures increase the rate of evapotranspiration, thereby reducing discharge. Cold temperatures may freeze the ground, restrict infiltration, increase overland flow and increase river discharge.
- Vegetation — this varies with the season. In summer there are more leaves on deciduous trees, so interception is higher and peak discharge lower. Plantations of conifers will have a less variable effect.
- Land use — water runs more quickly over impermeable surfaces such as caravan parks or agricultural land which has been trampled by cattle. Lag time is reduced and peak discharge increased.

Figure 1.9
The shape of the
storm hydrograph
before and after
urbanisation

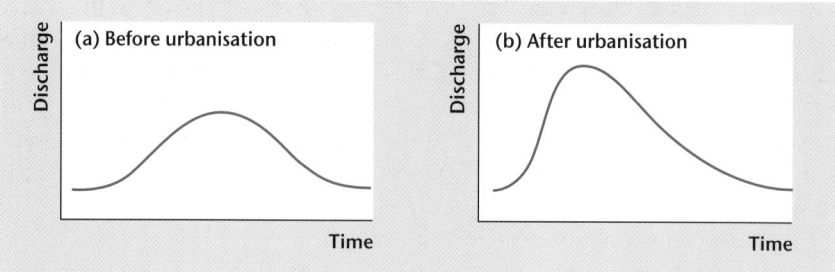

- Urbanisation — this is the main human impact on a storm hydrograph. The following processes combine to alter the shape of the hydrograph by reducing the lag time and increasing peak discharge (Figure 1.9):
  1 removal of topsoil and compaction of the ground with earth-moving machinery during building work
  2 building of roads which increase the impermeable surface area
  3 building of drains and sewers that transport water rapidly to river channels, reducing the lag time
  4 straightening of river channels and lining with concrete. This also leads to the faster delivery of water downstream of the urban area and increases the risk of flooding in downstream areas
  Less water reaches the channel by throughflow and base flow, and more gets there by overland flow. Local authorities and water companies may need to respond to these changes to prevent damaging flooding in their areas

Geomorphological variations within drainage basins

River processes

The work of a river involves three main processes — **erosion**, **transportation** and **deposition**. Erosion is the wearing away of the banks and bed of the river. The eroded material is transported as the river's load before being deposited.

Erosion

Rivers erode because they possess energy. Their total energy depends on three factors:

- the weight of the water — the greater the mass of the water the more energy it will possess due to the influence of gravity on its movement
- the height of the river above its base level (usually sea level) — this gives it a source of potential energy, and the higher the source of the river the more such energy it has
- the steepness of the channel — this controls the speed of the river which determines how much kinetic energy it has

Much of this energy is lost through friction, either internally through turbulence within the flow of the river, or externally through contact with the bed and banks of the river channel. Energy loss through friction can be great in all parts of a river, but it is perhaps easier to understand in the context of an upland river channel. Here, the channel often has a rocky bed with many large boulders.

The rough shape of the channel means that the **wetted perimeter** — the overall length of the bed and banks that the river is in contact with — is large. More energy is thus lost through friction, and the river's velocity, and therefore its energy level, is reduced. Hence, in normal conditions, the river is unable to perform much erosion. However, when the river contains large quantities of water, following heavy rain or snowmelt, it does possess the energy to perform great amounts of erosion.

There are four main processes of river erosion:

- **Abrasion** is the scraping, scouring and rubbing action of materials carried along by a river (the **load**). Rivers carry rock fragments in the flow of the water or drag them along the bed, and in doing so wear away the banks and bed of the river channel. Abrasion is most effective in short turbulent periods when the river is at bankfull or in flood. Larger fragments in the river's load produce more rapid erosion while smaller particles tend to smooth the surface.
- **Hydraulic action** is the movement of loose unconsolidated material due to the frictional drag of the moving water on sediment lying on the channel bed. As velocity increases, turbulent flow lifts a larger number of grains, particularly sand-sized particles, from the floor of the channel. Hydraulic action is particularly effective at removing loose material in the banks of meanders, which can lead to undercutting and collapse. It can be locally strong within rapids or below waterfalls where it may cause the rocks to fragment along joints and bedding planes or other lines of weakness.
- **Corrosion** is most active on rocks that contain carbonates, such as limestone and chalk. The minerals in the rock are dissolved in the water and carried away in solution.
- **Attrition** is the reduction in the size of fragments and particles within a river due to the processes described above. The fragments strike one another as well as the river bed. They therefore become smoother, smaller and more rounded as they move along the river channel. Consequently larger, more angular fragments tend to be found upstream while smaller, more rounded fragments are found downstream.

River erosion may be vertical or lateral. **Vertical erosion** is a characteristic of fast-flowing rivers with a large bedload of large angular particles. These erode by abrasion and hydraulic action and lower the bed relatively rapidly. Steep-sided valleys are often produced by such rivers. **Lateral erosion** is caused by a river meandering across its valley. The strongest current flows around the outside of the bend and hydraulic action causes the bank to be undermined and to collapse.

Transport

River energy not used for erosion or not lost through friction can be used to transport the river's load. A river obtains its load from two main sources:

- material that has been washed, or has fallen, into the river from the valley sides
- material that has been eroded by the river itself from the bed or banks

A river transports material in three ways:

- **Solution** — minerals that are dissolved within the mass of the moving water.

- **Suspension** — smaller particles of silt and clay are carried along within the flow of the moving water. Such material is not only carried but also picked up mainly through the turbulence that exists within the water in the river. The suspended load is usually the largest proportion of the load, and is the main cause of the brown appearance of many rivers and streams.
- **Movement** along the bed of the river — either by rolling (traction) or by bouncing (saltation). Larger particles are rolled along the bed of the river during times of very high discharge (and consequent high levels of energy). Slightly smaller particles may be thrust up from the bed of the river by turbulence only to fall back to the bottom again further downstream. As these particles land on the surface they in turn dislodge other particles upwards, causing more such bouncing movements to take place.

Two other terms are often used in the context of river transport — capacity and competence. Both of these are influenced by the velocity, and therefore the discharge, of the river.

The **capacity** of a river is a measure of the amount of material it can carry, that is, the total volume of the load. Research has found that a river's capacity increases according to the third power of its velocity. For example, if a river's velocity doubles, then its capacity increases by eight times (2^3).

The **competence** of a river is the diameter of the largest particle that a river can carry for a given velocity. Again, research has shown that a river's competence increases according to the sixth power of its velocity. For example, if a river's velocity doubles, then its competence increases by 64 times (2^6). This is because fast-flowing rivers have greater turbulence and are therefore better able to lift particles from the river bed.

Deposition

A river **deposits** when, owing to a decrease in its level of energy, it is no longer competent to transport its load. Deposition usually occurs when:

- there is a reduction in the gradient of the river, for example when it enters a lake
- the discharge is reduced, such as during and after a dry spell of weather
- there is shallow water, for example on the inside of a meander
- there is an increase in the calibre (size) of the load. This may be due to a tributary bringing in larger particles, to increased erosion along the river's course, or to a landslide into the river
- the river floods and overtops its banks, resulting in a reduced velocity outside the main channel

In general, the largest fragments are the first to be deposited, followed by successively smaller particles, although the finest particles may never be deposited (Figure 1.10). This pattern of deposition is reflected in the sediments found along the course of a river. The channels of upland rivers are often filled with large boulders. Gravels, sands and silts can be carried further and are often deposited further downstream. Sands and silts are deposited on the flat floodplains either side of the river in its lower course.

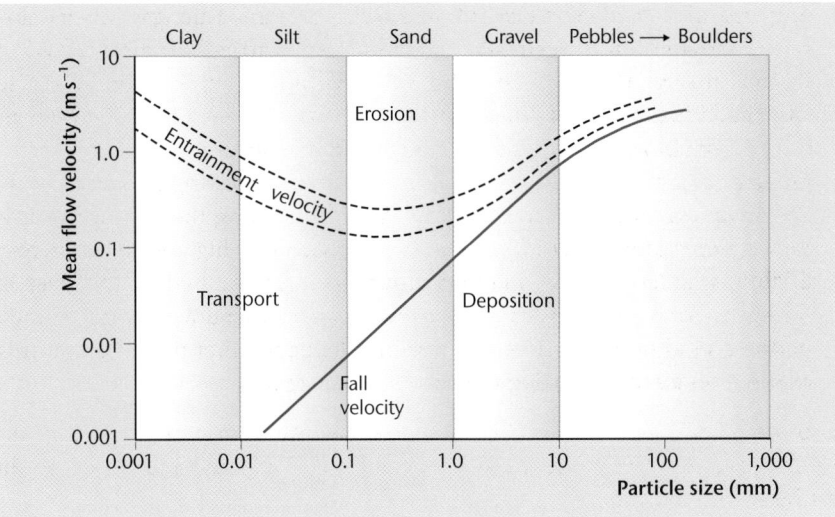

Figure 1.10
Hjulström's curve

How are erosion, transport and deposition related to changes in discharge?

The Hjulström graph (Figure 1.10) shows the relationship between the velocity of a river and the size of particles that can be eroded, transported or deposited. Velocity increases as discharge rises and generally this enables a river to pick up larger particles from the bed or banks of the channel. Similarly, as velocity and discharge reduce, then particles are generally deposited according to their size, largest first. However, Hjulström's research showed three interesting relationships:

- Sand particles are moved by lower pick up or critical erosion velocities than *smaller* silts and clays or *larger* gravels. The small clay and silt particles are difficult to pick up (entrain) because they tend to stick together. They lie on the river bed and offer less resistance to water flow than larger particles. Much more powerful flows of water are required to lift them into the water.
- Once entrained (picked up), particles can be carried at lower velocities than those required to pick them up. However, for larger particles there is only a small difference between the critical erosion velocity and the settling velocity. Such particles will be deposited soon after they have been entrained.
- The smallest particles, clays and silts, are only deposited at very low velocities. Indeed, some clay particles may never be deposited on the river bed and can be carried almost indefinitely. This explains why such deposits occur in river estuaries. Here the fresh water of the river meets the salt water of the sea, causing chemical settling of the clays and silts to occur and creating extensive areas of mudflats.

Figure 1.11 provides a summary of the processes of erosion, transportation and deposition from a river's source to its mouth.

Channel morphology

As a river flows from its source to its mouth, a number of changes take place in its morphology. These predominantly affect the shape of the channel, and its

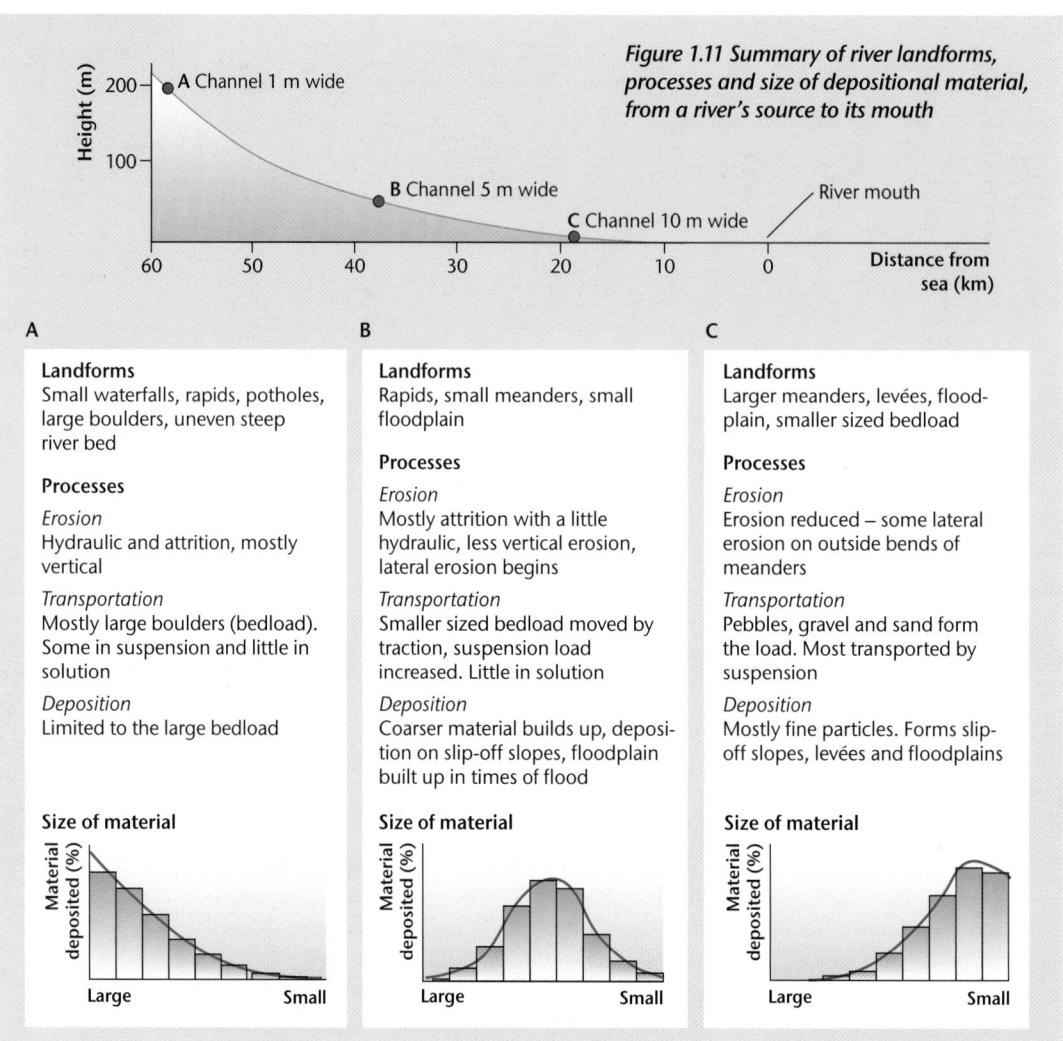

Figure 1.11 Summary of river landforms, processes and size of depositional material, from a river's source to its mouth

size, and result in distinctive landforms along its course. Five such landforms are waterfalls, meanders, oxbow lakes, floodplains and deltas.

Waterfalls

Waterfalls are steep falls of water where a river's course is markedly and suddenly interrupted. This may be the result of:

- a resistant rock occurring across the course of the river
- the edge of a plateau
- the rejuvenation of the area, giving the river renewed erosional power as sea level falls (see page 22)

The river falls over an edge into a deep plunge pool at the foot of the fall where the layers of weaker rock are excavated more quickly than the layers of resistant rock above. The force of the swirling water around the rocks and boulders enlarges

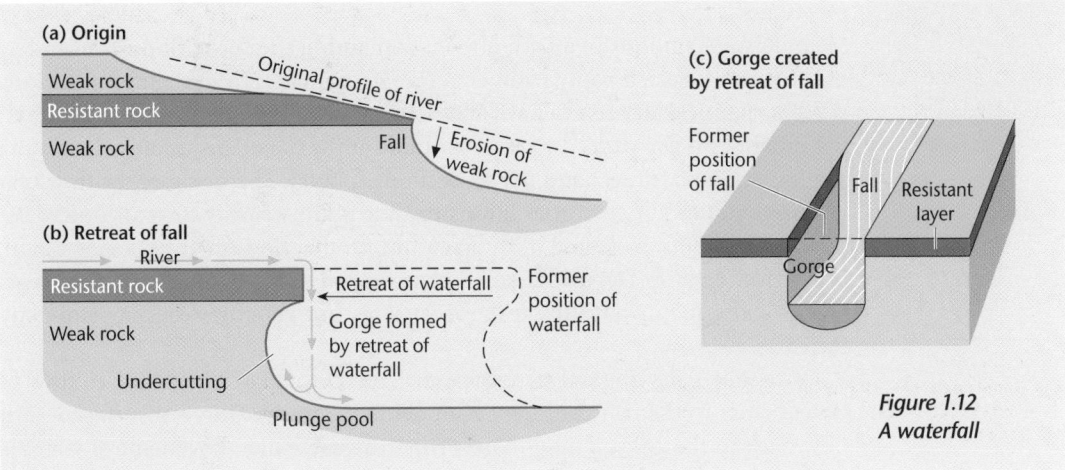

(a) Origin

Weak rock

Resistant rock

Weak rock

Original profile of river

Fall — Erosion of weak rock

(b) Retreat of fall

River

Resistant rock

Weak rock

Undercutting

Retreat of waterfall

Gorge formed by retreat of waterfall

Former position of waterfall

Plunge pool

(c) Gorge created by retreat of fall

Former position of fall

Fall — Resistant layer

Gorge

Figure 1.12
A waterfall

the plunge pool, undercutting the resistant rock (cap rock) above. Eventually the cap rock collapses and the waterfall retreats upstream, leaving a gorge ahead of it (Figure 1.12).

Case study *High Force in upper Teesdale*

In upper Teesdale an outcrop of an igneous rock called the Whin Sill causes the formation of the High Force waterfall. The Whin Sill is the resistant cap rock which overlies softer sandstone, limestone, shales and coal seams. These are eroded more quickly, leaving the overhang of High Force. The waterfall created is 22 m high — the tallest in England. Ahead of it lies a gorge stretching over 500 m downstream.

John Pallister

Photograph 1.3
High Force waterfall

Meanders

Meanders are sinuous bends in a river. Explaining the formation of meanders in a river has caused some problems for geographers. In low flow conditions straight channels are seen to have alternating bars of sediment on their beds and the moving water is forced to weave around these bars. This creates alternating shallow sections (riffles) and deeper sections (pools). The swing of the flow that has been induced by the riffles directs the maximum velocity towards one of the banks, and results in erosion by undercutting on that side. An outer concave bank is therefore created. Deposition takes place on the inside of the bend, the convex bank. Consequently, although the river does not get any wider, its sinuosity increases.

Once they have been created, meanders are perpetuated by a surface flow of water across to the concave outer bank with a compensatory subsurface return flow back to the convex inner bank. This corkscrew-like movement of water is called **helical flow**. In this way, eroded material from the outer bank is transported away and deposited on the inner bank. However, the zone of greatest erosion is downstream of the midpoint in the meander bend, because the flow of the strongest current does not perfectly match the shape of the meander. As erosion continues on the outer bank, the whole feature begins to migrate slowly, both laterally and downstream.

The cross section of a meander is therefore asymmetrical (Figure 1.13). The outer bank forms a river cliff with a deep pool close to the bank, the inner bank is a gently-

Figure 1.13 A meander

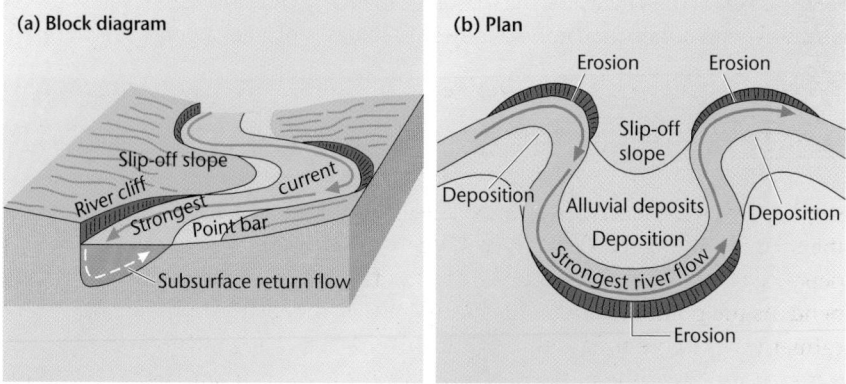

Figure 1.14 A point bar

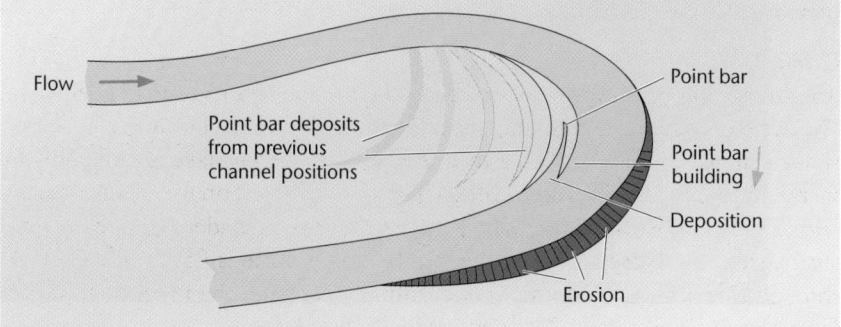

Kitchenham

*Photograph 1.4
Meanders on the
River Avon in
Hampshire*

sloping deposit of sands and gravels, called a **point bar** (Figure 1.14). The water in the river is moving slowly here, so it deposits material. The coarser particles are deposited first, at a point just downstream of the steepest part of the inside of the bend. As the point bar grows, finer sediments are deposited in the shallower and calmer water between it and the inside of the bend. Vegetation may encroach, trapping more sediment. When the meander migrates, point bar deposits from previous channel positions can be observed.

Oxbow lakes

An oxbow lake is a horseshoe-shaped lake separated from an adjacent river. The water is stagnant, and in time the lake gradually silts up, becoming a crescent-shaped stretch of marsh called a **meander scar**. An oxbow lake is formed by the increasing sinuosity of a river meander. Erosion is greatest on the outer bank, and with deposition on the inner bank, the neck of the meander becomes progressively narrower. During times of higher discharge, such as floods, the river cuts through this neck, and the new cut eventually becomes the main channel. The former channel is sealed off by deposition (Figure 1.15).

Figure 1.15
The development
of an oxbow lake

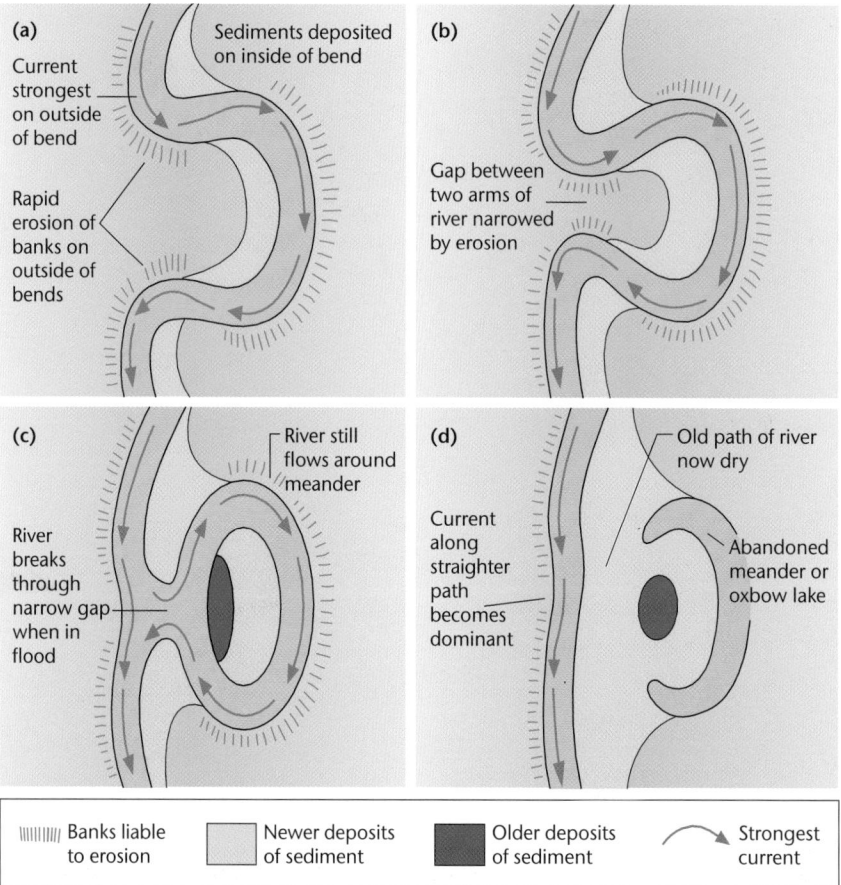

Floodplains

Floodplains are the most common depositional feature of a river. They are the relatively flat areas of land on either side of the river which form the valley floor. They are composed of alluvium — river-deposited silts and clays. The width of the floodplain is determined by the amount of meander migration and lateral erosion that has taken place, while the depth of the alluvium depends on the amount of flooding in the past. Over time point bar deposits and meander scars are incorporated into the floodplain. They become stabilised by vegetation as the meanders migrate and abandon their former courses.

When the river floods, the velocity of the water falls as it overflows the banks, and deposition takes place. Larger particles are deposited closer to the river channel, and smaller deposits further away. On some floodplains this has created high natural banks of sands and silts, called **levées,** which lie parallel to the river channel. In many areas levées have been artificially heightened and strengthened to act as flood defences.

Deltas

Deltas are formed by sediments which are deposited as a river slows when entering a lake or the sea. At this point, bedload and suspended material are dumped.

(a) Structure of a simple delta

Figure 1.16
Deltas

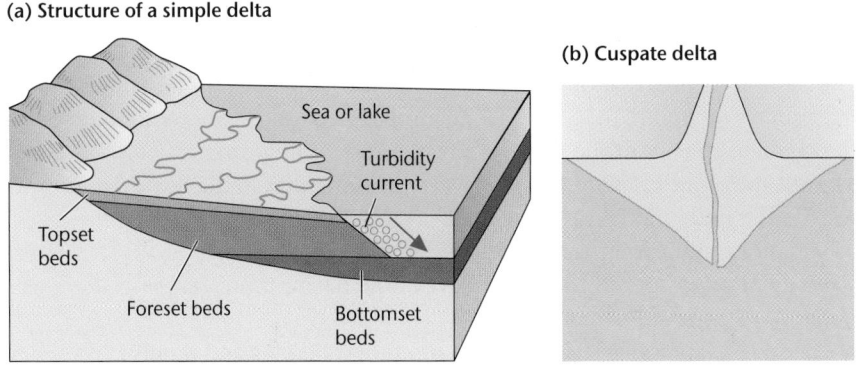

(b) Cuspate delta

(c) Arcuate delta

(d) Bird's foot delta

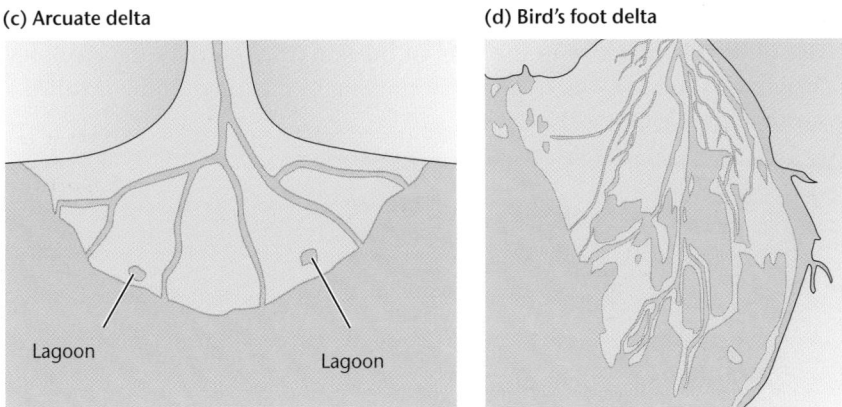

In addition, clay particles in suspension coagulate on contact with the sodium chloride in seawater. This process, known as **flocculation**, means that they sink more rapidly and settle out. Deltas are most readily formed when a river is heavily laden with sediment and the current and tidal effects are negligible. The rate of deposition then exceeds the rate of sediment removal.

Material deposited in a delta forms three types of bed (Figure 1.16a):

- **bottomset beds** — the lowest parts, composed of fine material built outwards by turbidity currents (sinking masses of sediment-laden water)
- **foreset beds** — inclined/sloping layers of coarse material lying over the bottomset beds; each bed is deposited above and in front of the previous one, building the delta seawards
- **topset beds** — composed of fine material, mainly silts and sands, essentially forming an extension of the river's floodplain

There are several delta shapes, but the three main types are (Figure 1.16):

- **cuspate** — a pointed shape produced by regular but opposing gentle water movements (e.g. Tiber)
- **arcuate** — a fan-shaped delta, where longshore drift or coastal currents keep the shape smooth (e.g. Nile)
- **bird's foot** — deposition pushes out into the sea along what are effectively extensions of the levées (e.g. Mississippi)

River valley profiles

Valley long profile

The valley long profile shows the changes in the altitude of a river along its course from source to mouth. In general a long profile is smoothly concave in shape, with the gradient steeper in the upper course and becoming progressively less steep towards the mouth (Figure 1.11). However, irregularities frequently exist, such as waterfalls, rapids and lakes.

Variations in the long profile can be explained in terms of:

- gradient — steeper in the upper part of the basin, more gentle in the lower part
- varying rock types — resistant rocks produce waterfalls and rapids
- natural lakes/artificial reservoirs which flatten out the long profile
- rejuvenation — the fall in sea level relative to the level of the land, or a rise of the land relative to the level of the sea, which revives the erosional activity of the river. The resultant steepening in the long profile is called a **knickpoint** (Figure 1.17)

Figure 1.17
Rejuvenation

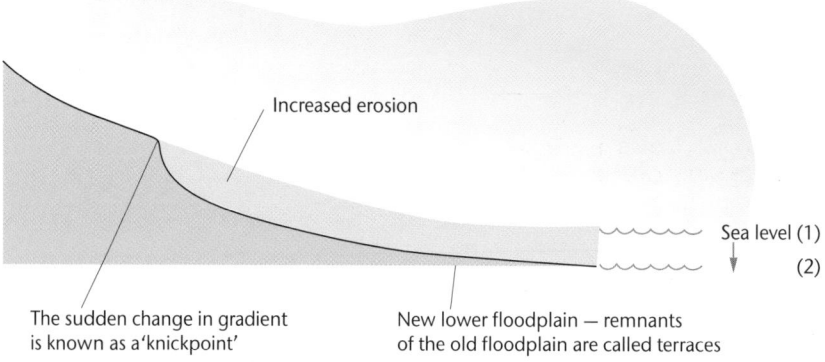

Increased erosion

Sea level (1)

(2)

The sudden change in gradient is known as a 'knickpoint'

New lower floodplain — remnants of the old floodplain are called terraces

Valley cross profile

The valley cross profile is the view of the valley from one side to another. For example, the valley cross profile of a river in an upland area typically has a V-shape, with steep sides and a narrow bottom.

Variations in the cross profile can be described and explained as follows (and as illustrated in Figure 1.18):

- in the upper course — a narrow steep-sided valley where the river occupies all of the valley floor. This is the result of dominant vertical erosion by the river
- in the middle course — a wider valley with distinct valley bluffs, and a flat floodplain. This is the result of lateral erosion, which widens the valley floor
- in the lower course — a very wide, flat floodplain in which the valley sides are difficult to locate. Here there is a lack of erosion, and reduced competence of the river, which results in large-scale deposition

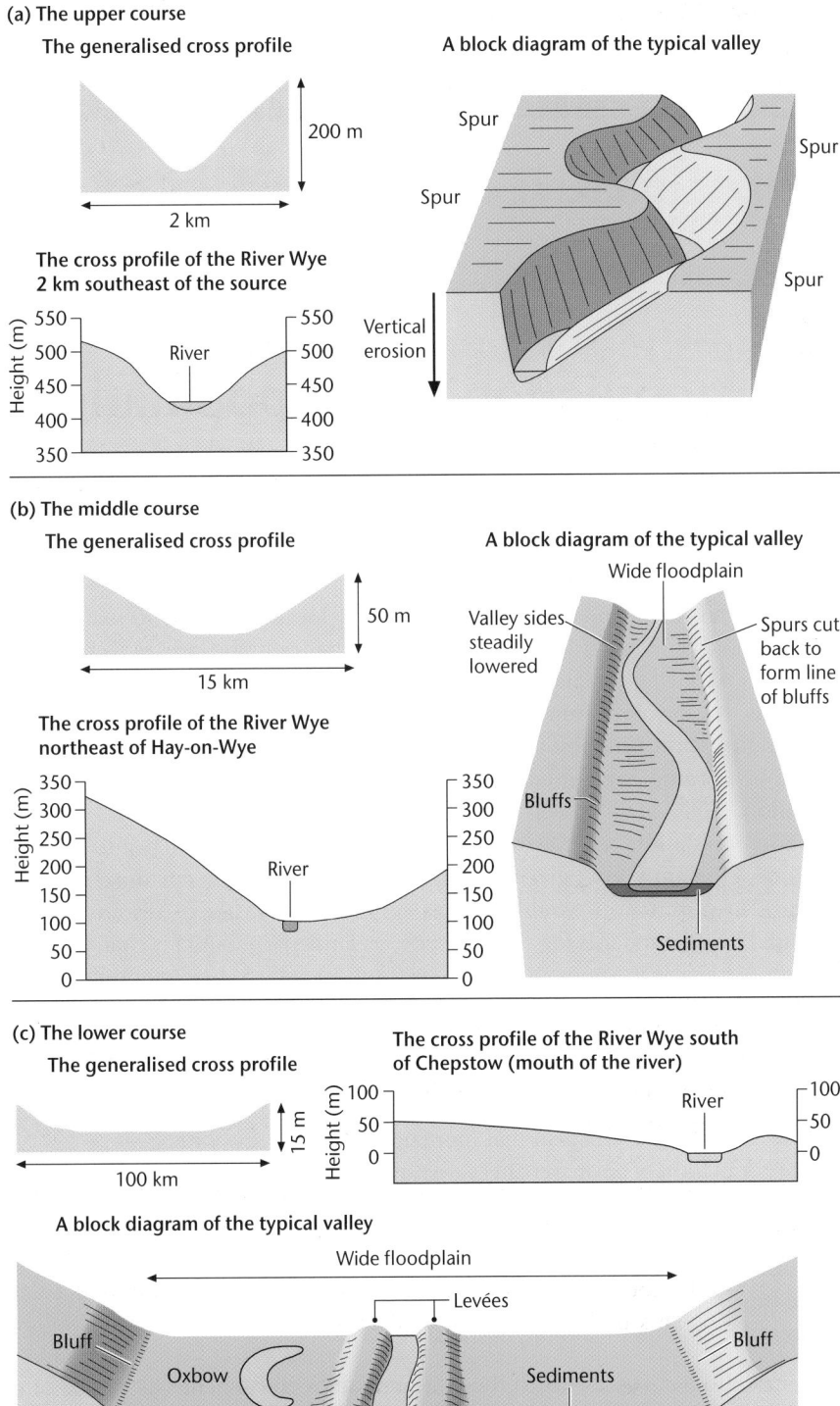

(a) The upper course

The generalised cross profile

The cross profile of the River Wye
2 km southeast of the source

A block diagram of the typical valley

(b) The middle course

The generalised cross profile

The cross profile of the River Wye
northeast of Hay-on-Wye

A block diagram of the typical valley

(c) The lower course

The generalised cross profile

The cross profile of the River Wye south
of Chepstow (mouth of the river)

A block diagram of the typical valley

*Figure 1.18
Valley cross profile
characteristics*

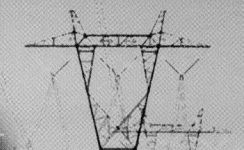
People and the environment
Population and resources

Overpopulation, underpopulation and optimum population

Overpopulation exists when there are too many people in an area relative to the amount of resources and the level of technology locally available to maintain a high standard of living. It implies that, with no change in the level of technology or natural resources, a reduction in a population would result in a rise in living standards. The absolute number or density of people need not be high if the level of technology or natural resources is low. Overpopulation is characterised by low per capita income, high unemployment and underemployment, and outward migration.

Underpopulation occurs when there are too few people in an area to use the resources efficiently for a given level of technology. In these circumstances an increase in population would mean a more effective use of resources and increased living standards for all of the people. Underpopulation is characterised by high per capita incomes (but not maximum incomes), low unemployment, and inward migration.

Optimum population is the theoretical population which, working with all the available resources, will produce the highest standard of living for the people of that area. This concept is dynamic — when technology improves, new resources become available which mean that more people can be supported.

The population of an area changes over time as demographic factors vary. These factors are:

- birth rates and fertility rates
- death rates, including infant mortality rates
- life expectancy and longevity
- rates of migration in and out of the area

The most effective way in which these factors can be illustrated is by means of a population pyramid. Figure 2.1 shows the main features of a population pyramid for a typical LEDC and a typical MEDC.

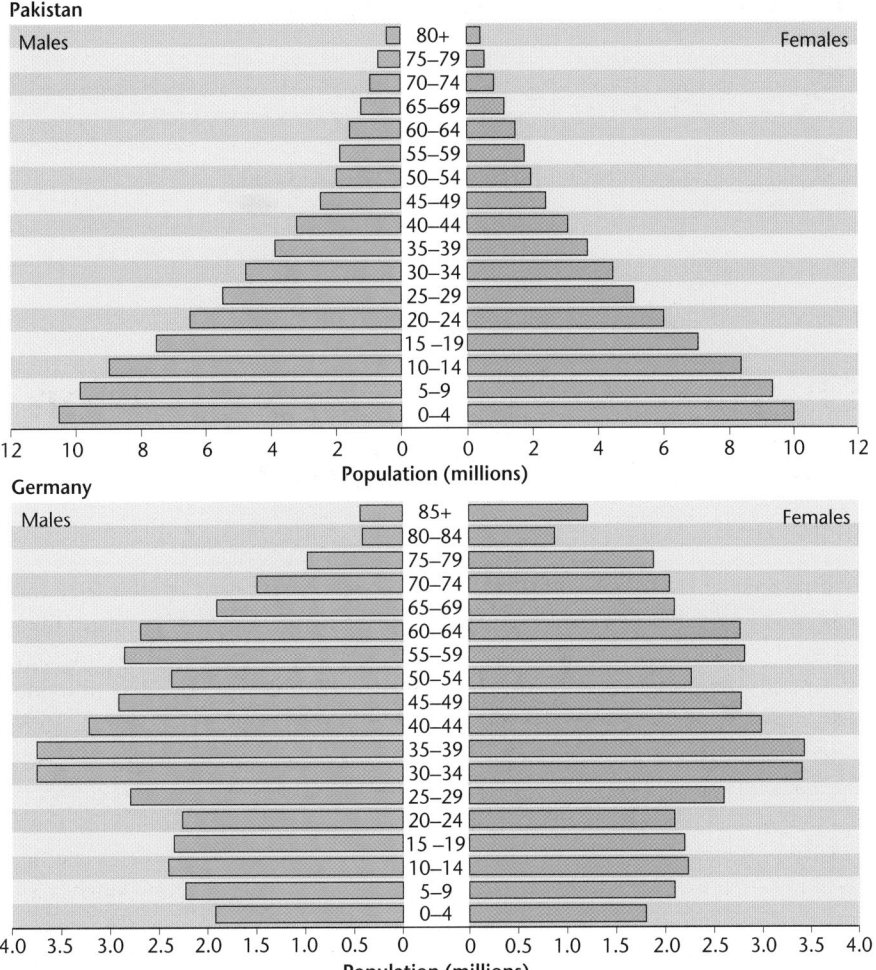

Figure 2.1 Population pyramids for Pakistan (a typical LEDC) and Germany (a typical MEDC)

Optimistic and pessimistic approaches

An optimistic approach

Ester Boserup, in *The Conditions of Agricultural Change: The Economics of Agrarian Change under Population Pressure* (1965), stated that environments have limits that restrict activity. However, these limits can be altered by the use of appropriate technologies which offer the possibility of resource development or creation. People have an underlying freedom to make a difference to their lives.

Boserup stated that food resources are created by population pressure. With demand, farm systems become more intensive, for example by making use of shorter fallow periods. She cited certain groups who reduced the fallow period from 20 years, to annual cropping with only 2–3 months fallow, to a system of multi-cropping in which the same plot bore two or three crops in the same year.

Key terms

Annual population change Cumulative change in the size of a population after both natural change and migration have been taken into account.

Birth rate A measure of an area's fertility. It is expressed as the number of live births per 1,000 people in 1 year.

Death rate The number of deaths per 1,000 people in 1 year.

Fertility The average number of children each woman in a population will bear. (Usually women between the ages of 15 and 50.) If fertility is 2.1, a population will replace itself.

Infant mortality The number of deaths of children under the age of 1 year expressed per 1,000 live births per year.

Life expectancy The average number of years from birth that a person can expect to live.

Longevity The increase in life expectancy over a period of time. It is a direct result of improved medical provision and increased levels of economic development. People live longer and this creates an older population.

Migration Any permanent change of residence by a person. Emigration is the movement of people away from an area. Immigration is the movement of people into an area.

Natural change The change in size of a population caused by the interrelationship between birth rate and death rate. If birth rate exceeds death rate, a population will increase. If death rate exceeds birth rate, a population will decline.

Population structure The make-up of a population of an area, usually in the form of age and sex distributions.

Photograph 2.1
School children collecting water in Zimbabwe. Water is a vital resource which is often in short supply in LEDCs

Neil Cooper/Still Pictures

The pressure to change comes from the demand for increased food production. As the fallow period contracts, the farmer is compelled to adopt new strategies to maintain yields. Thus necessity is the mother of invention.

Evidence to support this approach

The following two changes in agricultural practice support this view:

- The increasing intensity of shifting cultivation systems in various parts of the world. These move from 'slash and burn' systems in areas of very low rural population density, to systems making use of irrigation in areas of higher rural population density. People are adapting to their changing circumstances by adopting more intensive forms of agriculture.
- The Green Revolution — the widespread introduction of high-yielding varieties of grains, along with the use of fertilisers and pesticides, water control and mechanisation. The increased yields from these processes allow more people to be fed.

More recently, other writers, notably Julian Simon and Bjørn Lomborg, have contributed to these optimistic views. They refer to a number of so-called environmental scares of recent years. In the 1960s it was pesticides, carcinogens and the population explosion. In the 1970s there was the oil crisis, the imminent failure of the world's food supply and the fear of nuclear power. In the 1980s the deserts were advancing, acid rain was killing trees, the ozone layer was thinning and the elephant was on the point of extinction. The 1990s brought retreating rainforests, falling sperm counts, new diseases such as ebola, and genetically modified crops.

They argue that the alarmists were wrong. In their opinion none of these predictions has been fulfilled — there has been no rise in cancer caused by pesticides, population growth has slowed, oil reserves have increased, food production per head has increased even in the poorer countries of the world, nuclear accidents have been rare, deserts have not advanced, acid rain has killed no forests, the elephant has never been in danger of extinction, sperm counts are not falling and rainforests are still 80% intact. They think people are being made to indulge in environmental guilt when technology should in fact be encouraged to improve living standards throughout the world, rather than just for a rich minority.

A pessimistic approach

In *An Essay on the Principle of Population as it Affects the Future Improvement of Society* (1798), Thomas Malthus suggested that the environment dominates or determines patterns of human life and behaviour. Our lives are constrained by physical, economic and social factors.

His argument was that population increases faster than the supporting food resources. If each generation produces more children, population grows geometrically (1, 2, 4, 8 etc.) while food resources only develop arithmetically (1, 2, 3, 4 etc.) and cannot keep pace (Table 2.1). He believed the population/resource balance was maintained by various checks:

- increased levels of misery through war, famine and disease

- increased levels of moral restraint such as celibacy and later marriages
- increased incidence of activities such as abortion, infanticide and sexual perversions

Time periods (25 year intervals)	1	2	3	4	5	6	7
Population	1	2	4	8	16	32	64
Food supply	1	2	3	4	5	6	7

Table 2.1 Changes in population and resources (food supply) over time

Malthus asserted that the power of a population to increase its numbers was greater than that of the Earth to sustain it. This view is still held by so-called neo-malthusians. For example, in 1972 the Club of Rome (an international team of economists and scientists) predicted in a book entitled *The Limits to Growth* that a sudden decline in population growth could occur within 100 years if present-day trends continued. They argued that environmental degradation and resource depletion were not only related to population growth, but were also a function of the technologies and consumption patterns of greater numbers of people. They suggested greater control and planning of both population and resource use to create more stability.

Evidence to support this approach

Neo-malthusians believe that a number of recent issues support their views:

- They believe the war and famine in Ethiopia, Sudan and other countries of the Sahel region of Africa in recent decades suggest that population growth has outstripped food supplies. On a global scale, the Food and Agriculture Organization (FAO) suggests that over 800 million people are chronically malnourished, while 2 billion lack food security.
- Population growth accelerated rapidly in less economically developed countries (LEDCs) after their mortality rates began to fall. Rapid population growth impedes development and brings about a number of social and economic problems. In recent decades, however, population growth has slowed. In 2000, the population growth rate was 1.4% per annum compared with 2.4% in 1960.
- Water scarcity is predicted to be a major resource issue this century. The UN predicts that by 2050, 4.2 billion people (45% of the world's population) will be living in areas that cannot provide the required 50 litres of water a day to meet basic needs.

One of the most prominent neo-malthusians in recent years has been the American writer Paul Ehrlich. In the 1960s he suggested that India should not receive Western emergency aid because of its environmental state at the time. He said then that 'sober analysis shows a hopeless imbalance between food production and population'. However, optimists have since pointed out that India now more than feeds its population due to the advances of the Green Revolution.

The most recent scare from the pessimists is global warming, and one of the key features of this is that it cannot be proven either right or wrong within our lifetime. In response to this threat, at the 1997 Kyoto conference on the environment, the industrialised countries agreed to cut their carbon dioxide emissions by 30% by 2010. In the UK this was to be achieved by a switch away from coal-fired power stations to alternative sources, increases in public transport and taxes on fuel consumption. However, the USA, under President Bush, has refused to comply with the agreement.

In 2002, at the World Summit on Sustainable Development held in Johannesburg, key issues were sustainable management of the global resource base, poverty eradication and better health care. The last two were seen as ways in which population growth could be reduced. Clearly, the population–resources debate continues.

Managing the balance between population and resources

Overpopulation and underpopulation refer to an imbalance in the relationship between the level of population and the use of resources. Several countries have attempted to introduce policies aimed at managing or balancing this relationship.

Social policies of population control include:

- the limiting of population growth in China (the Chinese one-child policy — see case study below)
- the encouragement of population growth in Romania in the 1980s
- the use of immigration controls in Australia and the USA

Technological policies aimed at the improvement of living standards and levels of education include:

- the use of intermediate technology such as the introduction of windpumps to provide better water supplies in the Sahel countries of Africa
- the Green Revolution which introduced high-yielding varieties of seed, requiring the use of fertilisers, pesticides, mechanisation and irrigation
- the development of genetically modified crops (see case study on next page)
- educational improvements to increase literacy and numeracy levels and attract industries to the country, for example in Mauritius in the 1980s

Case study *The Chinese one-child policy*

One of the most documented population-control policies has been the Chinese one-child policy. During the second half of the twentieth century the Chinese government became concerned about population growth. There were two main reasons for this:

- the Chinese wanted to avoid a malthusian-type disaster in the future
- they realised that Chinese people could only have a rising living standard if the population was controlled

Chinese population policies have gone through a number of stages:

1950–59 The philosophy of the government under Chairman Mao was that 'a large population gives a strong nation'. The government encouraged people to have children for the good of the country. In 1959, there was a serious famine and 20 million people died.

1960–73 After the famine there was a population boom. The population increased by 55 million

(equivalent to the population of the UK) every year. Nothing was done to reduce the spiralling birth rate.

1974–79 There was a policy change and people were encouraged to reduce the birth rate by the slogan *'wan-xi-shao'* (later, longer, fewer):

- later marriages
- longer gaps between children
- fewer children

1979–90 The *wan-xi-shao* policy did not work well and the population went on increasing. In 1979 the government introduced the one-child policy which set very strict limits on who was allowed to have children, and when. Strong pressure was put on women to use contraception. Special family-planning workers in every workplace, and 'granny police' in housing areas, were instructed to make sure women were practising contraception, and to report on any suspicious or unauthorised pregnancies. Enforced abortions and sterilisation became common. The policy was very successful in urban areas, but less so in rural areas where disobedience

was more common. A disturbing effect of the policy was the practice of female infanticide. Couples wanted sons and so many baby girls were killed or 'disappeared'. The dominance of male babies also led to the spoilt 'little emperor' syndrome — the one child had no brothers or sisters and the attention of the extended family fell on him.

1990 onwards The one-child policy has been relaxed slightly. This is partly because it was so difficult to enforce, and because the Chinese government was concerned about the economic implications of a population in which there were far more older people than younger ones. In addition the revolution in global communication systems (the internet, satellite phones) has opened up the country to much greater social influence from the West. In more remote parts of the country the policy is still being encouraged. For example, the authorities in Guangdong, the state capital, ordered 20,000 abortions and sterilisations by the end of 2001 in the mountainous region of Huaiji.

Case study *The genetic modification of crops*

The latest revolution in plant breeding is a result of genetic modification (GM) of seeds. All living things contain DNA, a complex molecule that holds a genetic code for each plant or animal. DNA contains the instructions, inherited from the previous generation, for building the new organism. Genetic modification involves taking some of the DNA from one species and adding it to that of another species. When a plant is genetically modified, one or more characteristics of the donor species are transferred to the new plant.

How GM works

Some examples of the methods by which new varieties can be developed include:

- Adding the appropriate genes of a herbicide-resistant weed to a wheat seed to produce a type of wheat that is not harmed by herbicides. A field of wheat can then be sprayed to kill all the weeds without affecting the crop.

- Adding the genes of a species resistant to a particular pest to soya bean seed, so that the plant is not damaged by that pest.
- Adding a gene from a plant that grows well in an arid environment to the DNA of a rice plant. This would produce a plant that could grow in drier areas than traditional types of rice plant.

Arguments in favour

Those in favour of GM crops claim that the newly-engineered crops could solve many food shortages around the world, and also reduce the input of chemicals into farming. Trials of GM soya beans and maize have done well in the USA. Much of the soya imported into the UK and used in animal feeds is GM.

China has also invested a great deal into research of GM rice and cotton crops. Rice is the staple diet for its huge population, and cotton is an essential raw material for its clothing industry. Such developments are therefore important both for feeding its population and for improving its level of development. It is no surprise therefore that there is little opposition to GM crops in China.

Arguments against

The same cannot be said for the UK. Trials have been conducted here since 1999, but they have been very controversial. Critics of GM have the following objections:

- the pollen from GM plants may pollinate nearby plants and crops, spreading the modifications in an uncontrolled way
- crops on organic farms might be contaminated by the pollen from GM crops, causing the farms to lose their organic status
- the long-term effects of GM on human health are unknown

Protestors have destroyed GM field trials in the UK because of these fears (Photograph 2.2). Campaigners want GM crops to be banned completely and point to other agricultural innovations that have had serious health effects, such as BSE. On the other hand, some farmers and companies see a great opportunity to make profits from GM crops. The UK government is encouraging further testing in laboratories, along with carefully-controlled field trials. In the meantime, any GM crops that are sold to the public have to be clearly labelled.

The future

At a global scale, GM production is continuing in countries such as the USA and China. The international seed companies and food manufacturers are unlikely to be influenced by protestors in one country. Similarly, within a free-trade environment such as the EU, it will be increasingly difficult for governments to regulate the import of GM seeds or products.

In LEDCs, farmers may well face similar problems as those caused by the Green Revolution. GM seeds will only be available from the large seed companies, and it will not be possible to save seeds from one year to the next because many of the crops have been designed to produce infertile seeds. Poor farmers will not be able to compete with their richer neighbours.

Adrian Arbib/Still Pictures

*Photograph 2.2
An anti-GM protest
in Oxfordshire*

Environmental policies aimed at either increasing the amount of land that can be cultivated or improving the quality of land already available include:

- land reclamation schemes in the Netherlands
- the de-rocking scheme on the island of Mauritius which greatly improved soil quality

Energy resource issues

Differences between renewable and non-renewable resources

Renewable resources are those which cannot be used up. Living things such as forests are renewable as long as they are replaced at the same rate they are used up. For energy generation, renewable resources include the wind, tides, waves and moving water, as well as solar, geothermal or biomass generation.

Non-renewable resources are those which are finite — supplies could be used up. Coal, oil and natural gas are non-renewable — once used they cannot be replaced. Nuclear power can be included in this group, although it will take much longer to deplete supplies of uranium.

Environmental impacts

One of the main differences between these two resources is their impact on the environment. Renewable sources of energy on the whole are 'cleaner' and less harmful to the atmosphere than non-renewables. However, they can have some environmental impact. Many people think wind farms are visually polluting, and the burning of biomass can have serious environmental consequences, in terms of both deforestation and release of carbon dioxide. In LEDCs clearance of trees for fuel often damages the ecological balance of an area and can lead to desertification.

Non-renewable resources release harmful pollutants, such as carbon and sulphur compounds, into the atmosphere when they are burnt. Transporting the fuels from an area of production to an area of consumption also has an environmental impact. An obvious example is the movement of crude oil from one part of the world to another by means of tanker or pipeline. There is the danger of an oil spillage contaminating the local environment where it occurs, and transport by tanker also uses fuel which releases carbon dioxide and other pollutants into the atmosphere.

Another area of difference between renewables and non-renewables lies in the scale at which they are used. Renewable resources can often be exploited on a smaller, more localised scale than non-renewable resources. An example would be the mini-hydro systems used for electricity generation in remote Himalayan villages in Nepal. However, this is a generalisation — clearly most hydroelectric power (HEP) and tidal barrage schemes are as large scale as an oil-fired power station.

Examples of renewable and non-renewable energy systems

Renewable: biogas boilers in LEDCs

Biogas boilers are used to obtain methane gas from animal dung, human excreta and crop residues. The gas can then be used either directly as a fuel for cooking or, in some high-tech applications, to generate electricity using high-efficiency gas turbines.

The boiler is buried in the ground (Figure 2.2). Dung and crop residues are fed into it from above. An air pocket in the underground chamber allows methane gas to be produced as the waste matter decays. The methane is fed off and piped to houses or turbines. The residue and excess dung are used as high-quality fertiliser.

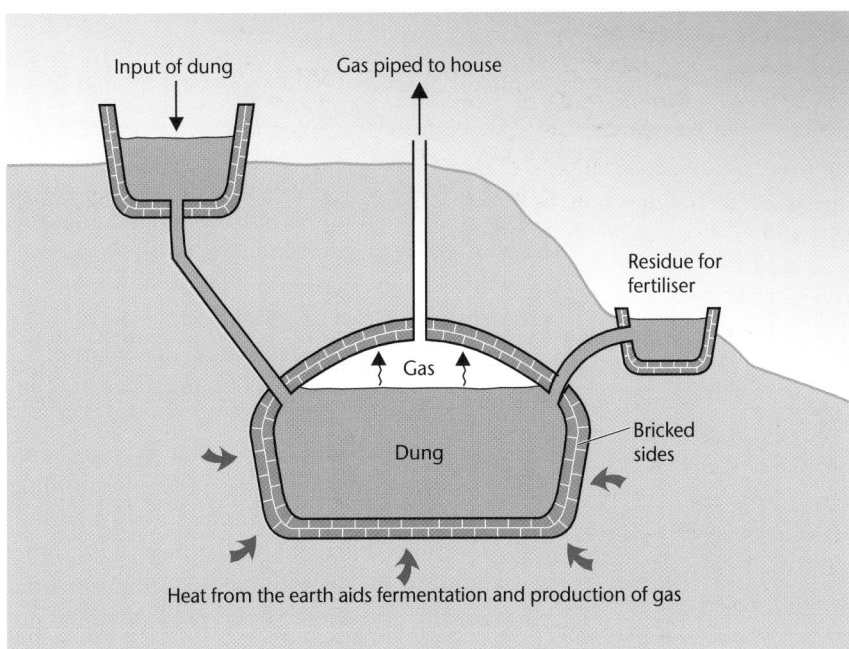

Figure 2.2
A biogas boiler

Input of dung

Gas piped to house

Residue for fertiliser

Gas

Bricked sides

Dung

Heat from the earth aids fermentation and production of gas

Pros and cons

The main positive features of this system are that it provides a cheap source of energy, and it allows the safe disposal of potentially unhealthy substances.

There are also negative sides to the system. It uses up a potential source of fuel and fertiliser, in particular those materials which have a soil-nutrient fixing role. The boiler gives off an unpleasant smell, and it also needs careful (and skilled) management to operate safely.

Renewable: wind energy in MEDCs

Wind power is the fastest growing renewable energy source. Many countries, particularly in Europe and North America, are seeking ways of developing wind power as one of their major sources of renewable energy.

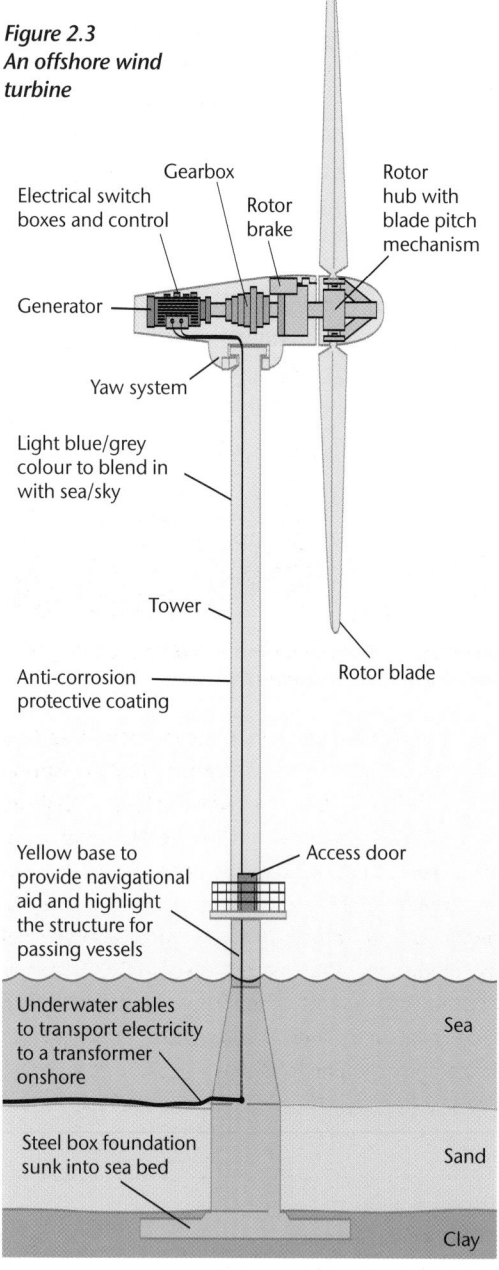

Figure 2.3
An offshore wind
turbine

Gearbox

Electrical switch
boxes and control

Rotor
brake

Rotor
hub with
blade pitch
mechanism

Generator

Yaw system

Light blue/grey
colour to blend in
with sea/sky

Tower

Anti-corrosion
protective coating

Rotor blade

Yellow base to
provide navigational
aid and highlight
the structure for
passing vessels

Access door

Underwater cables
to transport electricity
to a transformer
onshore

Sea

Steel box foundation
sunk into sea bed

Sand

Clay

Electricity is generated from the wind using one
or more wind turbines. It is more economical
if several turbines are sited in the same place to
create a **wind farm**. As objections have been raised
to building wind farms in upland areas, wind
energy companies have started to look for offshore
sites, where larger capacity turbines can be used
(Figure 2.3).

Pros and cons

Wind energy is pollution-free and does not
contribute to global warming. In Europe and North
America, winds tend to blow strongly in winter
when demand for electricity is at its highest. Wind
farms do not take up a lot of space (only 1% of
the land on which they are sited), which allows
farmland or natural habitats to exist around them.
Electricity generation by wind energy is becoming
increasingly competitive with coal-fired power plants
and is cheaper than nuclear fuel; it is still not as
cheap as gas-fired power stations, but wind energy
costs are likely to go down in the future. Supporters
of wind energy maintain that it represents an
excellent example of sustainability.

Opponents of wind farms claim that many of the
windiest sites are also areas of natural beauty. They
argue that wind turbines are an unwelcome
intrusion into the landscape and an eyesore. Some
people are worried about the noise the turbines
create, particularly as wind farms are often sited in
quiet locations; the damage that they could inflict
on wildlife (especially birds); and the potential
effect on property prices. Many critics also point out
that wind farms require large areas to produce only
small amounts of energy. It can take over 7,000 wind
turbines to produce the same amount of energy as
one nuclear power station. If wind energy is to be
viable, a lot of wind turbines will have to be built.

Case study *Wind energy production in the UK*

The UK is the windiest country in Europe. In 2004,
renewable energy sources in the UK generated just
over 3% of the total electricity supply, 30% of which
was derived from wind energy. By the end of 2005,
wind energy should generate 1.5% of the total UK
electricity supply — enough to meet the annual electricity needs of just under 1 million UK households.
The government's target is to generate 10% of the
UK's total electricity supply from renewable sources
by 2010 and it has announced that it intends to
increase this to 15% by 2015. Wind energy is
probably best placed to meet these targets.

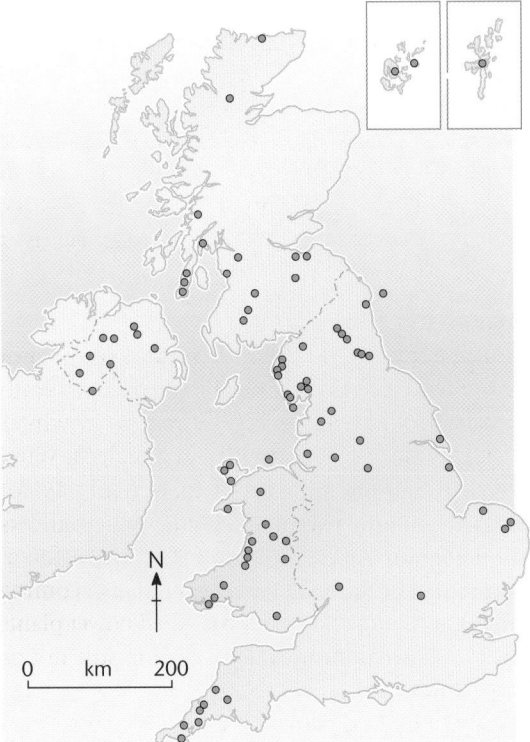

Photograph 2.3 Beinn An Tuirc wind farm on the Kintyre peninsula in Scotland

Figure 2.4 Wind energy sites in the UK, 2003

Although many wind farms are well established in the UK (Figure 2.4), there have been objections to both existing and planned developments. Even though he supports wind power, Jonathon Porritt, former director of Friends of the Earth, has his reservations: 'the real problem is that people building the things have been insensitive. They have put some of them in the wrong places and have not consulted local people or involved them in the benefits. The result is a growing anti-wind power lobby'.

As a result of these objections, the industry is starting to locate some of its new developments offshore. In November 2003, the North Hoyle Offshore Wind Farm was connected to the National Grid. This development, off the coast of north Wales, has 30 turbines which generate enough electricity to power 50,000 homes. It has been estimated that this station will offset the release of 160,000 tonnes of carbon dioxide into the atmosphere every year. David Bellamy, the well-known naturalist, is however campaigning against offshore wind farms, warning of 'plans that will make the British coastline ugly and impossible for birdlife'.

Non-renewable: gas-fired power stations in the UK

The so-called 'dash for gas' in the 1990s resulted in a number of gas-fired power stations being established around the UK. The gas they use is pumped from the offshore fields of the North Sea. Over 30 stations either have been built or are planned.

A number of factors have encouraged this:

- the development of the combined cycle gas turbine (CCGT) technology, which uses the gas more efficiently
- the increasing availability of natural gas and reduction in its price relative to that of other fuels

*Figure 2.5
Fuels used for
electricity
generation in the
UK, 1988 and 1997*

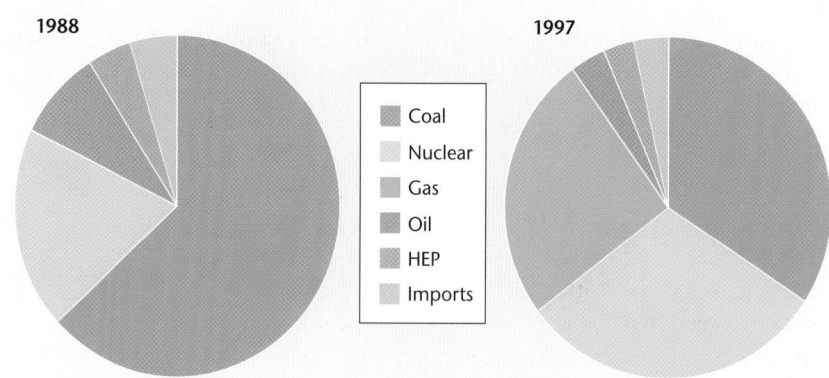

- the privatisation of the electricity supply industry, which has allowed independent companies to compete with the two main existing power station operators: PowerGen and National Power. An example of this is the proposed CCGT station at Avonmouth, promoted by Seabank Power (a company formed by Midland Electricity and British Gas)

Pros and cons

Gas is seen as being relatively clean and pollution-free. Modern gas-fired power stations produce very little sulphur dioxide and only a fifth of the nitrogen oxides produced by coal-fired stations. They also release 40% less carbon dioxide than coal-fired stations.

One major objection to the use of gas is that it can only be short-lived. Gas reserves in the UK may run out within 30 years, whereas there are proven coal reserves for 300 years. However, the use of coal-fired stations has been greatly reduced, and the coal industry has declined as a result (Figure 2.5). The proposed CCGT station at Avonmouth may have to supplement its gas supply by burning oil for up to 40 days a year as 100% gas supplies are not guaranteed.

Issues associated with the harnessing of energy

Acid deposition

Acid deposition consists of the dry deposition of sulphur dioxide, nitrogen oxides and nitric acid, and the wet deposition of sulphuric acid, nitric acid and compounds of ammonia from precipitation, mist and clouds (acid rain).

Effects

Acid deposition causes direct damage to trees, particularly conifers. It produces a yellowing of the needles and strange branching patterns. It also leads to the leaching of toxic metals (aluminium) from soils, and their accumulation in rivers and lakes. This in turn kills fish.

Acid deposition is blamed for damage to buildings, particularly those built of limestones, and to health problems in people, for example bronchitis and other respiratory complaints.

Causes

The major causes of acid deposition are the burning of fossil fuels in power stations, the smelting of metals in older industrial plants, and exhaust fumes from motor vehicles.

Solutions

Various solutions to acid deposition have been suggested:

- the use of catalytic converters on cars to reduce the amount of nitrogen oxides emitted in exhaust fumes
- burning fossil fuels with a lower sulphur content
- replacing coal-fired power stations with nuclear power stations
- the use of flue-gas desulphurisation schemes, and other methods of removing sulphur either before or after coal is burned
- reducing the overall demand for electricity and car travel

Case study *Scandinavia*

Scandinavia is one area of the world where forests and lakes have been severely damaged by acid rain. Scientists found that most of the pollution came from outside the region, mainly from the UK (Figure 2.6). The UK has now recognised its responsibility and has made significant moves to reduce sulphur emissions.

By 1999 total acid deposition on Scandinavia had fallen by 30%, but the residual effects mean that it will take time to repair the environmental damage. Today, much of the pollution affecting this area comes from Eastern Europe where countries have weak economies and cannot afford anti-pollution measures.

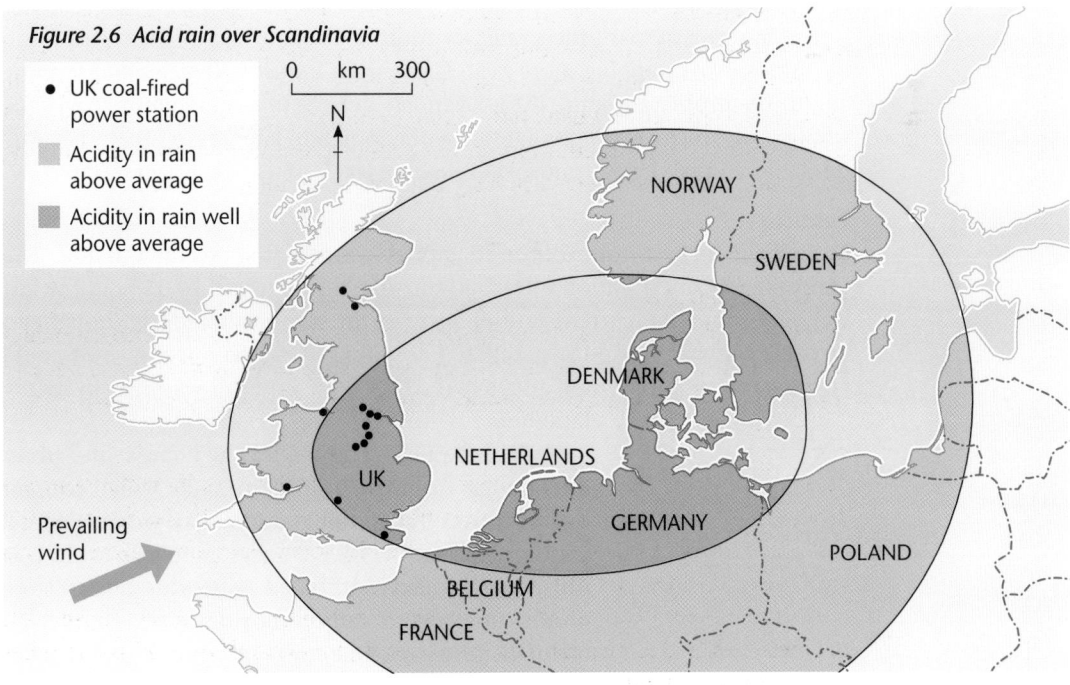

Figure 2.6 Acid rain over Scandinavia

- • UK coal-fired power station
- Acidity in rain above average
- Acidity in rain well above average

*Figure 2.7
The greenhouse
effect*

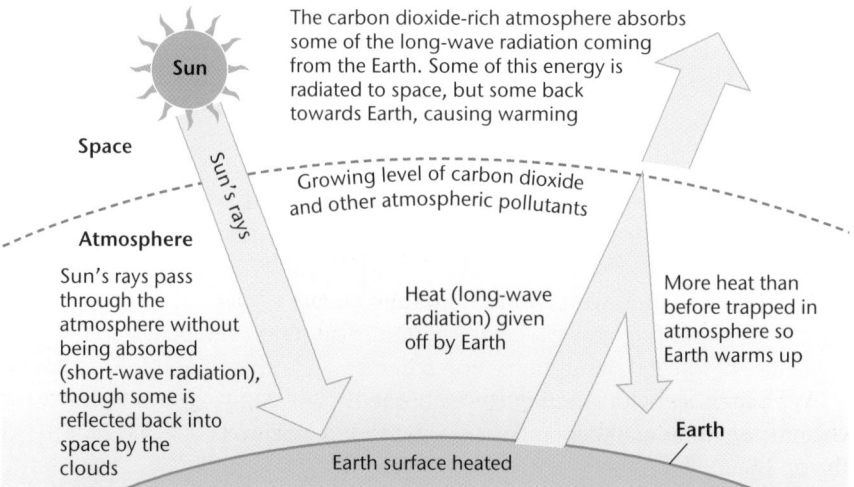

Sun

The carbon dioxide-rich atmosphere absorbs some of the long-wave radiation coming from the Earth. Some of this energy is radiated to space, but some back towards Earth, causing warming

Space

Sun's rays

Growing level of carbon dioxide and other atmospheric pollutants

Atmosphere

Sun's rays pass through the atmosphere without being absorbed (short-wave radiation), though some is reflected back into space by the clouds

Heat (long-wave radiation) given off by Earth

More heat than before trapped in atmosphere so Earth warms up

Earth

Earth surface heated

The emission of greenhouse gases

The sun's energy passes through the Earth's atmosphere as short-wave radiation and warms the surface of the Earth. Heat is then re-radiated back out again as long-wave radiation. Some of this heat is trapped by the atmosphere and causes it to become warmer (Figure 2.7). This so-called **greenhouse effect** is no bad thing for life on Earth. It is estimated that without it the average temperature on Earth would be about 30°C colder. The gases that are mainly responsible for trapping heat are termed **greenhouse gases** and include carbon dioxide, chlorofluorocarbons (CFCs), methane, nitrous oxide and ozone.

The burning of fossil fuels releases additional quantities of these gases into the atmosphere, trapping more heat. Gases from industry, vehicles and farming add to the problem (Figure 2.8). Scientists believe that temperatures over the whole world are increasing as a result, a phenomenon known as **global warming**. It is believed that this warming will cause sea levels to rise, as shown in Table 2.2.

Causes

The single biggest contributor to global warming is carbon dioxide. The atmospheric concentration of carbon dioxide has increased by 15% in the last 100 years, and the current rate of increase in carbon dioxide is thought to be 0.4% per year. It is estimated that a doubling of carbon dioxide levels could raise average surface temperatures by 2–3°C, with a greater warming in higher latitudes.

One of the main reasons for the increase in carbon dioxide has been the burning of fuels which contain hydrocarbons (coal, oil, natural gas) in developed countries such as the USA. Deforestation in areas such as Amazonia has also contributed, as trees are a major store of non-atmospheric carbon dioxide. More recently, the use of inefficient cooking stoves in many parts of Asia has been identified as a contributor to global carbon dioxide levels, but emissions by MEDCs are much greater.

| Year | Sea-level rise (cm) | |
	Best estimate	Worst estimate
2020	5	8
2040	12	20
2060	25	40
2080	35	60
2100	48	85

Table 2.2 Predicted sea-level rises due to global warming

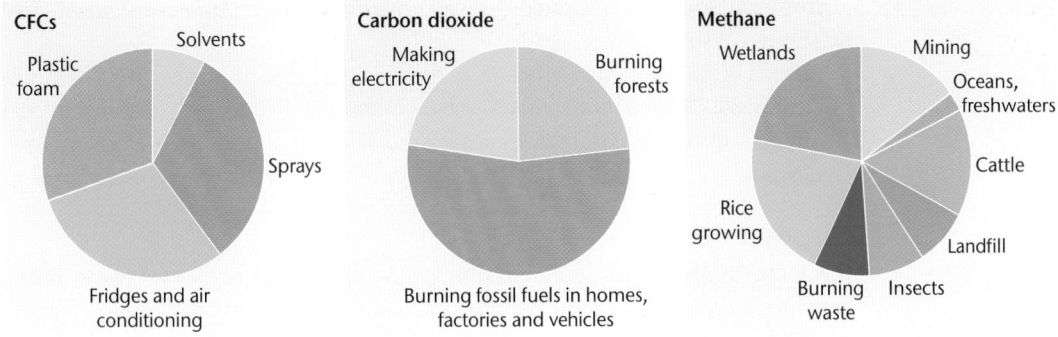

Figure 2.8
Where greenhouse gases come from

As LEDCs begin to develop, they too are beginning to generate energy as cheaply as possible and to consume quantities of fossil fuels, thereby adding to the problem.

Solutions

At the Kyoto conference on the environment in 1997 the following measures were agreed by the countries present:

- global greenhouse gas emissions were to be reduced by 5% from 1990 levels
- each country would have its own target for reduction — for example, the target for the USA was 7%

Just 4 years later, President Bush withdrew the USA from its Kyoto commitments, while insisting that developing nations should reduce their (much smaller) carbon dioxide emissions.

The hazard of nuclear waste

Types of nuclear waste

Nuclear power stations produce **high-level** radioactive waste in the form of used fuel rods which have been removed from reactors. These are taken to the THORP nuclear reprocessing plant at Sellafield in Cumbria (Figure 2.9). Here, re-usable uranium and plutonium are separated out to leave unusable radioactive waste. This is currently stored at Sellafield in steel-clad or lead-lined glass containers.

Low-level radioactive waste includes materials from hospital X-ray departments which may be contaminated. It is usually disposed of by burial.

Problems involved in disposal

Nuclear waste has a long half-life (the measure of how long it takes to lose half its radioactivity). The half-life of uranium is measured in millions of years. The material will therefore remain highly radioactive for a very long time, and this has to be borne in mind when disposing of it safely. Transport of the waste from one part of the country to another is also

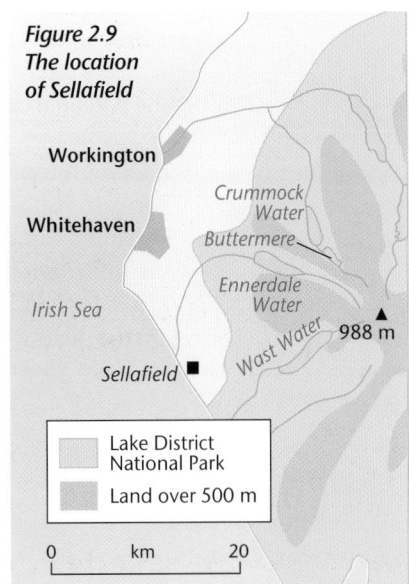

Figure 2.9
The location of Sellafield

39

a problem. Specially-designed railway containers have been constructed and extensively tested.

In the UK, the company NIREX has responsibility for disposing of all forms of nuclear waste, some of which does not originate in this country. It has tried unsuccessfully to find enough suitable sites for the disposal of both high-level and low-level waste.

Disposal sites

It is most difficult to find suitable sites for the disposal of high-level waste. When examining the potential of a site it is necessary to consider the following factors:

- the geology of the area — the ground must be geologically stable so that there is little chance of underground displacement
- unemployment figures of the area — jobs will be created by the activity
- the availability of land which has to be bought for the site
- transport links to the site, both locally and for transporting the waste long distances from nuclear power stations and ports
- the potential strength of local pressure groups
- the design features that will be necessary to make the site safe for many years
- the technology that will be necessary to ensure safe transport, storage and security

The construction of a site for the safe storage of nuclear waste raises a number of issues, which include:

- noise and disruption during construction
- short-term and longer-term safety concerns, with particular emphasis on leukaemia and cancer
- potential contamination of water supplies, again both in the short and longer term
- the effect on farming activities — will crops be safe to consume in the area and will animals become contaminated by grazing on grass which may be affected by the waste?
- the potential risk of accidents, and the worry that the site may become the target of terrorism
- effects on tourism — will the site destroy the tourist industry of the area or increase visitor potential?
- the 'hiding and forgetting' syndrome — what future problems may arise, which are difficult to predict and plan ahead for?
- if it is located in Cumbria, will the area become too economically dependent on the nuclear industry?

Chapter 3

Human geography
Changes in the UK in the last 30 years

Changes in manufacturing industry

Since the mid-1970s the UK, in common with other industrialised countries, has seen massive changes in its manufacturing industry. Some industries have undergone major decline, whereas others have grown markedly. Many of the areas of growth have been stimulated by investment from overseas.

Manufacturing industry has declined in its relative importance both in terms of employment and in its contribution to the national economy. Over 7 million people were employed in manufacturing in the mid-1970s, but by 2002 the number had fallen to less than 4 million. This progressive decline has been called **deindustrialisation**.

The main aspects of manufacturing change can be summarised as follows:

- manufacturing in the UK has declined and this has caused job losses and severe economic problems for communities once based on traditional manufacturing industries
- industrial revival, from both private and public investment, has had an uneven impact, with some areas still suffering the social and economic effects of industrial closures
- massive investment from overseas, and technological developments from research and development institutions, have contributed to industrial revival
- manufacturing industry now employs fewer workers than the growing service-based industries
- areas of the UK not traditionally linked with manufacturing have gained jobs more rapidly than urban areas
- the use of high technology in manufacturing processes has created its own requirements for industrial location, and has had a significant effect on working practices within industry
- sensitivity towards the environment, both locally and nationally, is a major influence on decision making

Key terms

Manufacturing industry Companies that convert raw materials into finished goods or assemble components made by other manufacturing companies.

Service industry Companies that provide a service for other businesses or for the people of an area. A wide range of employment is classified as services, ranging from education, health, police and local government to public transport, retailing, banking and the armed forces.

Decline in traditional manufacturing

The main manufacturing industries that have declined in the last 30 years are those which were established in the nineteenth century. Their growth was based on the use of coal and imported raw materials, such as iron ore and cotton. A key aspect of their development was the ability to export the finished products to other countries, particularly Britain's former colonies. For these reasons the main industrial areas were either on the major British coalfields or at coastal ports on deep-water estuaries (Figure 3.1). These industries have been given a variety of collective names, including 'smokestack' because of their link to coal, and 'sunset' because of their decline in importance.

Examples of such industries include:

- textiles — woollen cloth in West Yorkshire (Leeds, Bradford, Huddersfield) and cotton cloth in Lancashire (Bolton, Bury, Burnley)
- steel in Sheffield, Middlesbrough and south Wales
- shipbuilding in Newcastle, Sunderland, Belfast and Glasgow
- chemicals in the northeast (Middlesbrough) and the northwest (Widnes and Runcorn)
- the car industry and component suppliers in older locations, for example Birmingham and other parts of the West Midlands, and Luton (Photograph 3.1)

Figure 3.1
Traditional industrial areas in the UK

Source: Redfern, D. (2002) *Human Geography: Change in the UK in the Last 30 Years*, Hodder & Stoughton. Reproduced by permission.

The dynamics of change

Photograph 3.1
Car assembly at
British Motor
Corporation's
Longbridge works
near Birmingham
in the late 1960s.
Note the number of
workers on the line

- clothing, food processing, and other port industries in the East End of London
- pottery and other household goods in the area around Stoke

Reasons for the decline

There are many reasons for the decline in manufacturing. Some relate to changes within the UK, some to factors elsewhere in the world.

An increased use of **mechanisation**, such as automation, robotics and computerised production lines, has reduced the number of workers needed for the manufacturing process. In particular the tasks undertaken by unskilled and semi-skilled workers have been mechanised, reducing the number of people employed in manufacturing.

An inability to compete on price has caused UK industry to lose many of its overseas markets, as well as its home market. Competition from overseas countries producing similar products, particularly the newly industrialised countries (NICs) of the Pacific Rim (Hong Kong, Singapore, Taiwan, South Korea), has severely disadvantaged UK industry. In these countries, production costs are much lower, mainly because of lower labour costs. Outdated buildings and inefficient equipment have also added to production costs in the UK.

Working practices within the UK were traditional, or 'Fordist'. They involved the division of labour — the breaking down of a task into small repetitive fragments, each of which could be done at speed by workers with little specialist training. Such practices were characteristic of mass assembly lines which produced standardised products. In the newer industrial areas more flexible working practices, in terms of both production and the use of machinery and labour, have resulted in multi-tasking — one worker being able to do several jobs.

A world economic **recession** in the early 1980s combined with a rise in the value of sterling meant that the cost of British manufactured goods rose at a time

when worldwide demand was falling. The Conservative government at the time believed that manufacturing industry should strive to become competitive and not be protected by government assistance. A number of industries formerly owned by the government, such as British Steel, were privatised and forced to 'go it alone'. Subsidies and other support mechanisms were dropped. Rationalisation took place, concentrating production in a smaller number of highly-mechanised units so as to remain competitive. Uncompetitive industries closed. In most cases, rationalisation meant redundancy.

Political factors were also important during the period of manufacturing decline under the Conservative government which was in power from 1979 to 1997. This government wanted a less unionised labour force, particularly in the mining and manufacturing industries, and set out to 'defeat' many of the trade unions, and reduce their power. A previous Conservative government had suffered at the hands of trade unions in the coal mining and motor vehicle industries, and this one came to power soon after the 'winter of discontent' in 1979 during which the unions were thought by many to have overstepped their powers.

During the late 1990s and the early 2000s, the problems of **global overcapacity** previously suffered by the iron and steel and shipbuilding industries began to appear in the motor vehicle industry. The two 'giants' of the industry, Ford and General Motors, decided to cease car assembly at Dagenham and Luton respectively due to the falling demand for cars in Europe. The global strategies of these two firms do not allow for spare capacity, and there are worries that this will affect other manufacturers in the years to come, particularly as the major players in the industry continue to merge to share design and production costs.

Finally, although traditional manufacturing in the UK has declined (Figure 3.2 and Table 3.1), the overall output of manufacturing industry in the country has

*Figure 3.2
Standard regions
in Great Britain*

*Table 3.1
Changes in
manufacturing
employment by
standard region
in Great Britain,
1981–2001*

Standard region	1981 (thousands)	1995 (thousands)	2001 (thousands)	Percentage change 1981–2001 (%)
Southeast	1,683	883	845	−49.7
East Anglia	186	139	132	−29.0
Southwest	396	281	265	−33.0
West Midlands	801	511	486	−39.3
East Midlands	533	398	375	−29.6
Yorkshire and Humberside	579	398	345	−40.4
Northwest	800	469	438	−45.3
North	339	222	210	−38.1
Wales	238	227	223	−6.3
Scotland	502	319	298	−40.6
Great Britain	6,057	3,847	3,617	−40.3

Case study *The steel industry in Consett, County Durham*

The steel industry in Consett is an example of a declining industry in a declining manufacturing area.

In the mid-1960s the British Steel Corporation had a workforce of 250,000, producing nearly 25 million tonnes of steel at 23 locations in Britain. By the mid-1990s its workforce had been reduced to 55,000, with a production of 14 million tonnes at just four locations.

This reduction was a consequence of huge over-capacity around the world, with new steel-making facilities being developed in places such as India, South Korea and Taiwan. The response of the European nations was to rationalise production. Many high-cost inland locations such as those at Consett in northeast England (Figure 3.3) and at Bilston in the West Midlands were closed. Only the biggest and most efficient integrated works (Redcar-Lackenby, Scunthorpe, Llanwern and Port Talbot) survived in Britain. In 1999 British Steel merged with the Dutch company Hoogovens to form Corus, and this company announced the closure of the Llanwern steel-making plant in February 2001.

Steelworks are very much dependent on ease of transport of raw materials. They need access to wide, deep, sheltered estuaries through which coal and iron ore are imported from countries such as Australia, Brazil and Liberia. These same ports can be used for exporting the finished steel, and railway transport can be used for home markets.

Consett

Steel-making in Consett began in 1840, based around local deposits of coking coal and blackband iron ore. By the late 1880s the furnaces at Consett were producing 10% of the nation's steel. A company town grew up around the steelworks, which meant that most of the houses and shops were owned by the steel company.

The closure of Consett was announced in 1980, on economic grounds. Production was to be transferred to the more cost-efficient works on the coast at Redcar-Lackenby. The local raw materials had long

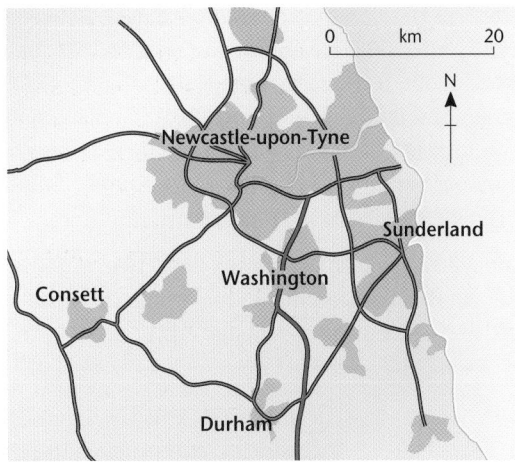

Figure 3.3 The location of Consett

since been exhausted, and losses were high due to the small capacity of the works. The main market for the Consett plant, shipbuilding on the Rivers Tyne and Wear and at Barrow-in-Furness, had also largely closed down. Closure resulted in the immediate loss of 4,000 jobs, and the town faced economic disaster. Male unemployment in the town was set to rise to 30%, adding to the unemployment caused by the decline of the surrounding coal industry. This was exacerbated by the subsequent closure of another large industry — a ball-bearing manufacturer.

Regeneration

Shortly after the closure of the steelworks a number of organisations (private businesses, local government agencies and the local authority) set up the Derwentside Industrial Development Agency. This encouraged various companies to move into the area, the most famous of which were Derwent Valley Foods (makers of Phileas Fogg snacks), which employs 80 people, and Blue Ridge Care, a manufacturer of disposable nappies. Over 200 companies, some computer-based, have been established in new purpose-built units, creating over 3,500 jobs in total. These companies were encouraged by grants and loans, but also by the substantial improvements made to the local environment, such as the greening over and landscaping of the spoil tips.

The area's economy changed considerably over the 20 years from 1980 to 2000. The early 1980s saw the area in a position of economic weakness. Between 1980 and 2000 Derwentside lost over 16,000 coal-mining jobs, 6,000 steel-related jobs and over 4,000 other jobs. Between 1977 and 1981 manufacturing employment declined by 66%, and in 1982 unemployment approached 30%, with a total of 8,900 people registered as unemployed.

During the same 20-year period (from 1980 to 2000), the industrial development programme successfully assisted 180 businesses to start up or grow in the Derwentside area, creating over 6,000 new jobs. Over 90% of these jobs were in manufacturing. The area's economy diversified and strengthened, but it still remains weak in comparison to that of the UK generally.

One of the main remaining problems is educational attainment. This is below both regional and national levels, with 18% and 40% of adults having low levels of literacy and numeracy respectively. Similarly, there are low attainment levels in both GCSE and A-level performances, at 25% and 31% below national levels respectively.

increased. This has been due to the expansion of new factories making motor vehicles, computers and related equipment, and other electrical goods. Clearly, deindustrialisation has been selective in both the industries and the areas it has affected.

The growth of new manufacturing industries and industrial areas

The majority of new manufacturing industries in the last 10–20 years in the UK have been high-technology industries, such as computers and computer-related equipment, telecommunications and microelectronics. In addition, many traditional industries, for example the car assembly industry, have advanced by adopting new technologies and working practices. A key feature of this type of industry is research and development (R&D). This is needed to develop new products and update existing products, as well as keeping 'ahead of the game' in product design.

For these 'new' industries, a highly skilled and qualified workforce is essential, while access to raw materials is less important. They have therefore become concentrated in areas where the workforce can be attracted or is available, or in places where the government encourages them to locate. They are often described as 'footloose' because they are not tied to certain locations, like traditional manufacturing, and as 'sunrise' because of their growth. At a local scale they are commonly located on new industrial estates on the edges of towns, or alongside motorways for efficient transport (Figure 3.4).

The following are the main areas of new industrial growth:

- 'silicon glen' in central Scotland
- the Cambridge area and along the M11
- the 'sunrise strip' of the M4 corridor
- the 'Honda valley' in south Wales, and the Honda assembly plant at Swindon
- the new car assembly plants in the northeast (Washington) and East Midlands (Burnaston)
- many small light industrial estates in 'rural' areas, e.g. East Anglia and Sussex

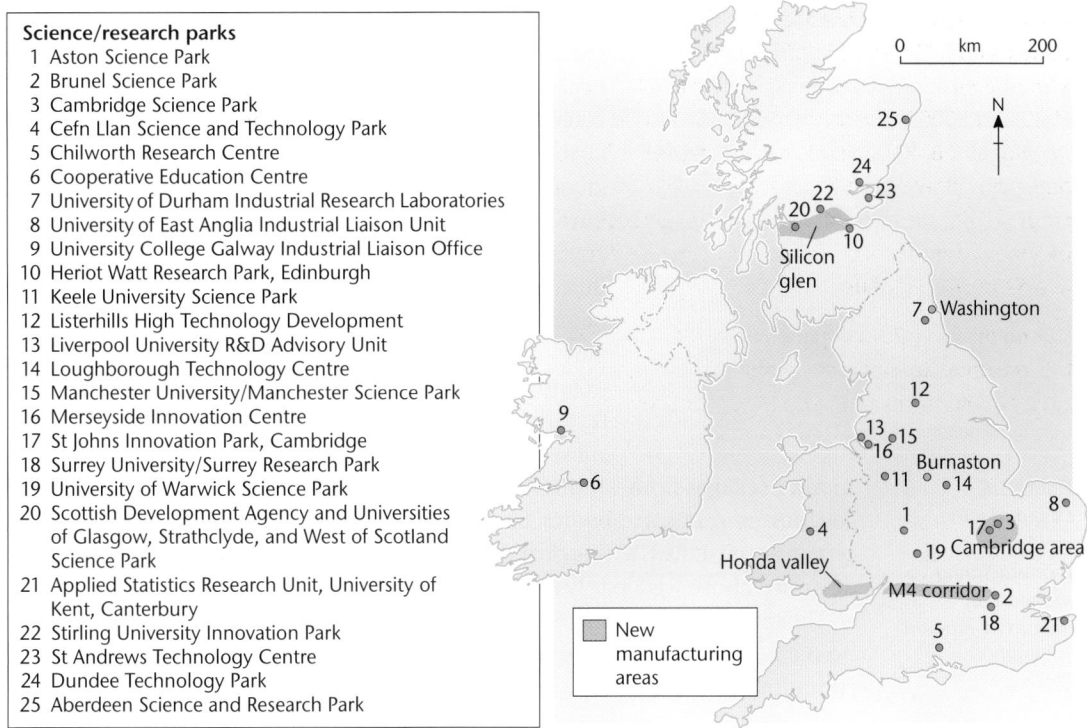

Science/research parks
1 Aston Science Park
2 Brunel Science Park
3 Cambridge Science Park
4 Cefn Llan Science and Technology Park
5 Chilworth Research Centre
6 Cooperative Education Centre
7 University of Durham Industrial Research Laboratories
8 University of East Anglia Industrial Liaison Unit
9 University College Galway Industrial Liaison Office
10 Heriot Watt Research Park, Edinburgh
11 Keele University Science Park
12 Listerhills High Technology Development
13 Liverpool University R&D Advisory Unit
14 Loughborough Technology Centre
15 Manchester University/Manchester Science Park
16 Merseyside Innovation Centre
17 St Johns Innovation Park, Cambridge
18 Surrey University/Surrey Research Park
19 University of Warwick Science Park
20 Scottish Development Agency and Universities of Glasgow, Strathclyde, and West of Scotland Science Park
21 Applied Statistics Research Unit, University of Kent, Canterbury
22 Stirling University Innovation Park
23 St Andrews Technology Centre
24 Dundee Technology Park
25 Aberdeen Science and Research Park

Figure 3.4
New manufacturing areas and science/research parks in the UK

Reasons for growth and location of new industry

The reasons for these changes in industrial growth and location reflect changes within the UK, and global economic factors.

Growth

There has been massive **inward investment by overseas transnational companies**, for example by Japanese, South Korean and German firms (Figures 3.5 and 3.6). In the case of the motor vehicle industry a number of Japanese car manufacturers

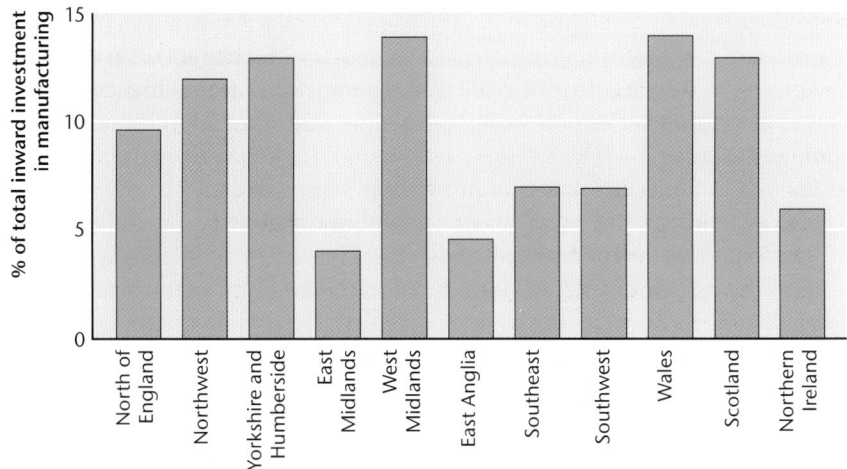

Figure 3.5
Location of inward investment in manufacturing in the UK by standard region, 1999

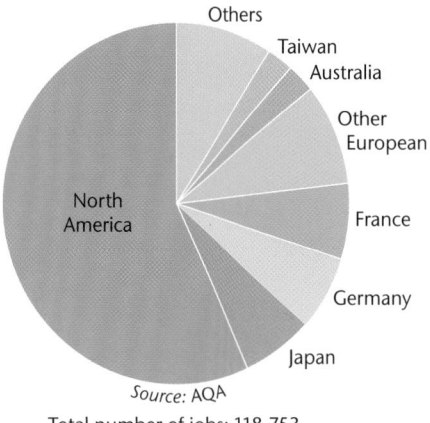

North America

Others

Taiwan

Australia

Other European

France

Germany

Japan

Source: AQA

Total number of jobs: 118,753

*Figure 3.6
Origin of jobs
created in
manufacturing
in the UK by
inward investment,
1998–2000*

have come to Britain and built huge new plants, the three largest being Nissan, Toyota and Honda. Nissan was the first. It began production in 1986 with an initial output of 100,000 cars per year, rising to over 270,000 in 1993. Its factory at Washington in the northeast of England was the largest single investment by a Japanese company in Europe. Other similar investments followed. A large Toyota plant was built next to the A38 trunk road at Burnaston in north Derbyshire, and a Honda assembly plant at Swindon on the M4. Inward investment in other parts of Britain has not always been as successful — examples are the Fujitsu (Japan) and Siemens (Germany) factories in the northeast of England which were both subsequently closed down.

Aid packages from various levels of government (local and national), or from government-sponsored bodies, have encouraged overseas and home investment in certain areas of the UK. It is clear that the major investments by Nissan and Toyota described above were partly attracted by the financial packages they were offered.

The revival of old industrial areas and the reindustrialisation of new areas has been a policy of most governments over the last 30 years, with varying degrees of commitment. A wide variety of regional policies has been set up to try to redress imbalances of economic and social development. These have included:

- Assisted Areas — carefully defined parts of the UK in which government grants may be given to persuade firms to locate. At various times in the last 30 years these have been called Special Development Areas, Development Areas and Intermediate Areas, and varying levels of incentives have been offered.
- Enterprise Zones — areas at a more local scale to which industry was attracted by the removal of certain taxes (local and national) and local authority planning controls. Typical Enterprise Zones were small, 100–200 hectares in size, sometimes built on greenfield sites and sometimes on areas within inner cities with 'development potential' (often a euphemism for empty and derelict).
- Urban Development Corporations and other urban regeneration schemes (for more details see chapter 7, pages 162–167).

The growth of new industries has allowed the **transfer of technology** to the UK. This is the movement of new working practices and other innovations into the country, including:

- The just-in-time (JIT) system of production. This is designed to minimise the costs of holding stocks of raw materials and components by carefully planned scheduling and flow of resources. It requires a very efficient ordering system and reliability of delivery. JIT was introduced to Britain by car manufacturers such as Nissan. In car plants there is hourly delivery of some parts, and many component manufacturers have been forced to relocate close to the assembly plants. Another requirement of JIT is zero defects and total quality control. Car manufacturers have close and strong links with their suppliers, which are monitored rigorously.

- The concept of teams of flexible multi-tasking workers who rotate jobs between them. This helps to increase skill levels. In car assembly, it means that production can be designed to fulfil the needs of the individual customer. Standardised products are fewer, and a greater range of specialised products is made. It also allows more flexible production which can meet even the smallest alteration in market demand at short notice. This links into the following two points.
- The use of 'envoys' — representatives from a 'buying' industry who are permanently based in the factories of component suppliers. They can pass on directly any required changes to design and specification at short notice.
- The use of 'milkmen' — representatives from a 'buying' industry who regularly visit the factories of component suppliers to inform them of required changes to design and specification.

Location

The development of numerous motorways, airports and high-speed rail links for ease of **communication** has been a major factor in the location of new manufacturing. For industries such as these, speed of access is important both for people and for raw materials and products.

Some industries have located close to universities which provide both **research expertise** and a pool of highly skilled/intelligent labour. An example of this is the Cambridge Science Park (see case study).

An **attractive environment**, such as a rural or landscaped out-of-town location, may also be an important locational factor. Many business and science parks have been built on greenfield sites where the relatively low cost of land has been an advantage. The high quality of the environment is thought to assist in the creative development required by such industries.

*Photograph 3.2
Robot assembly
in the body shop
at Nissan's
Washington factory*

Case study *The Cambridge Science Park*

The first and biggest science park in the UK is the Cambridge Science Park (Table 3.2), on the northern edge of this famous university city (Figure 3.7). There are well over 700 hi-tech companies within the Cambridge region (known as 'silicon fen') which employ over 20,000 people. The growth of the science park is clearly linked to the nearby university and its pool of highly educated and technologically qualified workers and scientists. These individuals have generated high personal incomes which have enabled entrepreneurship to develop even further. Research and development are encouraged by the university, which provides some of the facilities.

The building of the M11 and electrification of the railway have improved access to Cambridge and acted as growth factors. In 1999 an additional boost came when Microsoft chose Cambridge as the centre of its European operations and pledged £50 million to build its first European computer research centre there. This is bound to encourage more companies to move into the area, keen to take advantage of the **synergy** that will develop. Synergy is the intense localised interaction between different companies (research organisations, banks, entrepreneurs, service organisations) on the same site which creates benefits for all participants. It is sometimes summarised as '2 + 2 = 5'.

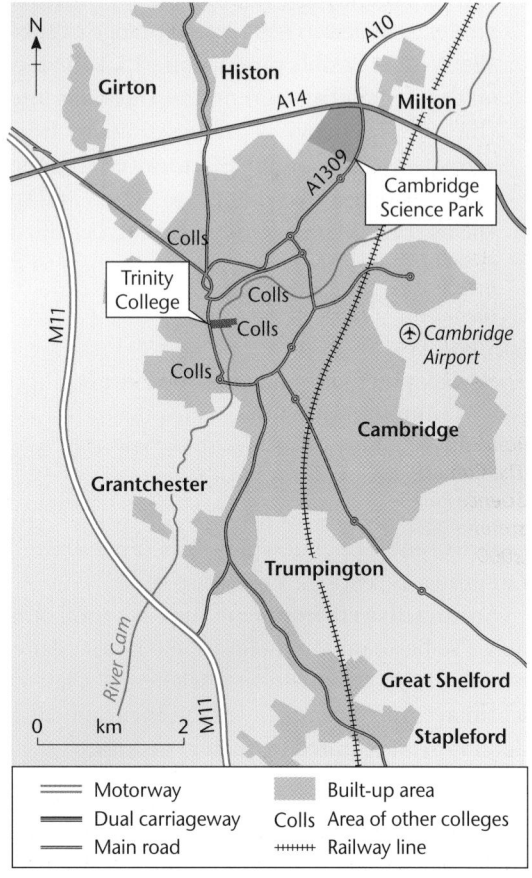

Figure 3.7 The location of the Cambridge Science Park

Photograph 3.3 The Cambridge Science Park

Timothy Soar

Economic activity	Companies Number	% of total
Computer systems and software	15	23
Research and development	10	16
Biotechnology	8	13
Pharmaceutical	6	9
Medical equipment	4	6
Associated services (e.g. licensing, patenting, travel)	4	6
Communications	3	5
Animal medicine	2	3
Health preparations	2	3
Investment/venture capital	2	3
Management/scientific consultancy	2	3
Others (e.g. lasers, structural engineers)	6	9

Note: Rounding means figures do not add up to 100.

Size of unit (square feet)	Number of companies
0–1,999	13
2,000–9,999	23
10,000–19,999	14
20,000–29,999	5
30,000–39,999	4
50,000–59,999	2
60,000–69,999	1
90,000–99,999	1
220,000–229,999	1

Number of employees	Number of companies
0–10	14
11–30	18
31–50	11
51–75	7
76–100	1
101–125	2
126–150	2
151–175	1
201–250	3
350	2
450	1
560	1

Table 3.2
The Cambridge Science Park — summary data, 2000

Years in science park	Number of companies
0–5	15
6–10	23
11–15	12
16–20	10
21–25	3
26+	1

Case study *The M4 corridor*

The M4 corridor to the west of London follows the route of the M4 motorway through Reading, Newbury, Swindon and Bristol and into south Wales (Figure 3.8). Many high-technology industries have located here, including information technology industries, computer-based industries, telecommunications and microelectronics. Research and development sites are also widespread, some connected to private industries and others to government institutions.

Advantages of the area

The location has a number of advantages:
- both the M4 (running west to east) and the A34 (running north to south) provide good road communication
- the electrified railway line west from Paddington provides a quick route into central London for business meetings
- Heathrow airport on the M4 provides easy overseas access
- the government research centres (Aldermaston and Harwell) already in the area have encouraged related industries to locate here
- inward migration of people from elsewhere in the UK seeking highly paid employment has created a skilled and motivated workforce
- nearby universities, such as Oxford, Brunel and Reading, have stimulated research projects and encouraged further development of expertise. Business parks and science parks similar to the one at Cambridge have become established across the area, encouraging the exchange of ideas and information
- the attractive environment of the Thames Valley and other nearby areas such as the Cotswolds, Mendips, Chilterns and Marlborough Downs provide homes for the highly paid workers

Swindon

Swindon lies halfway between London and Bristol, and owes its origin to the railway line between London and south Wales. It was the engineering centre of the Great Western Railway. Today it has attracted, like the rest of the corridor, a wide range of high-technology industries such as Intel, a leading microprocessing company. The percentage of people employed in manufacturing has gone down in Swindon (as it has in every other UK town), but the nature of that manufacturing industry has significantly changed.

An indication of the increasing wealth of Swindon is the fact that major service industries — the insurers Allied Dunbar and Commercial Union, and the Nationwide Building Society — have moved their administrative headquarters to the town.

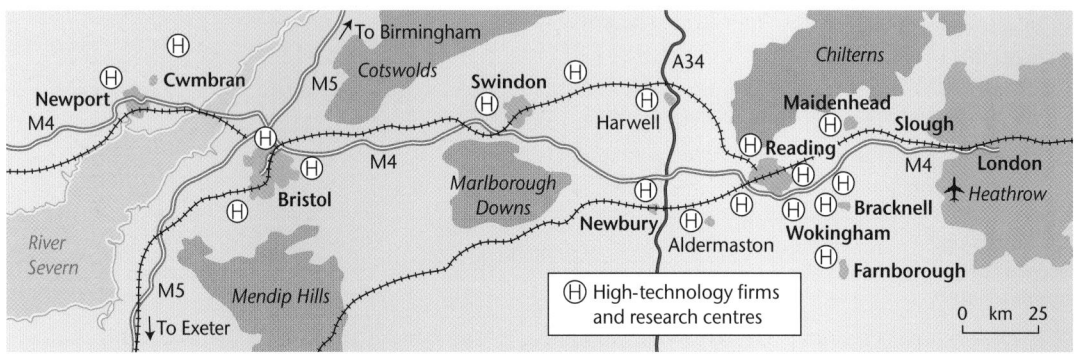

Figure 3.8 The M4 corridor

Source: Redfern, D. (2002) *Human Geography: Change in the UK in the Last 30 Years*, Hodder & Stoughton. Reproduced by permission.

Changes in the service sector

In the 1990s over 70% of the UK workforce was employed in the service sector. Employment in this sector can be divided into three groups: producer services, consumer services and public services.

Producer services exist to serve other organisations. For example, a computer consultancy will advise other organisations on the most suitable computer systems for their accounting. Similarly, a bank will provide a range of financial services to manufacturing industry. It is now common for many organisations, of varying sizes, to subcontract out services that once would have been an internal part of the organisation.

Consumer services are provided by retailers, hotels and leisure/entertainment organisations.

Public services provide for both the producer and the consumer. They cater for social needs (for example health, education and social services), infrastructure needs (for example road building, railway services), administration of local and national governments, regulation of public and private conduct (for example monitoring of environmental pollution or monitoring of flood levels), and national security (for example the armed forces).

Defining the service sector

As in manufacturing industry, rapid technological change is having a major effect on service industries, and it is becoming increasingly difficult to distinguish between manufacturing and service industries. In an economic environment increasingly dominated by computers, global communications and companies which constantly invest in research and development, the categorisation of industry is becoming more difficult. Computer-based industries on a science park may make computers (manufacturing), but will also distribute them, maintain them and seek to update them (services). Some writers have categorised this type of industry as **quaternary** or even **quinary**. Here we consider the key aspects of the 'traditional' service industries as described above.

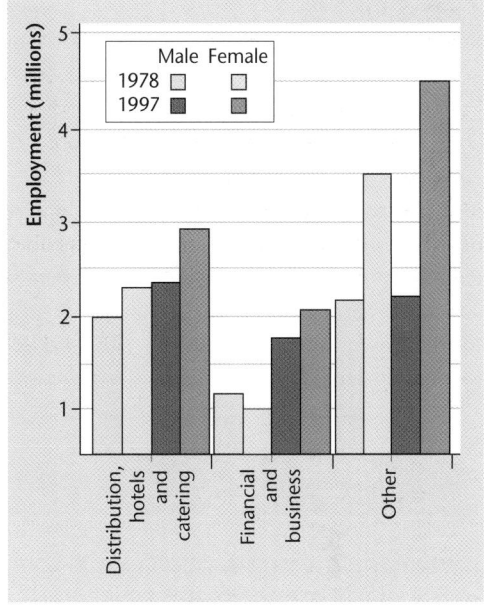

Figure 3.9 Employment in service industries in the UK — changes between 1978 and 1997

Employment in the service sector

Service-sector employment has increased in all areas of the country (Figure 3.9), but concentrations tend to be more notable in certain areas:

- London — 5.5% of the UK workforce is located in central London and the vast majority work in service-related activities. London is one of the world's main centres for finance, business and commerce. Employment has fallen slightly in recent years, mainly because some firms have relocated to other areas in the southeast of England
- the southeast of England — predominantly in business and finance
- the southwest of England — largely based around tourism and the leisure industry
- East Anglia — with significant growth in transport and communication-related activities

Case study *Financial services in the City of London*

Most of the office space in the City of London — the small area on the north bank of the Thames around St Paul's Cathedral — is given over to financial, legal and accountancy-based activity. The causes of this concentration are largely historical — central London has long been a key financial player in world economics. Employment in this area has been increased by the globalisation of services, and the need for international banking, accountancy and legal services.

Continued growth may cause some problems:

- finding enough highly-qualified and techno-logically-skilled people to supply a growing workforce
- providing and extending suitable office space — the costs per unit area of land are some of the highest in the world
- devising and introducing improvements in public transport systems sufficient to cope with the influx of workers into the City

Case study *Growth in the service sector on the south coast*

Coastal resorts, which have traditionally provided leisure and recreation facilities, have expanded into other service activities over recent years.

Bournemouth, for example, is a major conference centre out of season. It has also experienced rapid expansion of its office-based employment. The national headquarters of five companies have moved to the town, including Abbey Life and Chase Manhattan Bank. The latter has built a new purpose-

designed operation on the outskirts of the town. Over 1,000 jobs have been created, half of which were relocated from London.

New housing areas have been built, with a subsequent increase in local purchasing power. Chase Manhattan Bank is seeking to expand its offices in the town, and has received planning permission to do so. Bournemouth is now the fifth most important financial centre in the country.

Case study *Cornwall's tourist industry*

Tourism is by far the dominant industry in Cornwall. Over 4 million people visit the county each year, spending over £900 million. As Cornwall has a remote location, most visitors stay for periods of a week or more, which adds greatly to the economic wealth of the county. Over 30,000 people are employed in the industry, with many more at the peak of the tourist season. However, much of the investment originates from outside the county, so only about one-third of tourist spending is retained within Cornwall.

Employment

Much of the industry operates at low wage rates, and over time these have fallen behind national rates. For men the shortfall is as great as 25%, whereas for women it is 20%. Many of the jobs in tourism and in food and catering are semi or unskilled. In addition much of the work is seasonal. There is a significant fall in business between September and March, when unemployment rates increase. Unemployment is usually 2% less in July than in January, and typically 2–3% higher overall than in the rest of the UK.

Location

The majority of hotels and restuarants are concentrated on the coast at resorts such as Newquay,

Falmouth and Penzance. However, there are local concentrations at some inland tourist sites, such as the Eden Project (a major attraction), the Lost Gardens of Heligan (which attracts 10,000 people per week in summer) and the River Camel trail (Figure 3.10).

The tourist industry is expanding and taking advantage of events such as the solar eclipse in August 1999, which brought in an additional £50 million. Self-catering holidays are expanding, especially out-of-season.

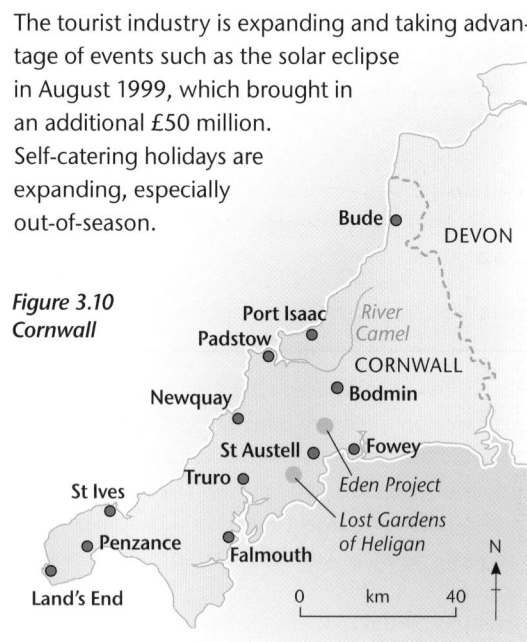

Figure 3.10
Cornwall

Socioeconomic changes in employment

The growth of service industries has had a significant effect on employment patterns (Figure 3.11).

Increasing numbers of **women** are employed in service industries, often making use of flexible working hours or part-time work to fit in with childcare (Tables 3.3 and 3.4). **Flexitime** is a common feature of service industries. It allows firms to respond easily and cheaply to peaks in demand, for example Christmas shopping at supermarkets, or hotels in the summer.

Year	Males	Females
1978	13.4	9.4
1991	14.4	11.2
1997	11.5	11.3

Table 3.3 Gender changes in total employment in the UK (millions)

Region	Male		Female		Table 3.4
	Change in numbers employed (thousands)	Gain/loss (%)	Change in numbers employed (thousands)	Gain/loss (%)	*Gender changes in regional employment (standard UK regions), 1976–94*
North	−127	−14	+69	+13	
Yorkshire and Humberside	−111	−8	+193	+13	
East Midlands	+20	+2	+203	+32	
East Anglia	+74	+15	+164	+57	
Southeast	−2	0	+729	+23	
Southwest	+123	+11	+325	+48	
West Midlands	−135	−7	+156	+17	
Northwest	−254	−14	+95	+8	
Wales	−37	−5	+193	+21	
Northern Ireland	+37	+9	+84	+37	

Note: Percentages have been rounded to the nearest whole number, so a fall of 2,000 male employees is rounded to zero — no change.

Source: Regional Trends, Office for National Statistics.

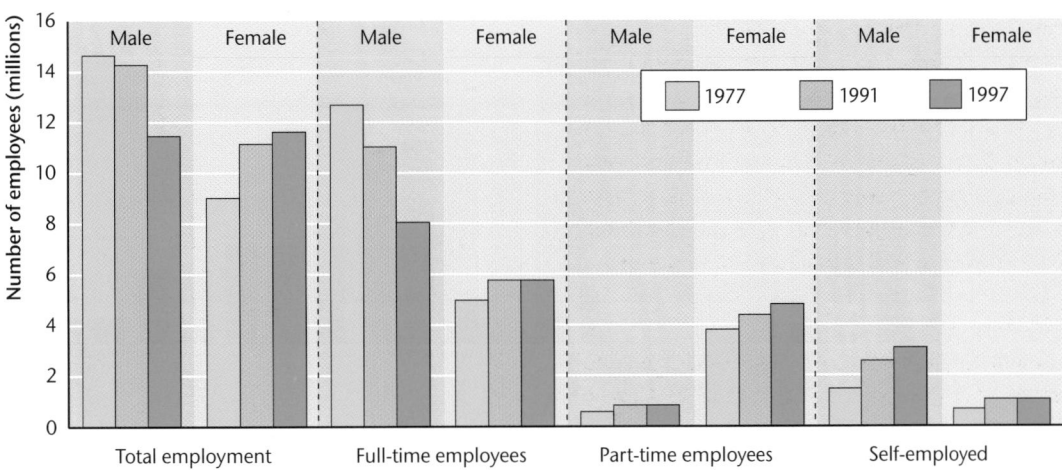

Figure 3.11 Changes in the nature of employment in the UK, 1977–97

Catering and hotel work are obvious examples of services that employ more women than men, but others are following suit, for example teaching and work in science and business parks. Extra security systems are being put in place by many such firms to enable female staff to come and go safely at night. In some service industries, men are beginning to take on what were once thought of as 'female' roles, for example cleaning services in schools and colleges, and nursing.

Worksharing is a feature of many service industries. This involves two people sharing one job. It is ideally suited to working mothers who wish to retain some working life and income as well as spending time at home with young children. Jobsharing is easier to carry out in some industries than others. Those that require continuity of interpersonal relationships, such as doctors and teachers, find it harder than, say, administrative office workers.

Homeworking and **teleworking** are also increasing with the development of telecommunications technology, including e-mail and the internet. A worker can be based at home and use this technology to communicate with colleagues and customers. Industries which make use of homeworking include design-based work, journalism, advertising and call-centre work (Figure 3.12).

The rise in **self-employment** is perhaps the most significant change in employment patterns in recent years (Figure 3.11). Self-employment now accounts for 12% of the UK workforce. For many years the majority of self-employed workers were either domestic repairmen or workers in the construction industry such as plasterers, plumbers, painters and decorators. Today, self-employment has moved into many other areas, from haulage work (many lorry drivers are sole proprietors) to insurance and salespeople. As these people carry their work around with them, or base their paperwork activities at their home, they have little geographical impact. However, their place within the national economy is important.

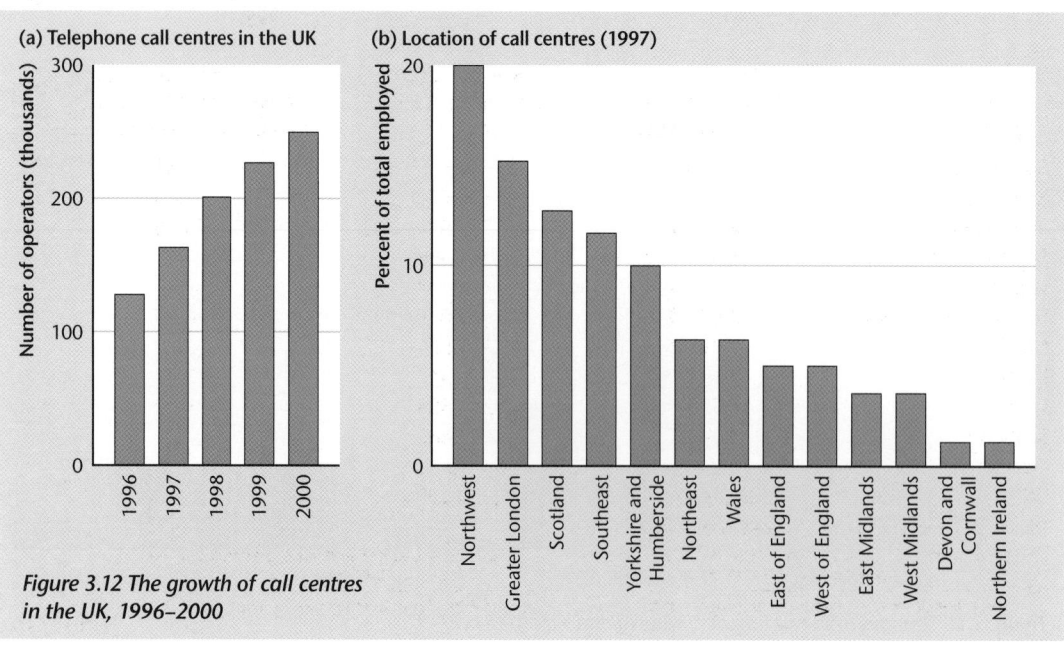

Figure 3.12 The growth of call centres in the UK, 1996–2000

Overseas investment and its effects

Overseas investment is the capital attracted to a region from beyond its boundaries. The UK has attracted such inward investment from a number of countries, including Japan, South Korea, the USA and other European countries.

The main effect has been to create new industries in many areas of the UK (Figure 3.13 and Table 3.5). Some of the larger investments include car assembly plants such as Nissan at Washington and Toyota at Burnaston. Other areas of major overseas investment have been in the Central Valley of Scotland, known as silicon glen, where transnationals based in the USA (for example Hewlett Packard) and Japan (for example Panasonic) have made large investments.

Who benefits from such investments?

The investors themselves benefit for the following reasons:

■ Access to relatively cheap, highly-skilled labour. Rates of pay in the UK are now below the European average, and this is a major consideration for any potential investor.

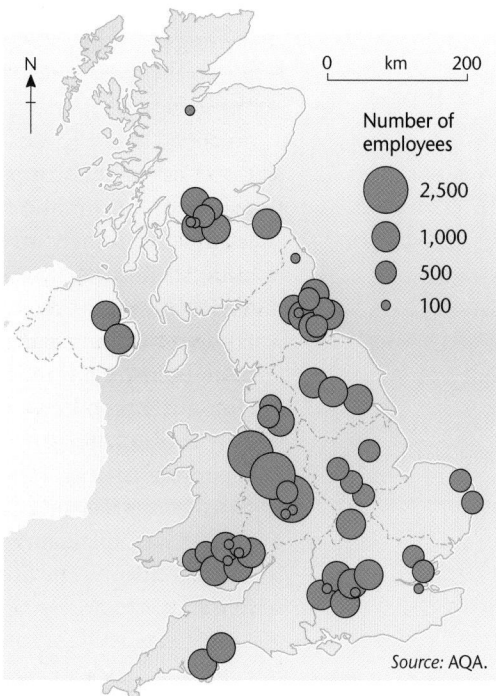

Figure 3.13 Location of employment in Japanese manufacturing companies in the UK, 1998

Table 3.5 Employment in Japanese manufacturing companies in the UK, 1998

Region	Number of employees in Japanese companies	Number of all manufacturing employees	Location quotient of employment by Japanese manufacturing employment
Southeast	6,380	1,417,778	0.41
East Anglia	923	184,600	0.45
Southwest	2,502	390,936	0.58
East Midlands	3,048	507,999	0.54
West Midlands	8,553	641,578	1.20
Yorkshire/Humberside	3,020	487,006	0.56
Northwest	1,998	665,999	0.27
North	9,369	289,167	2.92
Wales	15,763	248,384	5.72
Scotland	5,747	428,880	1.21
Northern Ireland	2,241	101,863	1.98
UK	59,544	5,364,190	–

The location quotient (LQ) is calculated as follows:

$$LQ = \frac{\text{Number of employees in Japanese manufacturing companies in a given region}}{\text{Number of all manufacturing employees in that region}} \div \frac{\text{Number of employees in Japanese manufacturing companies in the UK}}{\text{Number of all manufacturing employees in the UK}}$$

Source: AQA.

- Limited financial controls on profit. Recent UK governments (both Conservative and Labour) have become increasingly business friendly, and are keen to allow foreign transnationals fair systems of profit management.
- Sterling is a world currency with a strong exchange rate, so investments are not a risky prospect in this country.
- The government gives packages of incentives to peripheral economic areas as described above (page 48).
- Good access to European markets. The European Union alone has a potential market of 450 million people, many with high personal incomes. This market is attractive to new investors from overseas who can foresee huge growth in sales. There is also the potential new market of Eastern Europe and the former Soviet Union.
- European Union regulations stipulate that a proportion of the finished product of an industry has to be manufactured within the countries of the European Union if it is to be sold there. This has forced many transnationals, particularly those from the newly industrialised countries of the Pacific Rim, to support local component suppliers, or to encourage branches of their home-based component suppliers to relocate in Britain.
- The UK has a pleasant environment, with a mild climate, attractive scenery and a strong history. The excellence of UK golf courses is even cited as a key factor in influencing foreign investors.

Governments (national and local) benefit for the following reasons:
- Inward investment by a transnational replaces the shrinking traditional manufacturing base of a region, thus stimulating regional regeneration.
- Such investment creates a multiplier effect locally. As people's incomes increase there is more spending on local goods and services, and the general levels of wealth and investment in infrastructure increase.
- The export base of the country is strengthened, improving the balance of trade. Exports of manufactured goods increase, creating a more balanced relationship with imports and allowing some repayment of international debt.
- Transfer of technology and product development are more effective (page 48).
- For national government a clear benefit is the reduction in payment of unemployment benefits to former workers and their dependants.

Unemployed people in the area of investment benefit for the following reasons:
- Jobs are created, not only by the company locating in the area, but also by its suppliers — component suppliers, subcontractors, service companies, transport, hotels and catering.
- Living standards are raised as employment increases personal incomes.

Who loses from such investments?

Investment in an area by a large foreign company does not always have a beneficial effect, either for the local area or for the country as a whole. Some of the problems are described below:

- Transnationals may decide to leave an area once initial aid packages no longer apply. This has been the case in the northeast of England where initial investments by Fujitsu in 1991 and Siemens in 1996 have now closed down. Siemens received a very large subsidy on arrival, but this appeared to have little long-term effect (Figure 3.14).

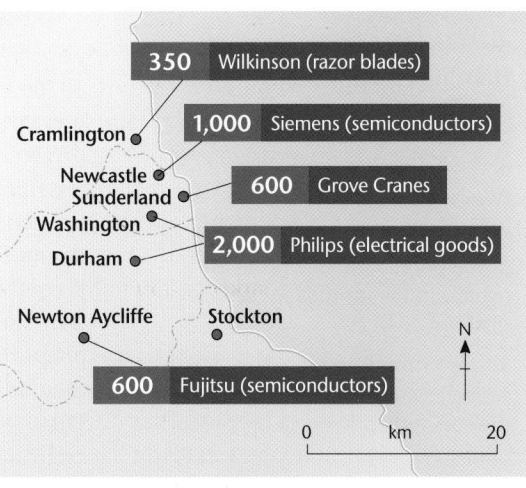

- Home-based industries producing similar products come under increased competition, in terms of both relative costs and product design. The motor vehicle industry illustrates this well. The overcapacity of car production within the country forced the sale of the Rover group by the German car manufacturer BMW. Rover was subject to a management buy-out but stopped producing vehicles in 2005. However, it is clear that the inroads into market share caused by the arrival of both Toyota and Nissan were key factors in its demise.

*Figure 3.14
Job losses from transnational companies in northeast England, 1998*

- Wages are sometimes lower than in domestic industries. Much of the employment created is part-time and female-orientated, with high annual turnover rates of staff. In the northeast of England the mobile telephone company Orange employs over 3,000 people in customer call centres. The firm has a high turnover of staff, and unions have expressed concern about stressful working environments.

- The positions created are often merely 'screwdriver' jobs, with low skill development and few long-term career prospects. Managerial positions, on the other hand, are filled by staff from overseas or from other parts of the UK.

- The local area often has to bear the costs of training. Local Training and Enterprise Councils (TECs) and local governments are frequently contracted to cover the costs of training and retraining of former employees of traditional industries.

Changes in the social and demographic environments in urban areas

Demographic change

The population pyramid for the UK for the year 2001 (Figure 3.15) shows a relatively straight pyramidal shape, with some slight bulges and indentations. These slight variations can be explained by the circumstances at the time each age group was born, and by later factors.

The bulge of people in their 30s demonstrates that there were slightly higher birth rates during the 1960s than the 1970s. There may be two reasons for this. First, the 1960s was a period of rising national prosperity and of increases in personal incomes. In the 'swinging sixties' there was also a lessening of sexual taboos and an increase in freedom for women. Second, this was the time at which people who themselves had been born in a 'baby-boom' following the Second World War were entering their fertile years. As there were more fertile individuals, more babies were born. The people who were young parents in the 1960s are now in their 50s, and also feature as a slight bulge on the pyramid. These examples demonstrate that population growth is cyclical, and to some extent changes can be predicted, as long as social norms are retained.

Two further points illustrate the changing nature of a population structure. There is a relatively large number of people over the age of 80 in the pyramid. The reasons for this are again complicated. People born in the period 1910–20 were often part of large families. It was traditional as well as functional to have a large family. Many children died in infancy, and a large family would act both as a source of income and as a form of security in old age. As the century progressed, however, death rates fell (despite the devastating impact of the First World War) and improvements in medicine have meant that some of these people are living into their 80s or 90s, particularly the women.

The second point concerns the younger part of the pyramid. As mentioned earlier, a relative increase in birth rates took place in the 1960s. The people born then would have become fertile in the 1980s and 1990s, and so a cyclical increase in birth rates would be expected at that time. This has not happened to the extent predicted. Once again social norms have modified. In recent decades young adults have been less willing to have children. Various factors account for this — increased availability of contraception, abortion and sterilisation, a growing importance attached to material possessions (houses, cars, holidays), and the

*Figure 3.15
UK population
pyramid, 2001*

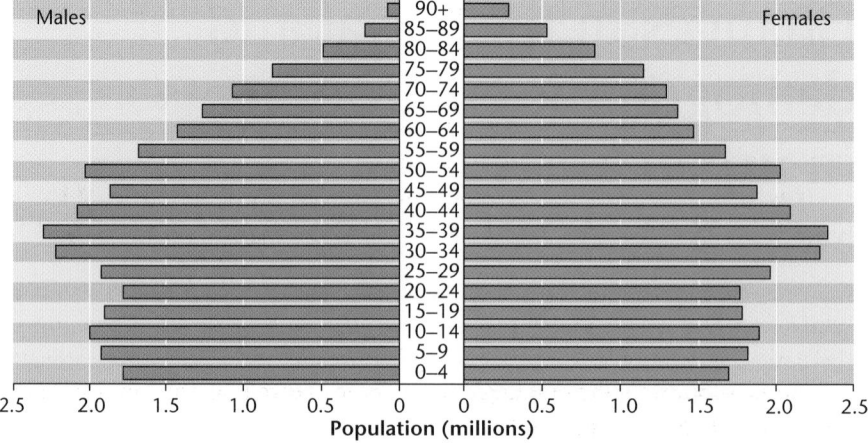

desire of women to have careers. With hindsight, it is possible to add another factor — the economic recession of the late 1980s and early 1990s which left many young adults financially insecure, and forced many women to become the main breadwinners as male employment in both mining and manufacturing industries fell.

The net result is that in the early part of the twenty-first century, the UK has an ageing population. The proportion of the population aged 50 and over has increased significantly since the start of the twentieth century. There has also been a rise in the 'very elderly' — people aged 80 and over. Population projections suggest that by 2021 there will be more than 3 million over 80s — 5% of the population. At the same time the number of people under 16 has been progressively falling. It is anticipated that in 2007 the number of people aged 65 and over will exceed those under 16 for the first time.

Changes in housing composition

The character, distribution and availability of housing stock are important elements of urban areas in the UK. There are two main providers — the private sector and the public sector, and individuals can now rent or buy within either of these sectors.

It has been government policy since the 1940s to improve the quality of housing for all groups in society. Until the late 1970s there were two main thrusts to this policy:

- The private sector was encouraged to provide dwellings for ownership by individuals or families. Private-sector rented accommodation was therefore progressively squeezed out. For example, between 1950 and 1980, private house ownership increased from 30% to 56%, and private rented housing decreased from 52% to 12% of all housing tenures.
- The public sector was encouraged to provide rented housing, mainly in the form of local authority council housing subsidised financially by central government. This resulted in the construction of large council estates on the outskirts of urban areas, and many council-owned high-rise flats in inner-city areas. Surprisingly, the proportion of council housing tenure did not increase greatly between 1950 and 1980 (from 18% to 32%). This can be accounted for by the fact that much of the existing council accommodation was demolished to make way for the new estates and flats.

After Margaret Thatcher's Conservative Party took power in 1979, four major changes in housing policy came into operation:

- a reduction in funding for local authority housing by central government
- restriction on using the money raised from council house sales to build new council dwellings or renovate existing housing stock
- the right to buy legislation of 1980 gave council tenants the opportunity to buy their homes at discounted prices
- Housing Associations were created — private bodies in receipt of government financial support for the building or renovation of dwellings for rent (for more detail see chapter 7, page 168).

These four policies caused an increase of more than 50% in annual building rates by private developers, while completions by the public sector fell by 60% (Figure 3.16). The changes were not evenly distributed across the country. Demand for private housing rocketed in the economically prosperous areas of the southeast. Here, the price of houses rose more rapidly than inflation, fuelled partly by rising land costs. In 1987, for example, the price of the land constituted 53% of the value of the dwelling price in the southeast, whereas in Yorkshire and Humberside it was only 12%. Consequently, the difference in housing prices widened greatly between the regions. The north–south divide was increased, with the popular view that there were significant problems in providing enough housing in the southeast. The construction of new private properties increased greatly in the Greater London and southeast areas, and in parts of East Anglia and the East Midlands.

Figure 3.16
Houses built in
England, 1975–2000

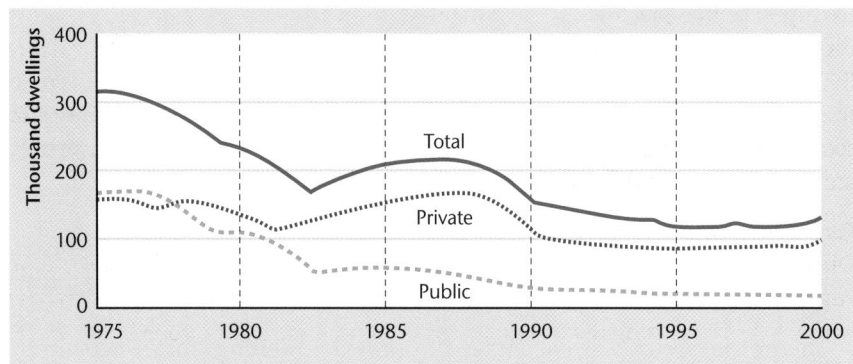

Changes in the supply of council housing

There are two ways in which a person can obtain a council house — either from the supply of newly-built houses or from existing properties which become vacant. During the 1980s the government steadily reduced the amount of money an authority could use to build new houses. In 1981, a local authority could use 50% of the funds raised by council house sales to build new homes, but by 1986 it had been reduced to 20%. In addition, by 1986 a further 650,000 council houses had been sold to their occupants, effectively removing them from the stock available.

Initially this did not cause too great a problem as many people who lived in council houses wanted to buy a house from the private sector. And an increasingly elderly population on the former council estates meant that deaths of tenants and movements into sheltered accommodation created sufficient vacancies for new tenants.

However, with the onset of the economic recession in the late 1980s demand for the falling numbers of council houses increased greatly. Applications for housing from homeless people in English local authorities increased from 220,000 in 1986 to 320,000 in 1996. This demand was not evenly spread: London was the area most severely affected.

The role of council housing then came under closer scrutiny. Was it increasingly to accommodate the vulnerable and the disadvantaged? New factors were coming into play — the rise in the number of divorces, the increase in the number of elderly people living alone, the growth of mortgage default, and the discharge of those with special needs from institutional care. As council house supply dwindled, local authorities increasingly concentrated on those they had a statutory duty to house, and those with greatest need. Council housing has become more and more 'welfare tenure'.

Waiting lists

All local authorities have more prospective tenants than vacant properties, which means there are waiting lists. Most waiting list schemes involve awarding points to establish priority based upon the urgency of the need for housing. Points schemes take into account a number of factors, each of which has a certain weight, and the points from each are added together to decide who should be housed first. Examples of these factors include whether or not children are involved; whether the existing dwelling is damp, or has no bathroom; whether a family is currently living with in-laws in an overcrowded house; and how long a name has been on the list. However, the weighting of factors varies between schemes — it would seem that the attitudes of local councillors and/or officials towards housing needs are more important than consistency and fairness.

Peripheral council housing estates

During the 1950s, 1960s and 1970s many local authorities built estates on the edges of their urban areas. These were to house overspill population and people who needed rehousing due to inner-city slum clearance. The estates consisted of typically designed council houses — semi-detached, uniform red brick with or without rendering, metal-framed windows, with gardens, limited garaging, and distinctive sequenced front door colours. Within these estates there were also system-built tower blocks and maisonettes made of prefabricated materials (Photograph 3.4).

Nigel Dickinson/Still Pictures

Photograph 3.4
A council estate
in Birmingham

These types of housing were relatively inexpensive to build and were a cheap way for local authorities to meet housing demand. Planning controls were limited, and construction was done in great haste. Over time evidence has emerged of corruption by local authority leaders in issuing lucrative contracts for both the design and the construction of these properties.

It seemed a good idea at the time to build these estates, using greenfield sites to provide decent homes, with open space and public amenities, for the poorer elements of society. However, the result was that communities the size of small towns were created on the outskirts of cities without proper facilities and without affordable transport links to the city centre or to places of work.

During the 1980s and 1990s the physical fabric of these estates deteriorated markedly, and the environmental quality of the streets and open spaces became poor. Maintenance costs escalated to the point where, for many estates, demolition has been the best option. The houses and flats have not proved popular under the right-to-buy legislation, and so many have remained in rental tenure. This means that such estates contain above average proportions of the more vulnerable groups in society — low-income households, the unemployed, and the elderly living in poverty. They have become the centre of a whole range of social and economic problems.

Case study *Housing estates in east Middlesbrough*

East Middlesbrough has 11 estates within an area of 5 km² containing over 10,000 homes, only 20% of which are privately owned. The housing is pre-dominantly two-storey terraced and semi-detached with gardens. Some houses have had central heating and double-glazing installed. Most of the three- and four-storey maisonettes have been removed, but the high-rise flats remain. These have had some improvement but there are still problems with damp and security.

The problems of the area are not all visible, and have more to do with the people than the property. The biggest problem is unemployment — male unemployment is 30–40%, caused by the decline of the traditional local industries of shipbuilding, chemicals and engineering. Unemployment means low income and, as there are also a lot of retired people living in the area, 70% of the council tenants are in receipt of housing benefit.

Social problems such as marriage breakdown, low academic attainment, drug abuse and petty crime are endemic. An additional problem is that of isolation —

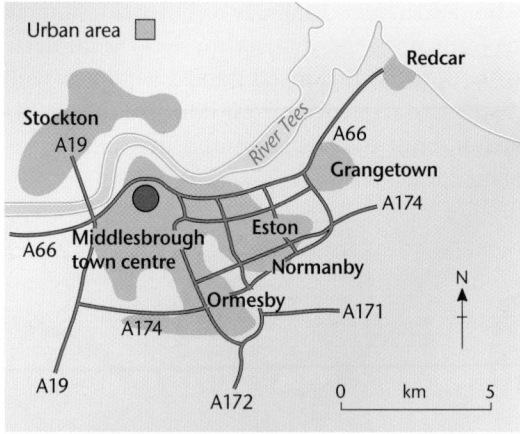

Figure 3.17 Housing estates in east Middlesbrough

many of the estates are 6–8 km from the city centre, and transport is an unaffordable luxury. Car ownership is low. Facilities are provided but they are restricted and choice is limited. Each estate tends to look inwards rather than outwards, and this means that its residents have low aspirations. Few people who do not live in these estates will have cause to visit them.

The boom in private housing construction

As described earlier in this chapter, there has been a significant increase in the numbers of high-tech companies, both manufacturing and services, in the southeast of England, in particular along the M4 corridor. The rise in employment in such industries has created a demand for private owner-occupied housing in counties such as Berkshire.

People attracted to the area by such employment tend to demand high-quality housing, and house builders have come under increasing pressure to build for the expensive 'trade-up' market. Many of these houses have been built on very large sites. Developers have been successful in pushing for land release in large blocks and central government has overridden the objections of local conservationist councils. During the 1980s, Berkshire County Council came under pressure to release land for up to 8,000 homes.

Three issues have arisen from this expansion of owner-occupied housing:

- house prices within the area have increased at rates faster than those in the rest of the southeast. These high prices probably reflect the high salaries being paid to the employees, rather than any absolute shortage of housing stock
- high house prices have made it difficult for people moving into the area from less wealthy areas, and the less affluent living locally, to afford adequate housing
- the drive for owner occupancy within the area has led to an increase in sales of council housing, further reducing the number of houses in that sector

The recent growth of private housing has tended to increase rather than decrease social tensions in the area. Campaigners who are against the construction of new housing on environmental grounds have become more vociferous. Their objective is to preserve the rural nature of the landscape as much as possible. On the other hand, industrialists point to the need to provide adequate and appropriate housing for the workforce moving into the area. Development control, in the form of green belt land, is increasingly being applied, but this tends just to move the problem to areas where opposition is less well organised, and away from areas of highest demand. For example, Slough has some of the county's most acute housing problems and has Berkshire's largest local authority house-building programme, yet its growth is restricted by the green belt.

Where and what shall we build in the future?

The government's population projections suggest that, between 1996 and 2021, the population of England will rise by 3.4 million to 52.5 million, an increase of 7%. At the same time, the number of households will rise by 3.8 million, an increase of 18%, giving a total of 24 million households. Most new households want their own housing unit, and this creates pressure on land.

There are two main reasons why there is likely to be a faster growth in households than in population:

- The increase in the adult population, which accounts for 77% of the growth in demand for households. This is due to the change in the age structure of the

population, with a fall in the proportion of people in younger age groups and a rise in the proportion of those in older age groups.

■ Changes in the way people choose to live — in particular more divorces and later marriages. Of the 3.8 million new households, 71% will be single-person households.

Some people argue that population predictions have been wrong before and we should not build new houses until we are certain that the demand is there. However, if we wait there is a danger that there will then be a shortfall of suitable housing. This could result in more crowded households and house-price inflation — a real problem for young people or those on low incomes who want to buy their own homes.

The following solutions to this issue have been suggested:

■ Increasing the number of people living in homes that already exist. There are 800,000 empty houses in England, over 80% of which are in the private sector. In the north of England many of these are abandoned houses that the owners cannot sell. Another tactic would be to provide tax incentives to encourage people to take in lodgers or share their homes. In some areas empty council housing is being improved. The advantages of all these practices are that they are cheap and do not use up new land.

■ Building new houses on both greenfield sites (in rural areas) and brownfield sites (in cities). The government now talks in terms of building houses on 'previously developed land', most of which is in urban areas. It excludes farmland and land in cities such as parks, allotments and playing fields. The government has set a target that, by 2008, 60% of new houses built each year will be on such land. Some say this is too low. However, there is still a mismatch between where land is available (the industrial areas of northern England) and where the pressure for housing is greatest (southeast England).

■ Building new homes near to work and leisure activities. They need to be within the 'pedshed' — 10 minutes walk from railway stations and bus stops. To achieve this, innovative developments will be required to create high-density housing that is socially and environmentally appealing, and affordable.

■ Allowing building in rural areas and small towns on land that is not 'previously developed'. Greenfield sites are cheaper to develop than brownfield, and there is still plenty of rural land available. In 1991, 11% of the land in England was in urban use. By 2016, this will rise to only 11.9%. Much agricultural land is not being used productively — for example over half a million hectares were 'set aside' in the 1990s under EU subsidy.

Of the 3.8 million new households it is estimated will be required by 2021, a considerable proportion will be built in rural areas and small towns in the southeast of England, where demand is greatest. The main proposed areas of development are the Thames gateway, Ashford in Kent, Milton Keynes and the Cambridge to Stansted corridor.

Urban segregation of social groups

The segregation of social groups means distinct groups of people living almost entirely separately. It produces clusters of people with similar characteristics in separate residential areas within an urban area. Segregation may occur according to wealth, age group or ethnicity.

The reasons for segregation are complex, but can be summarised in the following ways:

- Social groups of various types may choose to live in separate areas or they may be constrained to do so. The reality is often a combination of the two. For example, ethnic minorities often dominate certain areas of a town because these were the areas with the cheapest accommodation when the original migrants arrived and worked in low-income jobs. With time the areas developed characteristics and support mechanisms which made them more attractive to new migrants to the town. Within the areas people retain their own identity, and preserve their traditions and faiths.
- Once created, social divisions tend to be reinforced through lack of contact or interaction with other social groupings. Certain views are held about the elderly or the poor, and there is no real understanding of one group by the rest of society. Existing differences become exacerbated, and fear or suspicion add to the feeling of 'them' and 'us'. Segregation increases as a group feels more safe in its own area. Social tensions may arise, and sometimes manifest themselves in extreme ways, adding to the lack of integration.
- It is surprising how often a physical barrier — for example a main road or railway line — encourages social segregation. Such a feature may separate a council estate from a private housing estate, or a fence may separate a gentrified up-market housing area from its poorer surroundings. These barriers mark out relative territories and act to reduce interaction.

The effects of geographical segregation

Segregation based on wealth

Housing

Many urban areas are characterised by poorer and richer people living in very different neighbourhoods. The general pattern has been one of the poor living in the inner cities, and the rich living in the outskirts or suburbs. This arose partly because of supply factors — small building firms acquiring land would tend to build housing of a similar type to that surrounding the plot, and so large areas of similar housing developed. Demand factors also played a part — people usually want to live among other people who share their values.

In the second half of the twentieth century the pattern was complicated by the building of peripheral council estates with concentrations of poorer people. As described in the case study on east Middlesbrough (page 64), a great deal of council house development has occurred on the edges of cities, for example

Seacroft (Leeds) and Newall Green (Manchester). This has led to the spread of poorer people into the periphery of many towns and cities.

More recently the gentrification of some inner-city areas has inverted the original pattern. Now richer people live very near to poorer people in the inner cities, but the amount of interaction between them is still low. Gentrification has occurred in the riverside area in Leeds and the Castlefields area in Manchester, for example. Pockets of high-status housing exist in inner-city areas where there is a favourable environment or where redevelopment has made the area desirable to live in again. The London Docklands with its new marinas, luxury converted warehouses and spectacular river views is another well-documented example of this phenomenon. The following is a common description of the type of property in these areas:

> It is the open plan space and the original structure of the warehouse such as the high ceilings, large windows, exposed brickwork, highlighted by the well considered modern inserts of the designers, that make home life comfortable. With its own landscaped courtyard and canal inlet, the area has secluded and peaceful outdoor areas. An on-site gymnasium means that leisure time can be filled without leaving the gates of the site.

Despite these changes over the past few decades, areas on the edge of many cities are still occupied by the wealthy, for example Alwoodley in Leeds, and Bramhall, Cheadle Hulme and Alderley Edge in Greater Manchester. Alderley Edge houses commuters who work in Manchester and further afield, and who command high incomes for their work. Houses regularly sell for substantially more than houses in other areas of Manchester. Car ownership in the area is high at 86%, compared with the national average of 67%.

Similarly, many inner-city areas are still poor, for example Harehills in Leeds, and Ancoats in central Manchester. Ancoats lies to the north of Piccadilly station and consists mostly of council housing in the form of two-storey houses,

Figure 3.18
Examples of census data from Manchester (2001) to illustrate variations in wealth across a city. Each area represents a grouping of enumeration districts with common characteristics. An average for that area has been calculated

Census data reproduced with the permission of the Controller of HMSO © Crown Copyright

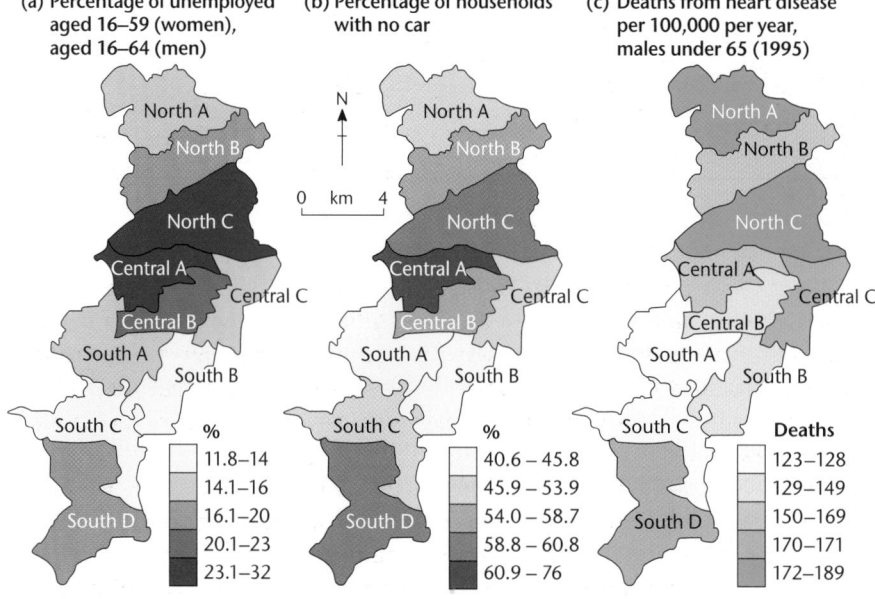

maisonettes and high-rise flats. Car ownership in the area is well below the national average at 16%, indicating the low levels of personal wealth. Unemployment is high and environmental quality is poor, with much derelict land. Restoration of the area is progressing, but it is still an example which fits the traditional view of a poor inner-city area (Figure 3.18).

Some prestigious inner-city areas have long been popular with certain wealthy groups. In London, the areas of Mayfair, Belgravia, Chelsea and Kensington have never lost their status. In recent years these have been added to by other fashionable locations within the city. Notting Hill, Islington and Primrose Hill are becoming increasingly popular with the rich and famous.

Education

Schools tend to be evenly spread, reflecting population coverage, although higher concentrations of older Victorian schools are still found in many poorer inner-city areas (Photograph 3.5). Most of these still operate as primary schools, while secondary schools in such areas are usually built from the same modern and cheaper building materials as the houses around them.

Photograph 3.5
A Victorian primary school in Nottingham

Maggie Murray/Photofusion

Out-of-town council estates usually have purpose-built comprehensives similar in design to secondary schools in the inner cities. They consist of rectangular buildings, up to three storeys in height, with a spread of flat-roofed single-storey buildings around them. Many of these schools were built in the 1950s and 1960s and now need a great deal of maintenance work, particularly in terms of flat-roof drainage and asbestos removal.

Each major new housing development tends to have its own primary school. However, this is not the case for new, small, private housing developments in the vicinity of existing housing areas. Such developments can place pressure on local primary schools. In some cases, it is the good reputation of the school that has

encouraged the building of the housing development. Access to a 'good school' is often an important factor when people are making decisions about where to live, and can actually influence house prices. This is a good example of the close inter-dependence between private sector development (housing) and public sector provision (education), for which it is difficult to have overall strategic planning.

The quality of educational facilities therefore varies within a town. Older schools tend to be found in poorer areas and are often in a poor physical state, with crumbling buildings and vandalism. Out-of-town comprehensives in low-income council estates face similar problems. There is a great danger that some of these areas are becoming educational backwaters with low educational attainment, low rates of attendance and high levels of truancy. Recent governments have adopted strategies to try to counteract these trends, for example making use of Single Regeneration Budget (SRB) funding, and Objective One funding. The full benefits of these have yet to be seen.

In those parts of urban areas with higher personal incomes the major problem appears to be demand for places, at both primary and secondary school level. More private schools and schools that have opted out of local authority control (formerly known as grant-maintained schools) are established in these areas. There is currently a range of initiatives that schools can subscribe to, such as technology college and language college status, which are more likely to be found in the more affluent parts of an urban area.

Health care

There is often a greater demand for medical facilities in poorer areas, possibly because there are more young children and elderly people, and a higher population density. Health centres in such areas are likely to be purpose-built, single-storey buildings linked to either a small shopping centre or a residential nursing home for the elderly. Some larger medical centres of this type may offer out-patient facilities from the local hospital, and carry out minor surgical operations. In wealthy areas medical facilities are less likely to be purpose-built and there is more conversion of private houses. It is common to find health-visitor and small-scale clinical treatments available on the ground floor or in basements, with the appointment-based general practitioner on the first or second floors.

Hospital location is more variable, but there is still a concentration of National Health Service (NHS) hospitals in the older, more central parts of a town or city. Here they are accessible to the poorer population (public transport access) and the more wealthy (car access) alike. Modern hospitals tend to be built on the edge of a town. Here they have the same benefits and problems as out-of-town shopping complexes. They are more accessible by road transport, and there is space for expansion and for car parking. However, they are less accessible to those with lower incomes who don't have cars. Ambulance provision is not normally affected by these developments and ambulance depots tend to be more evenly spread within an urban area.

Private hospitals have a variable locational pattern. Some are in pleasant wealthy suburbs, others are on major routeways out of town, and some are very close to NHS establishments because they need to share expertise and personnel.

Some private facilities are located well away from the urban area altogether. There is also a mixture of purpose-built provision in some areas and conversion and extension of existing property in others. A geographical pattern of this type of medical provision is difficult to see.

Segregation based on age

As the number of elderly and the age to which they live increases (Figures 3.19 and 3.20, and Table 3.6), so some degree of segregation has taken place, particularly in terms of housing. Many elderly people have to decide whether or not to leave the family home when they are left on their own or have difficulty caring for themselves. Old people living in council houses on their own have found that very often they do not have a choice. Housing departments move them out into sheltered accommodation or nursing homes because their houses are required for families.

Segregation based on age has manifested itself in a number of ways in towns in this country.

- On council estates, it is common to see clusters of purpose-built bungalows occupying one small part or parts of the estate. In some areas maisonettes with security access have been built. This type of housing for elderly people is provided in the belief

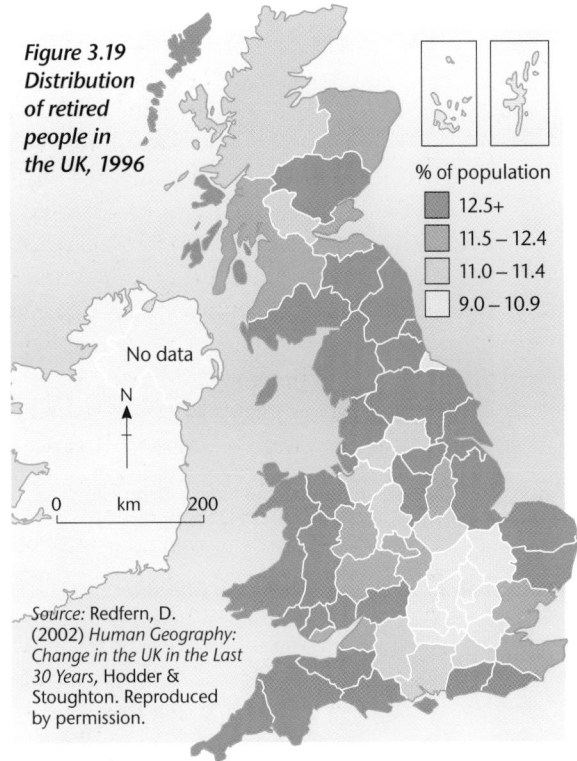

Figure 3.19 Distribution of retired people in the UK, 1996

% of population
- 12.5+
- 11.5 – 12.4
- 11.0 – 11.4
- 9.0 – 10.9

No data

N

0   km   200

Source: Redfern, D. (2002) *Human Geography: Change in the UK in the Last 30 Years,* Hodder & Stoughton. Reproduced by permission.

Age (years)	% of population
0–15	20
16–64	63
65+	17

Table 3.6 Age structure, Stony Stratford, Milton Keynes, 2001

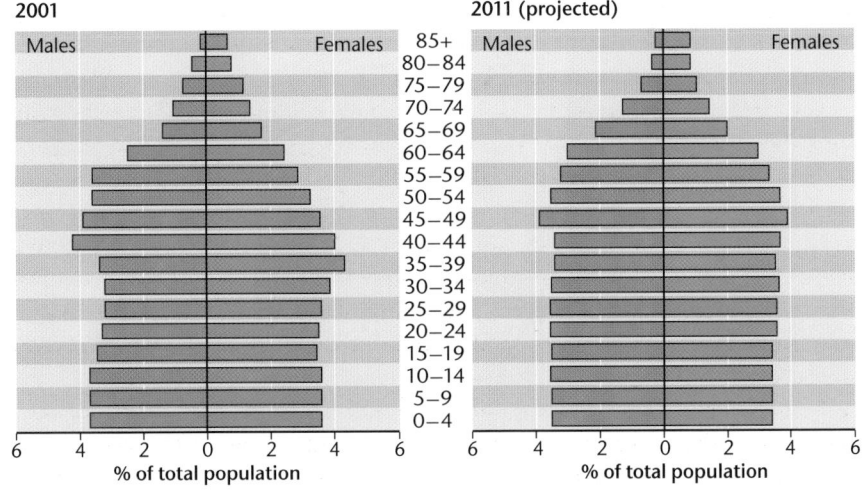

2001

Males   Females

2011 (projected)

Males   Females

85+
80–84
75–79
70–74
65–69
60–64
55–59
50–54
45–49
40–44
35–39
30–34
25–29
20–24
15–19
10–14
5–9
0–4

6   4   2   0   2   4   6
% of total population

6   4   2   0   2   4   6
% of total population

Figure 3.20 Population pyramids for Milton Keynes

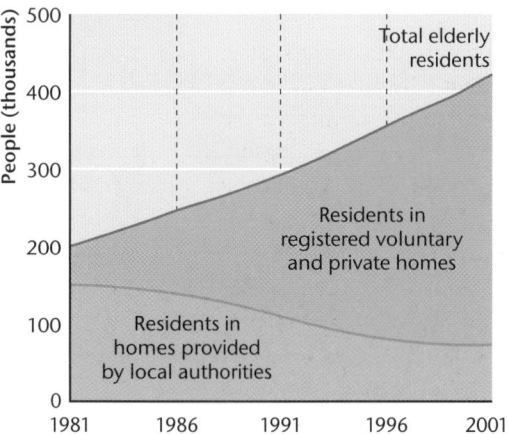

Figure 3.21 Changes in numbers of elderly people in residential accommodation, UK, 1981–2001

Photograph 3.6 a–c Nursing homes in Doncaster

that it is best for them to live in the community for as long as they are fit enough. They are often people who have lived in the area for many years. They have friends and relatives living locally and they are integrated into social functions such as the church or social clubs.

■ A more recent provision has been sheltered accommodation — a complex of flats or units with some shared facilities, overseen by a warden or manager. In some cases purpose-built blocks of flats, some for single people and some for married couples, have been constructed. A mobile warden may oversee a number of complexes. The location of these facilities is only just beginning to establish a pattern in some urban areas.

■ Nursing homes have been increasing in number to cater for the growing number of elderly people who have difficulty looking after themselves. Initially, both local authorities and private developers provided such housing, but local authorities have been cutting back their provision (Figure 3.21). In many urban areas, concentrations of nursing homes are becoming clear. They are often in both inner and outer suburbs, in areas where large Georgian and Victorian houses can either be converted or extended for this purpose (Photograph 3.6). Close links with medical provision are also a factor, and some of the most financially successful nursing homes are located on a main road within a town so as to facilitate the arrival of ambulances.

Segregation based on ethnicity

Ethnic segregation is the clustering together of people with similar ethnic or cultural characteristics into separate urban residential areas. There are numerous examples of this in the UK. The largest ethnic minority in the country is the Indian population, which forms 27% of the total ethnic minority population. The next largest is the Pakistani ethnic minority (17%), closely followed by the black Caribbean (15%). Smaller, but still significant, ethnic minorities of Bangladeshi, black African and Chinese people also exist in the country.

Ethnic minorities are concentrated in the major urban areas of the country, particularly London and the southeast, the West and East Midlands, Manchester and West Yorkshire. Over 50% of ethnic minorities live in the southeast, which has only 30% of the white population, so the concentration here is highest. A significant proportion of ethnic minorities consists of people born in the UK, descended from migrants who arrived from the former Commonwealth countries in the 1960s and 1970s.

There are some slight variations in the geographical distribution of ethnic groups which owe their origin to factors in the early days of immigration. For example, there are large concentrations of the Indian ethnic minority in the East and West Midlands (places like Leicester, Wolverhampton and Sandwell) and Greater Manchester (Blackburn, for example). The Pakistani minority is concentrated in parts of Bradford (Figure 3.22), Leeds and Birmingham, and there are large Bangladeshi communities in Luton, Oldham and Birmingham.

The 1991 census brought to light some geographical variations within urban areas between some ethnic minorities. Bangladeshi and black Caribbean groups are concentrated in high-density inner-city areas characterised by low levels of

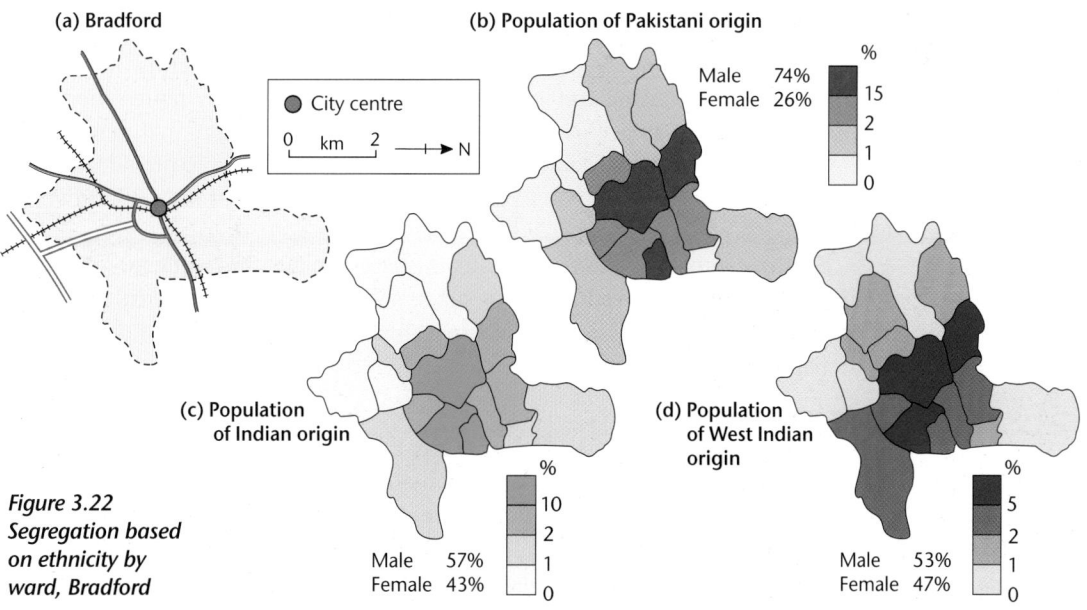

Figure 3.22 Segregation based on ethnicity by ward, Bradford

owner occupancy and high levels of unemployment. In contrast, people of Chinese origin are distributed across a wide range of area types — from deprived inner-city areas to more affluent suburbs. Some research suggests that ethnic segregation is more geographically pronounced in northern areas than in, say, London.

London has a very high proportion and diversity of ethnic minorities. Sixty per cent of the UK's black Caribbean population lives here, and 52% of the Bangladeshi population, but only 18% of the Pakistani population. The majority of these ethnic minority populations live in inner-city areas, and some areas are dominated by certain ethnic groups (for example, Bangladeshi people in Tower Hamlets), but there is wider diversity in other localities such as Brent and Newham. There is a strong concentration of black Caribbean people south of the River Thames in Lambeth and Southwark.

Even within these areas, there are some pronounced variations. For example, in the Northcote ward of Ealing, nearly 70% of the population is Indian, and over 90% of the population of that ward is non-white. The London Borough of Brent has the most diverse ethnic structure. Here there are large concentrations of black Caribbean, black African and Indian people, as well as a large subsection of the white community which is Irish-born.

Housing

In the initial phases of immigration, multiple occupancy in rented accommodation in inner-city areas (terraced houses) was widespread. Ethnic minorities have been much less successful in securing conventional mortgage loans and this has forced them to use less conventional and more expensive forms of financing which limit the price of housing they are able to afford. The prospect of an expensive mortgage on a sub-standard property in a deprived area contributed to the low rate of owner occupancy among the ethnic minority population.

Ethnic minorities have also been discriminated against in terms of access to local authority housing and as a result are disproportionately represented. In a number of urban areas this has led to the development of internal networks of housing provision, where landlords belonging to an ethnic group provide housing for members of that group — a process known as the 'racialisation of residential space'.

More recently there has been greater owner occupancy and some more wealthy individuals have moved out into more suburban areas. In addition, many individuals from ethnic minorities run a small business such as a shop, and live in part of the same building.

Education

Concentrations of minorities in inner-city areas have led to some schools being dominated by one ethnic group, which has affected education requirements. For example, special English lessons may be needed for children and their parents (mothers in particular), and bilingual reading schemes may be introduced. In some areas, special religious provision for minority groups has developed into separate schooling, but this is rare.

Variation in the educational attainment of different ethnic minorities is still being examined. There is some evidence to suggest that children from black Caribbean backgrounds are underachieving compared not only with the white population but also with other ethnic groups. Conversely, the performance of children from Indian, Pakistani and Chinese backgrounds appears on average to be better than that of white children. It has been said that the white 'working-class' male is currently the lowest achiever in schools.

Health care

In the past there have been problems associated with lack of resistance to childhood diseases among the children of newly-arrived immigrants, and fears about immunisation. Literature in ethnic minority languages aimed at informing parents has been produced. However, as literacy and educational standards have improved, particularly among second and subsequent generation ethnic minorities, there have been fewer concerns. Many ethnic minority groups continue to live in inner-city areas (which still tend to be more run down) and there remains a higher concentration of communicable and transmittable disease in such areas. But this is more a reflection of the living standards in these areas than of the people who live in them.

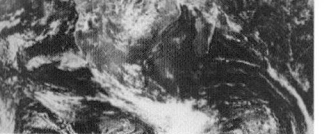

Assessment exercises

Physical geography: shorter term and local change

1 Study the weather chart for the British Isles at 0900 GMT on 23 May 2000 (Figure A).
  a The data from Station X are given in Table A. Construct a station plot to show the weather conditions at this location. (5 marks)
  b Describe how you would measure (i) precipitation and (ii) wind speed. (4 marks)

Table A

Weather	Station X
Temperature	11°C
Precipitation	Rain
Wind speed	14 knots
Wind direction	SSE
Cloud cover (oktas)	7/8

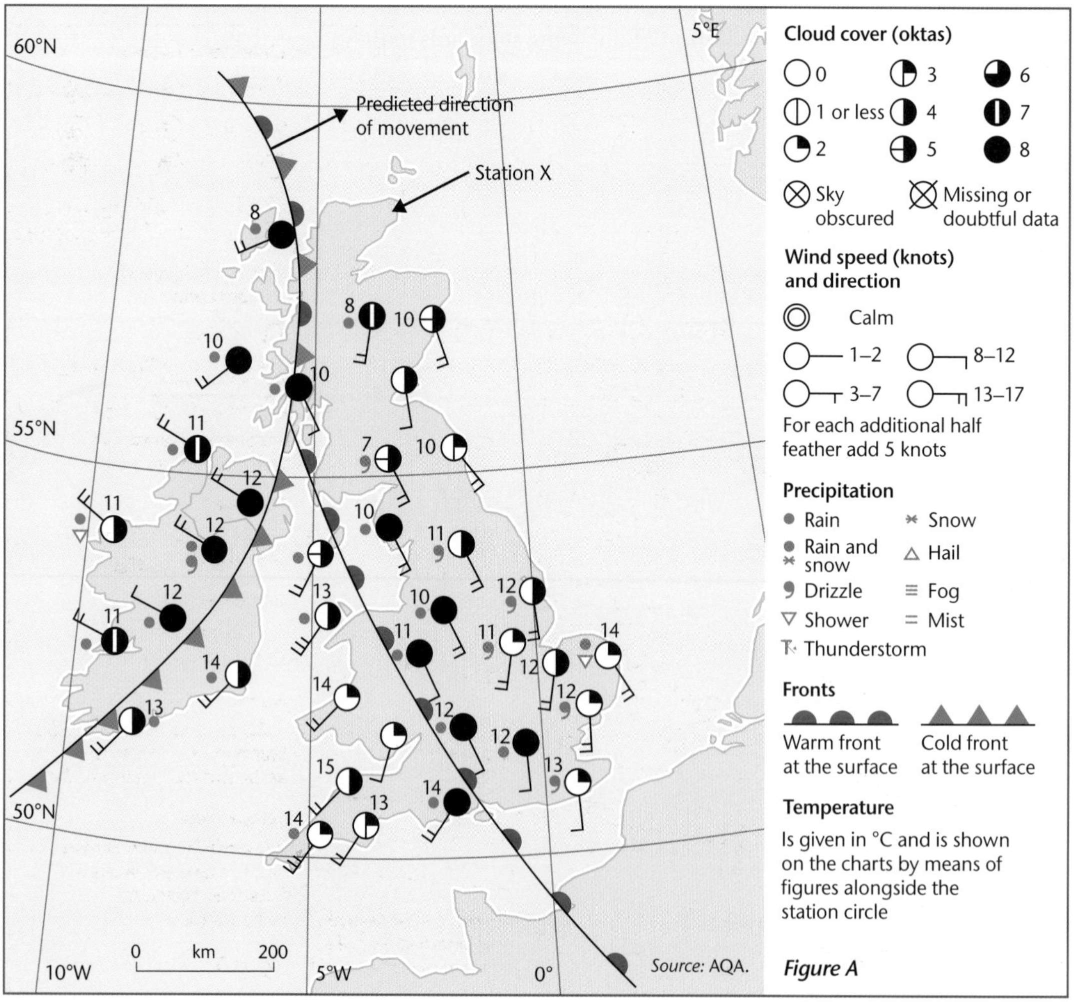

Cloud cover (oktas)
- 0
- 1 or less
- 2
- 3
- 4
- 5
- 6
- 7
- 8
- Sky obscured
- Missing or doubtful data

Wind speed (knots) and direction
- Calm
- 1–2
- 3–7
- 8–12
- 13–17

For each additional half feather add 5 knots

Precipitation
- • Rain
- ✳ Snow
- Rain and snow
- △ Hail
- ❵ Drizzle
- ≡ Fog
- ▽ Shower
- = Mist
- T Thunderstorm

Fronts

Warm front at the surface   Cold front at the surface

Temperature

Is given in °C and is shown on the charts by means of figures alongside the station circle

Source: AQA.   *Figure A*

c Give two pieces of evidence on the weather chart that could have been used by the meteorologist to locate the position of the warm front. (4 marks)

d Describe and suggest reasons for the differences in weather conditions between southwest England and Northern Ireland as shown on the weather chart. (8 marks)

e Explain why fogs commonly occur under anticyclonic conditions. (4 marks)

(25 marks)

2 Look at Figure B which shows a synoptic chart for the British Isles at 0900 GMT on 11 January 2001. An anticyclone is stationed over northwest Scotland.

a State two differences between the weather conditions at Station A and those at Station B. (2 marks)

b (i) Describe the variations in temperature and wind speed shown on Figure B. (3 marks)

(ii) Suggest reasons for the variations in temperature and wind speed. (6 marks)

c How is the weather associated with a depression different from that experienced under anticyclonic conditions? Suggest reasons for these differences. (9 marks)

(20 marks)

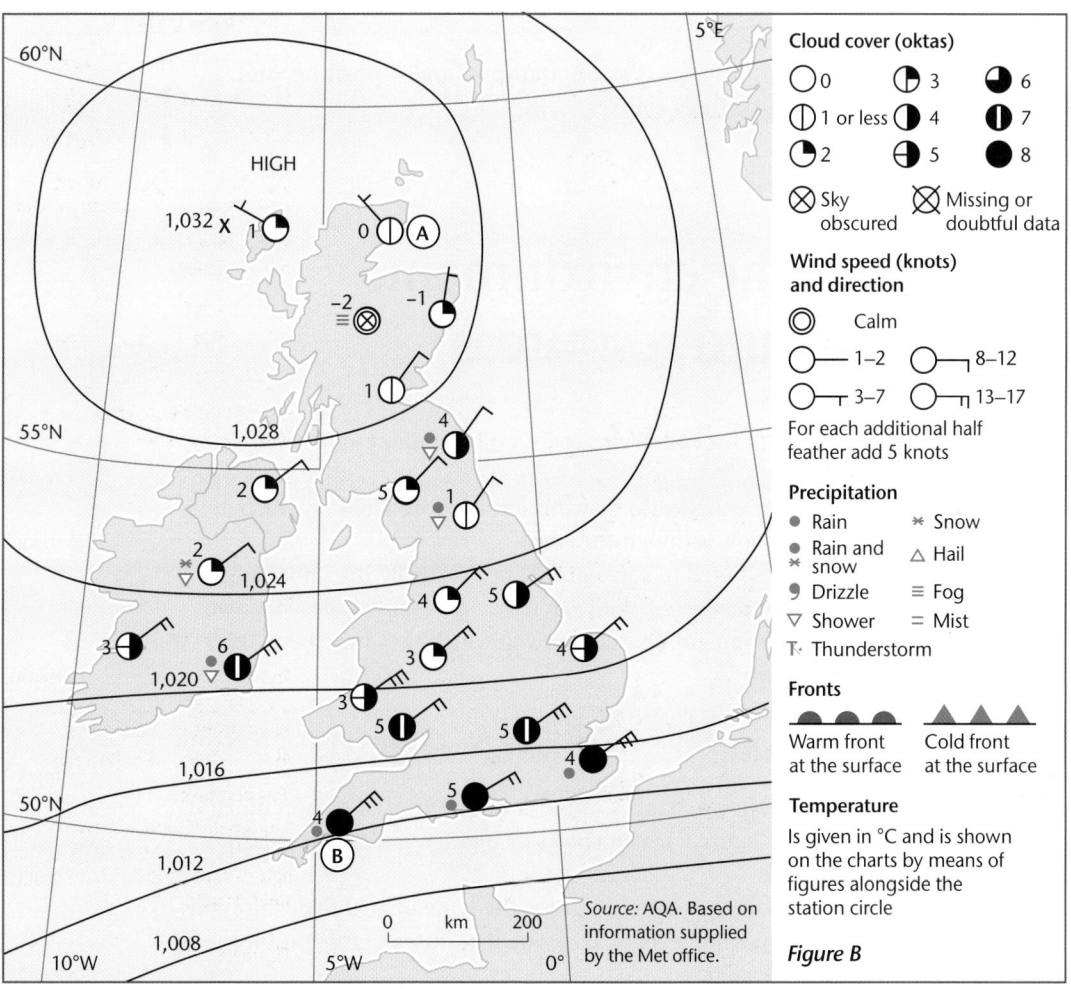

Source: AQA. Based on information supplied by the Met office.

Figure B

3 a What is meant by each of the following terms: (i) river abrasion;
(ii) hydraulic action? (4 marks)

   b (i) Draw and label two different valley cross profiles found along the course
of a river. (4 marks)
(ii) For one of these profiles, explain how processes of river erosion have
contributed to its formation. (4 marks)

   c Choose one depositional feature produced by a river. Name the feature and
explain how it has developed. (5 marks)

   d (i) For a river that drains a catchment area underlain by impermeable rock,
draw and label a typical storm hydrograph which would be produced by a
rainstorm following a period of predominantly wet weather. (4 marks)
(ii) Suggest reasons for the shape of the hydrograph. (4 marks)
(25 marks)

4 a What is meant by each of the following terms: (i) base flow; (ii) interception;
(iii) throughflow? (6 marks)

   b With the aid of a diagram, describe the characteristic features of a waterfall.
Explain how these features are created. (10 marks)

   c Explain the relationships between erosion, transport and deposition, and
discharge/velocity along a river's course. (9 marks)
(25 marks)

People and the environment: population and resources

1 a What is meant by each of the following terms: (i) optimum population;
(ii) overpopulation? (4 marks)

   b Look at Figure C. Describe the world distribution of those areas in which at
least 20% of the population is undernourished. (6 marks)

   c Many countries have attempted to solve problems caused by the imbalance
between population and resources. With reference to examples, describe two
different attempts and comment on the extent to which they have been successful
in changing the situation. (10 marks)
(20 marks)

2 a Look at Figure D. Compare the pattern of energy production in Country A with
that in Country B. (4 marks)

   b Look at Figure E which shows changes in the concentration of carbon dioxide
in the atmosphere, and Figure F which shows changes in global temperature.
Compare the two graphs. (4 marks)

   c In addition to the release of greenhouse gases, the issue of acid deposition has
arisen as a result of harnessing energy. Describe the causes of this problem and
suggest strategies to reduce it. (6 marks)

d Write an account of the advantages and disadvantages of having a nuclear waste disposal site built in an area. (6 marks)

(20 marks)

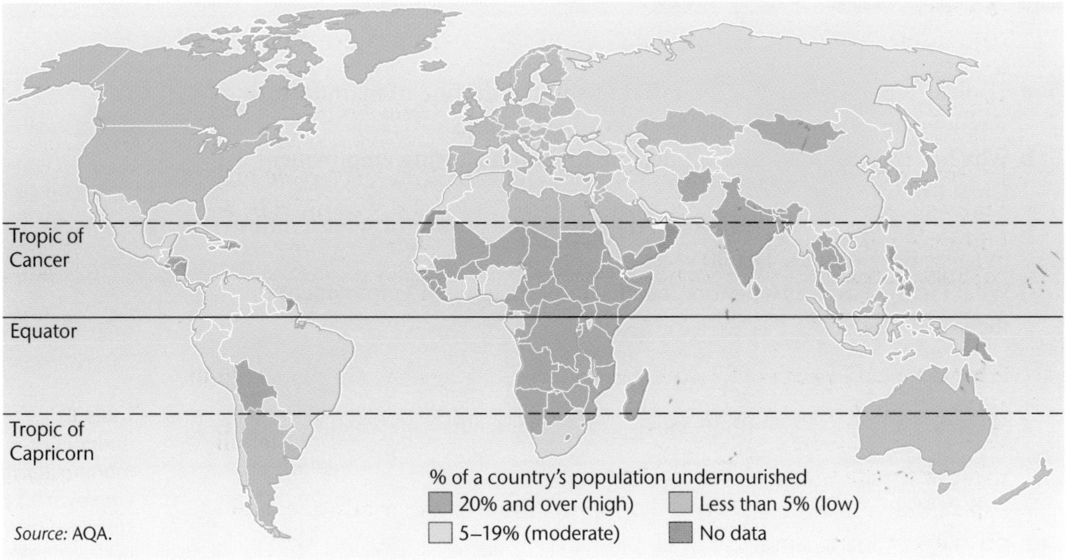

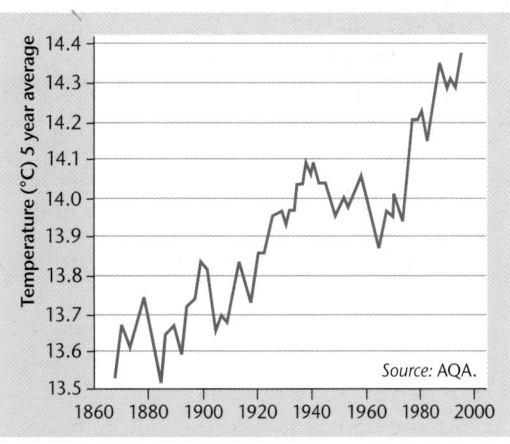

Tropic of Cancer
Equator
Tropic of Capricorn
% of a country's population undernourished
20% and over (high)
Less than 5% (low)
5–19% (moderate)
No data
Source: AQA.

Figure C Levels of undernourishment (UN Food and Agriculture Organization)

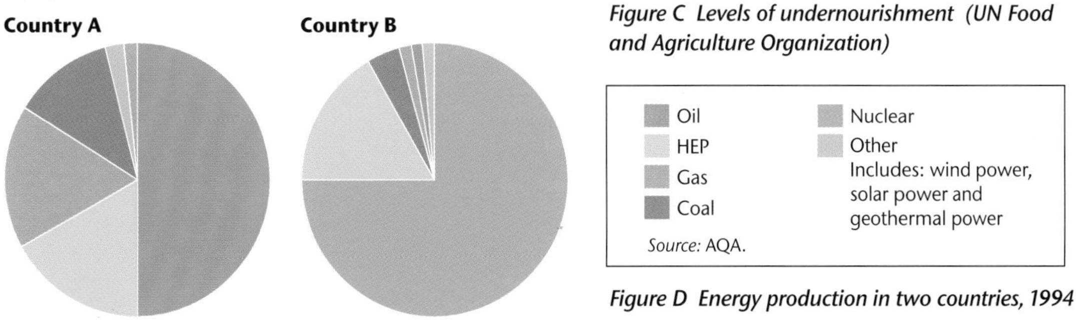

Country A
Country B
Oil
HEP
Gas
Coal
Nuclear
Other
Includes: wind power, solar power and geothermal power
Source: AQA.

Figure D Energy production in two countries, 1994

Carbon dioxide concentration (parts per million)
370
360
350
340
330
320
310
300
290
1860 1880 1900 1920 1940 1960 1980 2000
Source: AQA.

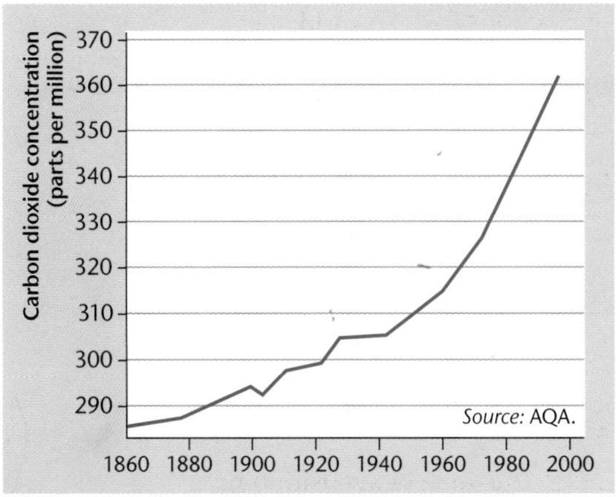

Figure E Changes in the concentration of carbon dioxide in the atmosphere over the last 150 years

Temperature (°C) 5 year average
14.4
14.3
14.2
14.1
14.0
13.9
13.8
13.7
13.6
13.5
1860 1880 1900 1920 1940 1960 1980 2000
Source: AQA.

Figure F Changes in global temperature over the last 150 years

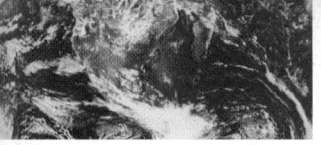

Human geography: changes in the UK in the last 30 years

1 a Look at Table B. Describe the variations in the decline of manufacturing employment in Great Britain as shown in the table. (5 marks)

  b Why has there been an overall decline in manufacturing employment in Great Britain? (5 marks)

  c Suggest reasons for the regional variations in the growth of service industries in Great Britain in the last 30 years. (7 marks)

  d What factors have caused more females to participate in employment during the last 30 years? (8 marks)

  (25 marks)

Standard region	1981 (thousands)	2001 (thousands)	Percentage change (%)
Southeast	1,683	845	−49.7
East Anglia	186	132	−29.0
Southwest	396	265	−33.0
West Midlands	801	486	−39.3
East Midlands	533	375	−29.6
Yorkshire and Humberside	579	345	−40.4
Northwest	800	438	−45.3
North	339	210	−38.1
Wales	238	223	−6.3
Scotland	502	298	−40.6
Great Britain	6,057	3,617	−40.3

*Table B
Changes in manufacturing employment by standard region in Great Britain, 1981 and 2001*

2 a The general pattern of distribution of social groups, according to wealth, is that the poor live in the inner-city area, and the wealthy live in the suburbs and on the urban fringe. Describe and explain the ways in which this pattern has been modified in the last 30 years. (7 marks)

  b How have these changes affected the type and size of housing which has been developed in the area around the edge of the CBD? (6 marks)

  c Suggest reasons for the distribution of the non-white ethnic population within urban areas in the UK. (7 marks)

  (20 marks)

AS
Module 2

The physical options

Chapter 4

Option P
Glacial environments

Ice formation and movement

When a climate starts to become colder, more precipitation in winter falls as snow. Summers also begin to shorten, so there is less time for the winter snow to melt. At first, this leads to permanent snow cover in upland areas, the lower edge of which is known as the **snow line**. As the climate continues to deteriorate, the snow line moves down the slope.

At present, the snow line is at sea level in Greenland but at 6,000 m on the equator. There is no permanent snow cover in the British Isles, so there is no snow line, but scientists estimate that if the Scottish mountains were 200–250 m higher, there would be. In the northern hemisphere, the snow line is found at a higher altitude on south-facing slopes than on those that face north, as the south-facing slopes receive more insolation (Figure 4.1).

Snow initially falls as flakes, which have an open, feathery structure that traps air. As the snow accumulates, compression by the upper layers gradually turns the lower snow into a more compact form known as **firn** or **neve**. Meltwater seeps into the gaps and then freezes, further compacting the mass. As more snow falls, air is progressively squeezed out of the lower snow by the weight of the upper layers and after a period of some years (most experts put it between 20 and 40), a mass of solid ice develops. Where there is no summer melting, this process takes longer. During this period, the mass changes colour from

Key terms

Fluvioglacial (glacifluvial) Processes and landforms associated with the action of glacial meltwater.

Glacier A tongue-shaped mass of ice moving slowly down a valley. Processes and landforms associated with such a feature are called glacial.

Ice ages The common term for the period when there were major cold phases known as glacials, and ice sheets covered large areas of the world. The last ice age lasted from about 2 million years ago to about 10,000 years BP (before present). It was also known as the Quaternary glaciation. During the Quaternary there were many episodes of glaciation, the last major period beginning around 120,000 BP and reaching its maximum in 18,000 BP. At that time about 30% of the Earth's surface was covered by ice (compared to 10% today).

Periglacial Processes and landforms associated with the fringe of, or the area near to, an ice sheet or glacier.

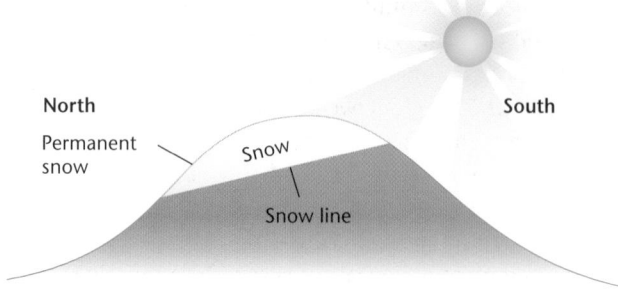

Figure 4.1 The snow line in the northern hemisphere

*Photograph 4.1
Franz-Joseph glacier
in New Zealand*

white, indicating the presence of air, to a bluish colour, indicating that the air has been largely expelled. This is the ice that begins to flow downhill as a glacier.

There are two types of glacier — temperate and polar. The characteristics of each affect ice movement and glacial processes.

Temperate (alpine) glaciers melt in summer, releasing huge amounts of meltwater. This acts as a lubricant, reducing friction. Temperate glaciers move by basal flow, extending/compressing flow, creep and surges. This type of glacier is more likely to erode, transport and deposit material.

Polar glaciers occur in areas where the temperature is permanently below 0°C, and therefore no melting occurs. Movement is slower than in temperate glaciers as they are frozen to their beds and thus move mainly by internal flow. Much less erosion, transportation and deposition occurs.

As ice moves downhill it does not always behave in the same way. It has great rigidity and strength, but under steady pressure it behaves as a plastic (mouldable) body. In contrast, when put under sudden compression or tension, it will break or shear apart. This gives two zones within the glacier (Figure 4.2):

- the upper zone where the ice is brittle, breaking apart to form crevasses
- the lower zone which has steady pressure. Here meltwater resulting from that pressure and from friction with the bedrock allows a more rapid, plastic flow. At depth in the glacier the melting point of the ice is raised slightly by the increased pressure. Basal ice is therefore more likely to melt at temperatures close to 0°C (pressure melting point)

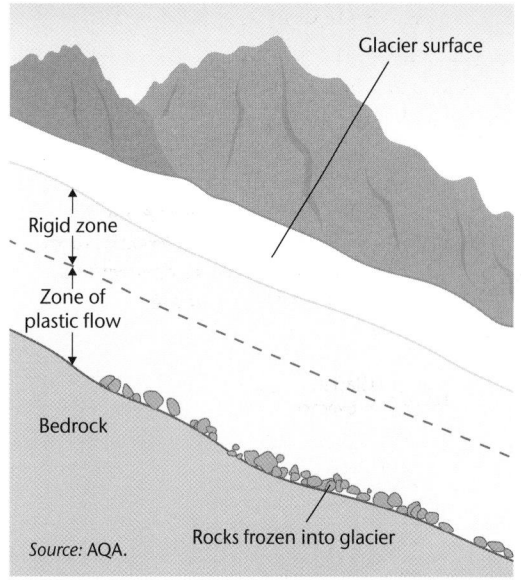

Glacier surface

Rigid zone

Zone of
plastic flow

Bedrock

Source: AQA.   Rocks frozen into glacier

**Figure 4.2
Zones within a
glacier**

Ice movement is generally very complex, and several types of movement can be identified. The main types are:

- **Compressing flow** which occurs where there is a reduction in the gradient of the valley floor leading to ice deceleration and a thickening of the ice mass. At such points ice erosion is at its maximum (Figure 4.3).
- **Extending flow** which occurs when the valley gradient becomes steeper. The ice accelerates and becomes thinner, leading to reduced erosion (Figure 4.3).
- **Basal flow (sliding/slippage)** — as the glacier moves over the bedrock, there is friction. The lower ice is also under a great deal of pressure and this, combined with the friction, results in some melting. The resulting meltwater acts as a lubricant, enabling the ice to flow more rapidly.

- **Surges** occur from time to time when an excessive build-up of meltwater under the glacier leads to the ice moving rapidly forward, perhaps by as much as 250–300 m in one day. Such surges represent a hazard to people living in the glacial valley below the **snout**.
- **Internal flow** occurs when ice crystals orientate themselves in the direction of the glacier's movement and slide past each other. As surface ice moves faster, crevasses develop. Internal flow is the main feature of the flow of polar glaciers as, without the presence of meltwater, they tend to be frozen to their beds.
- **Creep** occurs when stress builds up within a glacier, allowing the ice to behave like plastic and flow. It occurs particularly when obstacles are met.
- **Rotational flow** occurs within the corrie (cirque), the birthplace of many glaciers. Here ice moving downhill can pivot about a point, producing a rotational movement. This, combined with increased pressure within the rock hollow, leads to greater erosion and an over-deepening of the corrie floor.

**Figure 4.3
Extending and
compressing flow**

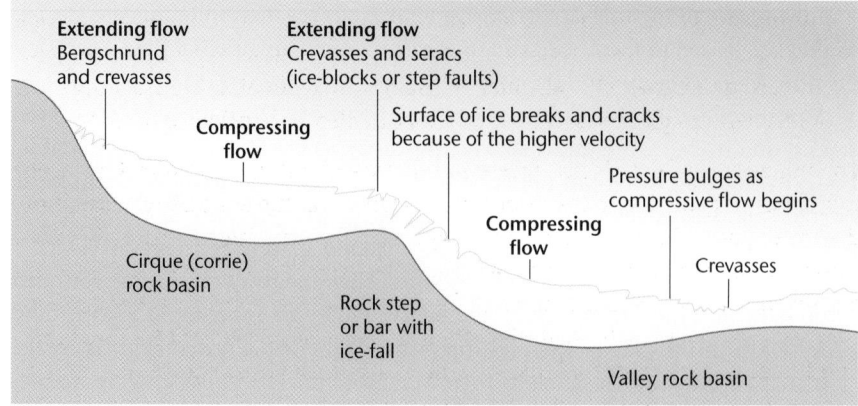

Extending flow
Bergschrund
and crevasses

Extending flow
Crevasses and seracs
(ice-blocks or step faults)

Compressing
flow

Surface of ice breaks and cracks
because of the higher velocity

Pressure bulges as
compressive flow begins

Cirque (corrie)
rock basin

Compressing
flow

Crevasses

Rock step
or bar with
ice-fall

Valley rock basin

Within a glacier there are different rates of movement. The sides and base of the glacier move at a slower rate than the centre surface ice (Figure 4.4). As a result, the ice cracks, producing crevasses on the surface. These also occur where extending flow speeds up the flow of ice and where the valley widens or the glacier flows from a valley on to a plain (piedmont glacier).

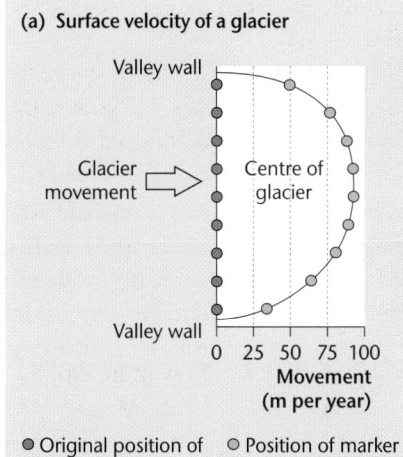 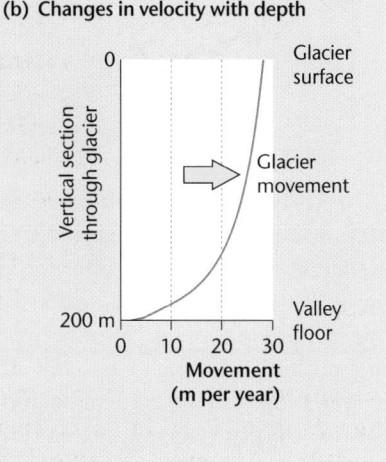

Figure 4.4
Differential rates of flow within a glacier

Glacial budgets

A glacier can be viewed as a system with inputs, stores, transfers and outputs:
- **inputs** are precipitation in the form of snow and ice, and avalanches which add snow, ice and debris from the valley side
- **storage** is represented by the glacier itself
- **transfer (throughput)** is the way that the ice moves (the various types of flow are described above)
- **outputs** are water vapour (from evaporation of water on the ice surface and sublimation — the direct change of state from ice to water vapour), calving (the formation of icebergs), and water in liquid form from **ablation** (melting). The debris deposited at the snout (moraine) can also be considered an output

The upper part of the glacier, where inputs exceed outputs and therefore where more mass is gained than lost over a year, is known as the **zone of accumulation**; the lower part, where outputs exceed inputs, and where mass is lost rather than gained, is known as the **zone of ablation**. Between the two zones is the line of **equilibrium** which separates net loss from net gain and represents the snow line on the glacier (Figure 4.5). A glacier that is characterised by large volumes of gains and losses will discharge a large volume of ice through its equilibrium line to replace mass lost at the snout and will therefore have a high erosive capacity.

Figure 4.5
Glacier budget

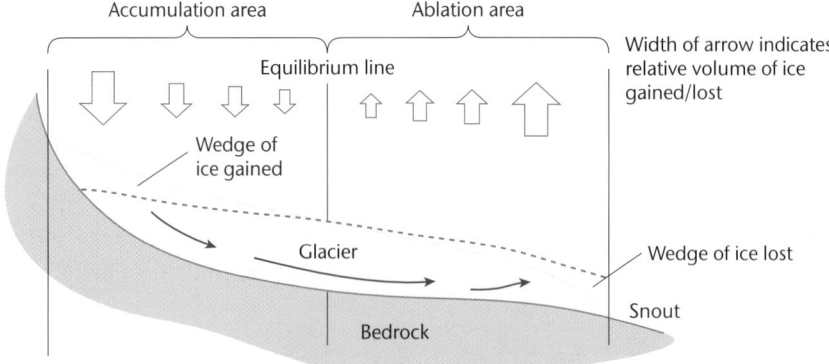

The **net balance** is the difference between the total accumulation and the total ablation during 1 year (Figure 4.6). In temperate glaciers, there is a negative balance in summer when ablation exceeds accumulation, and the reverse in winter. If the summer and winter budgets cancel each other out, the glacier appears to be stationary. In other words, the snout of the glacier remains in the same position, although ice is still advancing down the valley from the zone of accumulation into the zone of ablation.

If the 'supply' begins to exceed the losses, then the snout moves down the valley. This is known as glacial **advance**. When the reverse is true, the glacier begins to shrink in size and the snout moves its position up the valley. This is called glacial **retreat**. Even though the *position* of the snout is moving backwards (retreating), ice continues to move down from the upper parts of the system.

Figure 4.6
Net balance
in a northern
hemisphere
temperate glacier

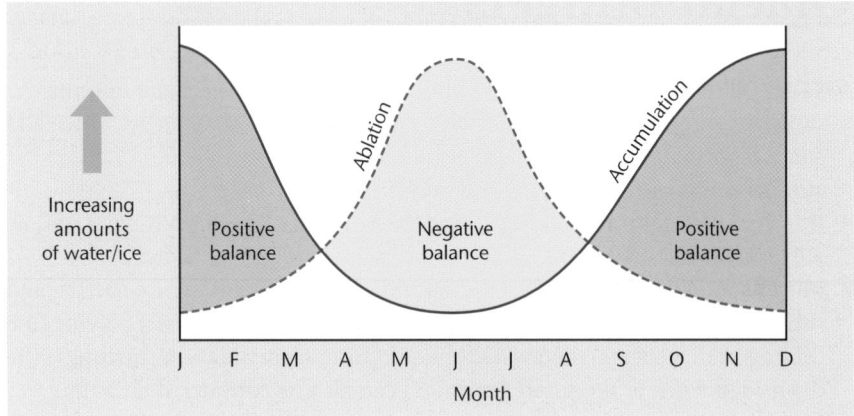

Glacial erosion

Glacial erosion tends to occur in upland regions and is carried out by two main processes:

■ **Abrasion** occurs where the material the glacier is carrying rubs away at the valley floor and sides. It can be likened to the effect of sandpaper or a giant file. The coarser material may leave scratches on the rock known as **striations**; the finer debris smoothes and polishes rock surfaces. The debris involved in abrasion is often worn down by the process into a very fine material known as **rock flour**.

■ **Plucking** involves the glacier freezing onto and into rock outcrops. As the ice moves forward, it pulls away masses of rock. Plucking is mainly found at the base of the glacier where pressure and friction often result in the melting of the ice. It is also marked in well-jointed rocks and in those where the surface has been weakened by freeze–thaw action (frost shattering). Plucking leaves a very jagged landscape.

There are two other processes associated with glacial action that produce the debris glaciers use in their erosive action. Both of these are weathering processes:

■ **Freeze–thaw action/frost shattering** occurs in areas where temperatures rise during the day but drop below freezing at night for a substantial part of the winter. Water which enters cracks in the rocks freezes overnight. Ice occupies more space than water (just under 10% more) and therefore exerts pressure on the crack. As the process continues, the crack widens, and eventually pieces of rock break off (Photograph 4.2). On steep slopes this leads to the collection of material at the base, known as **scree**. In a glacial valley, much of this material falls from the valley side onto the edges of the glacier and some finds its way to the base of the ice via the numerous crevasses which cross the glacier's surface.

■ **Nivation** is a series of processes that operate underneath a patch of snow, leading to the disintegration of the rock surface (see page 99 for details).

There are a number of major landforms which are mainly produced by glacial erosion. These include corries (cirques), arêtes, pyramidal peaks, glacial troughs, hanging valleys and truncated spurs.

Photograph 4.2 Frost-shattering: a limestone boulder in County Clare, Ireland, shattered into several pieces by freeze–thaw action

Alan Young

Corries and associated landforms

A **corrie** is an armchair-shaped rock hollow, with a steep back wall and an over-deepened basin with a rock lip. It often contains a small lake (or tarn). In the British Isles, corries are mainly found on north, northeast and east-facing slopes where reduced insolation allows more accumulation of snow.

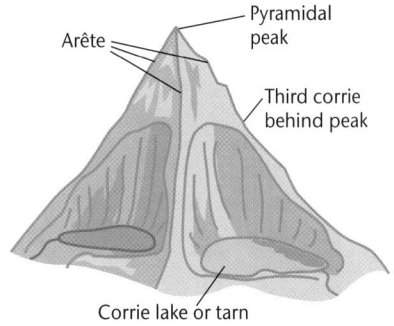

Figure 4.7
Arêtes and pyramidal peaks

If several corries develop in a highland region, they will jointly produce other erosional features. When two corries lie back to back or alongside each other, enlargement will often leave a narrow, steep-sided ridge between the two hollows, called an **arête**. An example is Striding Edge on Helvellyn in the Lake District. If more than two corries develop on a mountain, the central mass will survive as a **pyramidal peak**, which often takes on a very sharp appearance due to frost shattering (Figure 4.7). An example is the Matterhorn in the Alps.

Corrie formation is the result of several interacting processes (see Figure 4.8). The original process is believed to be **nivation** which acts upon a shallow, preglacial hollow and enlarges it into an embryo corrie (this may take a long time and be spread across several glacial periods within an ice age). As the hollow grows, the snow becomes thicker and is increasingly compressed to form firn and then ice. The back wall becomes steeper through the action of **plucking**. The **rotational movement** of the ice, together with the debris supplied by plucking and frost shattering on the back wall, abrades the floor of the hollow which over-deepens the corrie.

As the hollow deepens, the thinner ice at its edge does not produce the same amount of downcutting and a rock lip develops on the threshold of the feature. Some thresholds have their height increased by morainic deposits formed when the glacier's snout was in that position. After the last ice has melted, the corrie fills with meltwater and rainwater to form a small lake (tarn).

Glacial troughs and associated landforms

Figure 4.8
The formation of a corrie

Glaciers flow down pre-existing river valleys as they move from upland areas. They straighten, widen and deepen these valleys, changing the original V-shaped river feature into the U-shape typical of glacial erosion. The action of ice, combined with huge amounts of meltwater and sub-glacial debris, has a far greater erosive power than that of water.

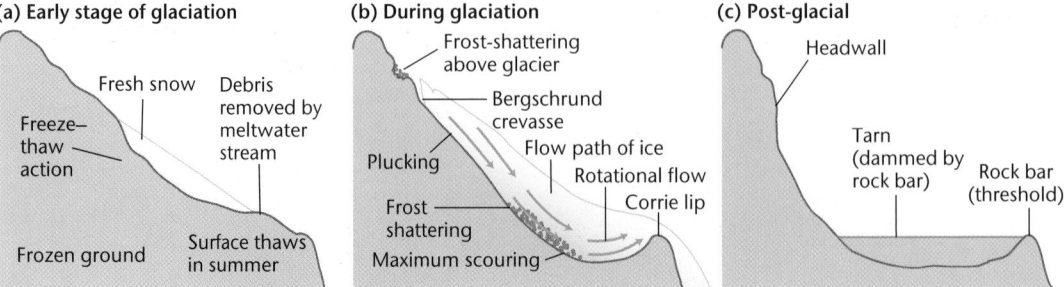

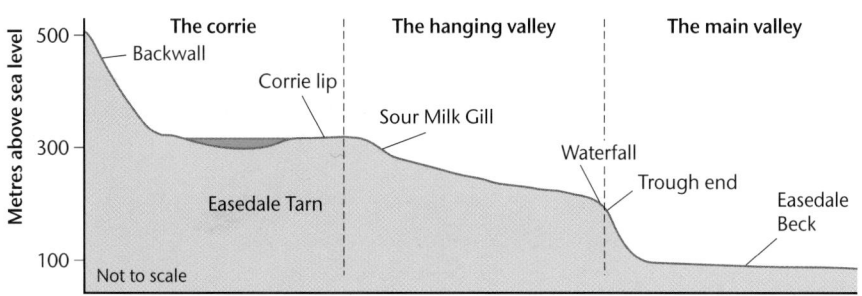

Figure 4.9
The long profile of a glacial valley: Easedale, Lake District

As both extending and compressing flow are present, the amount of erosion varies down the valley. Where compressing flow is present, the glacier will over-deepen parts of the valley floor, leading to the formation of **rock basins**. It is also suggested that over-deepening is caused by increased erosion at the confluence of glaciers, areas of weaker rocks or zones of well-jointed rocks.

The major features of glacial troughs are:

- usually fairly straight with a wide base and steep sides — a U-shape
- stepped long-profile with alternating steps and rock basins
- some glacial valleys end abruptly at their heads in a steep wall, known as a **trough end**, above which lie a number of corries (Figures 4.9 and 4.10)
- rock basins filled with **ribbon lakes**, e.g. Wastwater in the Lake District
- over-deepening below the present sea-level — this led to the formation of **fjords** when sea-levels rose after the ice ages and submerged the lower parts of glacial valleys, for example on the coasts of Norway and southwest New Zealand (Milford Sound)
- **hanging valleys** on the side of the main valley (e.g. the valley of Church Beck which flows down into Coniston Water in the Lake District). These are either pre-existing tributary river valleys which were not glaciated, or tributary glacial valleys (Figure 4.11). In tributary glacial valleys there would have been less ice and therefore less erosion than in the main valley. The tributary valley floor was therefore left higher than that of the main valley when the ice retreated.
- areas of land projecting from the river-valley side (spurs) have been removed by the glacier, producing **truncated spurs**

Figure 4.10
A trough end

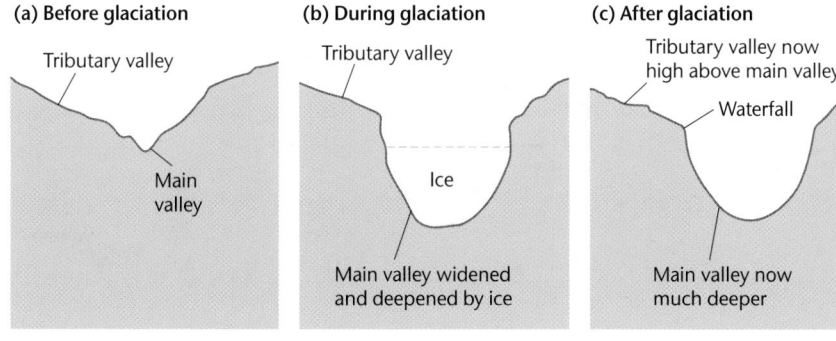

Figure 4.11
The formation of a hanging valley

Alan Young

Photograph 4.3
A roche moutonnée
in the Lake District.
Deep striations are
visible on the right
(upstream) side

- small areas of rock on the valley floor are not always completely removed and this leaves **roches moutonnées**. These have an upstream side polished by abrasion and a downstream side made jagged by plucking (Figure 4.12, Photograph 4.3)
- after ice retreat, many glacial troughs were filled with **shallow lakes** which were later infilled, and their sides were modified by frost shattering and the development of screes which altered the glacial U-shape (e.g. Great Langdale, Lake District) (Figure 4.13)

Figure 4.12
Formation of a
roche moutonnée

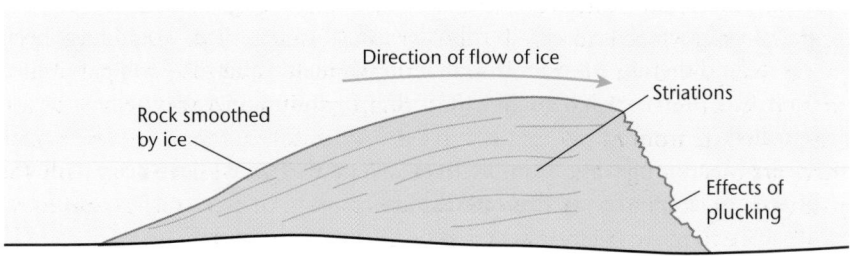

Figure 4.13
Formation of a
glacial valley

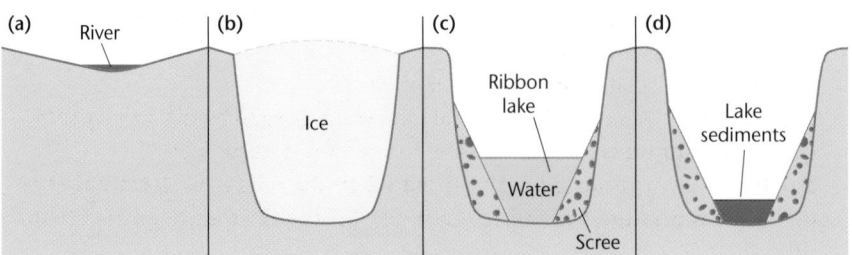

Glacial transportation and deposition

As well as eroding the rock over which it is flowing, a valley glacier is also capable of transporting large amounts of debris. Some of this may be derived from rockfalls on the valley side. It is then transported on the surface of the glacier (**supraglacial** debris) or buried within the ice (**englacial**). Material found at the base of the glacier is known as **subglacial** and may include rock fragments that have fallen down crevasses and material eroded at the base. Another way of describing the material carried by the glacier is shown in Figure 4.14. Strictly speaking, **moraine** applies to a type of landform, but many textbooks now use the term in this way.

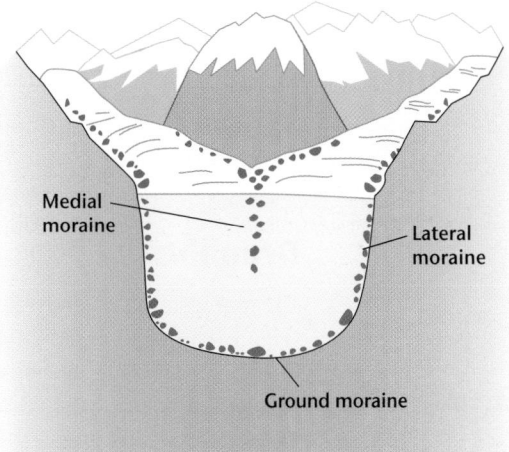

Figure 4.14
Transport of debris within a glacier

The huge amounts of material carried down by a glacier will eventually be deposited. The bulk of this will be the debris released by the melting of the ice at the snout. It is also possible for the ice to become overloaded with material, reducing its capacity. This may occur near to the snout, as the glacier melts, or in areas where the glacier changes between compressing and extending flow. Material that is deposited directly by the ice is known as **till** or **boulder clay**, although the latter term tends not to be used today.

Till is used to describe an unsorted mixture of rocks, clay and sand that was mainly transported as supraglacial or englacial debris and deposited when the ice melted. Individual stones tend to be angular to sub-angular, unlike river and beach material which is rounded. Till reflects the character of the rocks over which the ice has passed. In the till of south Lancashire, for example, it is possible to find rocks from the Lake District (e.g. Shap granite) and southern Scotland (e.g. riebeckite from Ailsa Craig in the Firth of Clyde). In the till of East Anglia, there are pieces of granite from southern Norway. This indicates not only the passage of the ice but the fact that the sea level must have been considerably lower to allow ice to move over the area that later became the North Sea.

Sometimes it is possible to find a large block of rock that has been moved from one area and deposited in another which has a very different geology. Such a feature is known as an **erratic**.

Two types of glacial deposit are recognised:

- **lodgement till** — subglacial material that was deposited by the actively moving glacier. A **drumlin** is a typical feature formed from this material
- **ablation till** — produced at the snout when the ice melts. **Terminal (end)**, **push** and **recessional moraines** are typical features produced from ablation till

Figure 4.15
Drumlins

Direction of movement

Maximum width

Blunted
end

Tapered
end

Long axis

Section

Plan

Drumlin swarm – 'basket of eggs' topography

Drumlins

The term drumlin is derived from the Gaelic word *druim*, meaning a rounded hill (Figure 4.15, Photograph 4.4). The main features of drumlins are:

■ they are smooth, oval-shaped small hills, often resembling the top half of an egg
■ they can be as long as 1.5 km (although most are much smaller) and up to 50–60 m in height
■ they have a steep end known as the **stoss** and a gently sloping end, the **lee**
■ they are elongated in the direction of ice advance with the stoss at the upstream end and the lee at the downstream end
■ they are often found in groups known as **swarms** and, given their shape, this is sometimes referred to as a 'basket of eggs topography'
■ they are formed from unsorted till
■ they are found on lowland plains such as the central lowlands of Scotland. A well-known swarm is at Hellifield in the Ribble Valley. Many are found at the lower end of glacial valleys

Photograph 4.4
Risebrigg Hill in
North Yorkshire is
a drumlin, lying
within a swarm of
drumlins known as
the Hills of Elslack

Alan Young

There is some controversy over the origin of drumlins, which are formed underneath the ice. The most widely-held view is that they are the result of the ice being overloaded with debris. This reduces its capacity to carry and deposition occurs at its base. Once this material has been deposited it is then streamlined by further ice advance. There could also be pre-existing sediment (older till from a previous glacial advance, for example) that is caught up in the streamlining process.

Moraines

Moraines are lines, or a series of mounds of material, mainly running across glacial valleys. The main type is the terminal or end moraine which is found at the snout of the glacier. **Terminal moraines** show the following features:

- they consist of a ridge of material (or several mounds) stretching across a glacial valley
- they are elongated at right angles to the direction of ice advance
- they are often steep-sided, particularly the ice-contact side, and reach heights of 50–60 m
- they are often crescent-shaped, moulded to the form of the snout
- they are formed from *unsorted* ablation material

Terminal moraines are formed when the ice melts and the material it has been carrying is deposited. This is why they contain a range of unsorted material, from clay to boulders.

Figure 4.16
Moraines

As the glacier retreats, it is possible for a series of moraines to be formed along the valley, marking points where the retreat halted for some time. These are known as **recessional moraines** (Figure 4.16).

If the climate cools for some time, leading to a glacial advance, previously deposited moraine may be shunted up into a mound known as a **push moraine**. Such features are recognised by the orientation of individual pieces of rock which may have been pushed upwards from their original horizontal position.

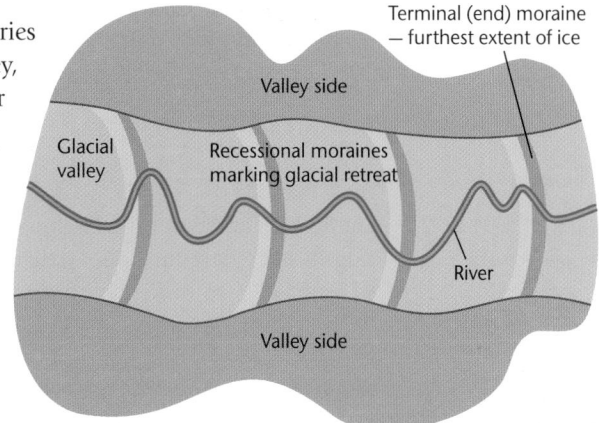

Fluvioglacial deposition

The melting of ice produces a great deal of water which has the capacity to carry much debris. As the water often flows under considerable pressure, it has a high velocity and is very turbulent. It can therefore pick up and transport a larger amount of material than a normal river of similar size. It is now believed that this water, with its load, is responsible for the creation of subglacial valleys that are often deep and riddled with potholes.

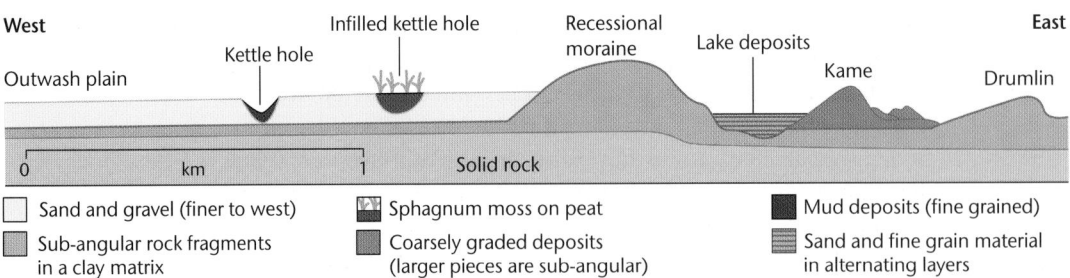

West ... Infilled kettle hole ... Recessional moraine ... Lake deposits ... East

Outwash plain ... Kettle hole ... Kame ... Drumlin

0 ... km ... 1 ... Solid rock

☐ Sand and gravel (finer to west)

☐ Sub-angular rock fragments in a clay matrix

▨ Sphagnum moss on peat

▨ Coarsely graded deposits (larger pieces are sub-angular)

■ Mud deposits (fine grained)

▤ Sand and fine grain material in alternating layers

Figure 4.17
Features of lowland glaciation

When the meltwater discharge decreases, the resultant loss of energy causes the material being carried by the meltwater to be deposited. As with all water deposition, the heavier particles will be dropped first, resulting in sorting of the material. Deposits may also be found in layers (stratified) as a result of seasonal variations in the meltwater flow. The main features produced by fluvioglacial deposition are eskers, kames and the outwash plain (Figure 4.17). Lakes on the outwash plain may have layered deposits in them called varves.

Eskers

Eskers have the following main features:

- they are long ridges of material running in the direction of ice advance
- they have a sinuous (winding) form, 5–20 m high
- they consist of sorted coarse material, usually coarse sands and gravel
- they are often stratified (layered)

Eskers are believed to be deposits made by subglacial streams. The channel of the stream will be restricted by ice walls, so there is considerable hydrostatic pressure which enables a large load to be carried and also allows the stream to flow uphill for short distances. This accounts for the fact that some eskers run up gentle gradients. The bed of the channel builds up above the surrounding land, and a ridge is left when the glacier retreats during deglaciation. In some areas, the ridge of an esker is combined with mounds of material, possibly kames (see below). Such a feature is known as a **beaded esker**.

Kames

Kames are mounds of fluvioglacial material (sorted, and often stratified, coarse sands and gravel). They are deltaic deposits left when meltwater flows into a lake dammed up in front of the glacial snout by recessional moraine deposits. When the ice retreats further, the delta kame often collapses. Kame terraces are frequently found along the side of a glacial valley and are the deposits of meltwater streams flowing between the ice and the valley side.

Outwash plains (sandur)

Outwash plains are found in front of the glacier's snout and are deposited by the meltwater streams issuing from the ice. They consist of material that was brought down by the glacier and then picked up, sorted and dropped by running water beyond the position of the ice front. The coarsest material travels the shortest distance and is therefore found near to the glacier; the fine material, such as clay, is carried some distance across the plain before being deposited. The deposits are also layered vertically, which reflects the seasonal flow of meltwater streams.

Meltwater streams that cross the outwash plain are **braided**. This happens as the channels become choked with coarse material because of marked seasonal variations in discharge. On the outwash plain there is often a series of small depressions filled with lakes or marshes. These are known as **kettle holes**. It is believed that they are formed when blocks of ice, washed onto the plain, melt and leave a gap in the sediments. Such holes then fill with water to form small lakes. Aquatic plants become established in the lakes and this leads over time to the development of a marshy area and then peat.

Lakes on the fringe of the ice are filled with deposits that show a distinct layering. A layer of silt lying on top of a layer of sand (Figure 4.18) represents 1 year's deposition in the lake and is known as a **varve**. The coarser, lighter layer is the spring and summer deposition when meltwater is at its peak and the meltwater streams are carrying maximum load. The thinner, darker-coloured and finer silt settles during autumn and through the winter as stream discharge decreases and the very fine sediment in the lake settles to the bottom. Varves are a good indicator of the age of lake sediments and of past climates as the thickness of each varve indicates warmer and colder periods.

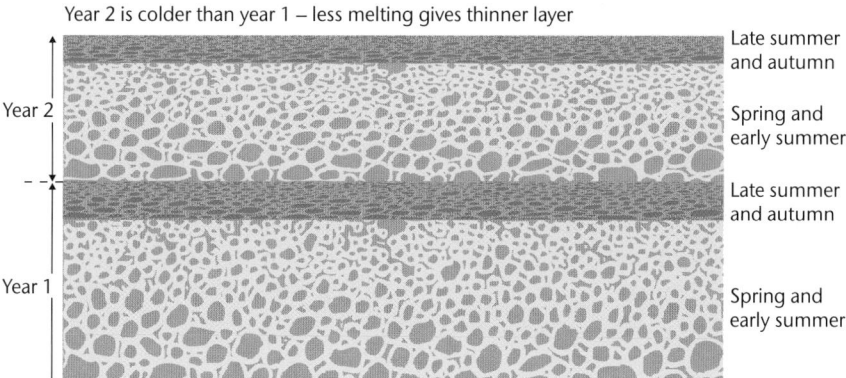

Year 2 is colder than year 1 – less melting gives thinner layer

Year 2 — Late summer and autumn / Spring and early summer

Year 1 — Late summer and autumn / Spring and early summer

Figure 4.18
Varves

Drainage diversion

Where ice sheets and glaciers exist it is possible for them to divert the course of the existing drainage systems (rivers and valleys). This happens as a result of direct glacial intervention (watershed breaching) or through erosion by meltwater.

Watershed breaching

Watershed breaching describes the situation where the glacier itself has been responsible for producing a new course for a river. One way in which this may be achieved is shown in Figure 4.19.

- **Stage 1** Two rivers flow in parallel valleys. On the watershed, there is a lower area known as a col.
- **Stage 2** During glaciation, valley glaciers occupy both valleys but the glacier in valley A is of such a large volume that pressure forces some of the ice over the col into valley B, cutting a side valley in the process.
- **Stage 3** On deglaciation, meltwater from glacier A runs down the valley. If point X has been cut lower than point Y (there may also be glacial debris at point Y), then the meltwater will run through the valley at X and join the meltwater from glacier B.
- **Stage 4** The new drainage pattern of the area is left. If, of course, point X had *not* been cut lower than point Y, then the drainage would have resumed its pre-glacial pattern as in stage 1.

Figure 4.19
Watershed breaching

(a) Stage 1

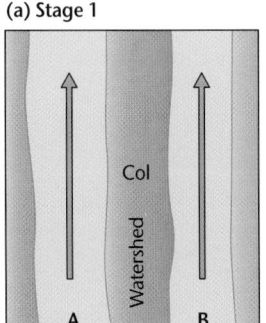

(b) Stage 2

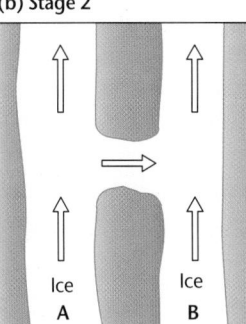

(c) Stage 3

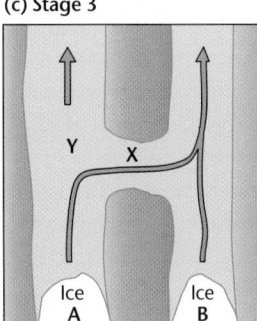

(d) Stage 4

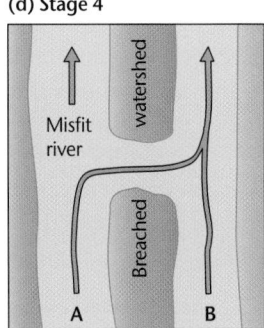

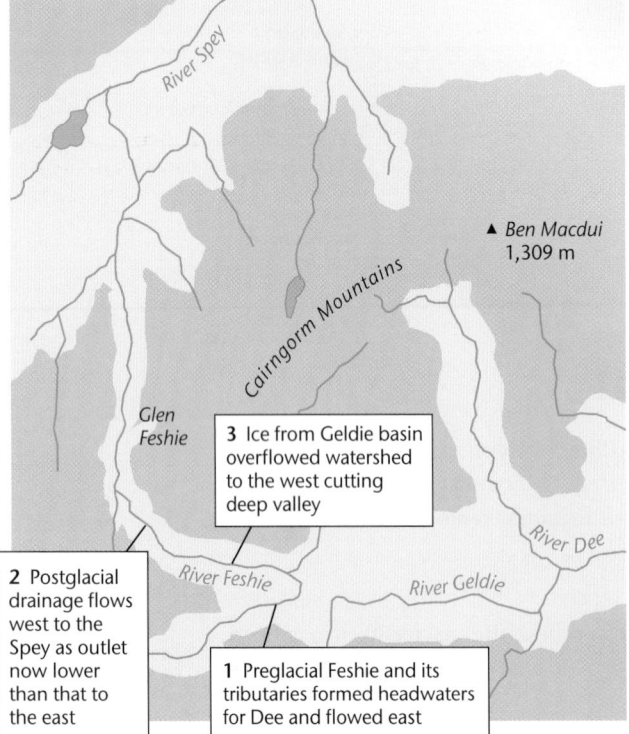

An example of watershed breaching causing a reversal of the preglacial drainage occurs in the Cairngorm Mountains in Scotland. The River Feshie and its tributaries formed part of the headwaters of the River Dee system that flowed towards the east. During the Quaternary glaciation large amounts of ice built up in the Geldie basin and some ice eventually overflowed the watershed, cutting a valley across it towards the west. This ice also overdeepened the existing valley on the northwestern side of the mountains. After the ice retreated this outlet to the northwest was lower than that to the east. The previous upper tributaries of the Geldie therefore joined and flowed west as the River Feshie to join the Spey system (Figure 4.20).

Figure 4.20 Map of the present-day southwest Cairngorms showing where watershed breaching occurred

Proglacial lakes and overflow channels

As has already been stated, glacial meltwater has great erosive power because of its volume and the large amounts of debris it contains. During deglaciation, lakes develop on the edges of the ice, some occupying large areas. Overflows from these lakes which cross the lowest points of watersheds will create new valleys. When the ice damming these meltwater lakes totally melts, many of the new valleys are left dry, as drainage patterns revert to the preglacial stage. In certain cases, however, the postglacial drainage adopts them, giving rise to new drainage patterns.

Large meltwater lakes of this kind occurred in the English Midlands (Lake Harrison), the Vale of Pickering in North Yorkshire (Lake Pickering) and the Welsh borders (Lake Lapworth, Figure 4.21) at the end of the last glaciation. The River Thames is thought to have followed a much more northerly course before the Quaternary glaciation — its modern course was formed when ice filled the northern part of its basin and forced it to take a different route.

The River Severn is also believed to have been diverted during the last glaciation. Figure 4.21 shows the stages of this process.

- **Stage 1 Pre-glacial** The River Severn flowed northwards to enter the Irish Sea in what is now the estuary of the River Dee. The present Lower Severn was a shorter river flowing from the Welsh borderlands to the Bristol Channel.
- **Stage 2 The last Ice Age** Ice coming down from the north blocked the River Severn valley to the north. The water from the blocked river formed a huge proglacial lake known as Lake Lapworth. The lake eventually overflowed the watershed to the south to join the original Lower Severn. In the process it cut through a solid rock area, creating the gorge at Ironbridge.
- **Stage 3 Deglaciation and the postglacial period** As the ice retreated to the north, the way should have been left open for the two rivers to return to the preglacial situation. The route north, however, was blocked with glacial deposits, and as the Ironbridge gorge had been cut very deep (lower than the exit to the north), the new drainage adopted this rather than its former route. The River Severn now flows from central Wales to the Bristol Channel.

*Figure 4.21
Theoretical stages
in the diversion of
the River Severn*

(a) Stage 1

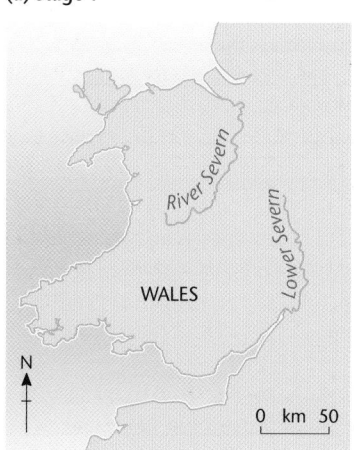

(b) Stage 2

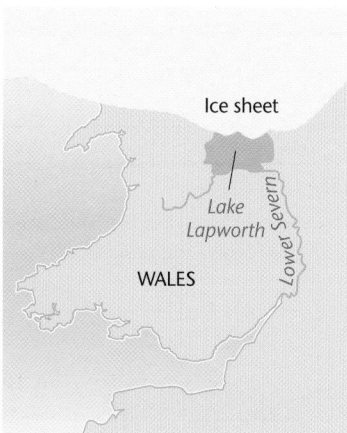

(c) Stage 3

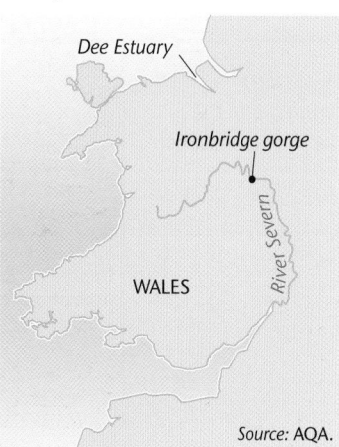

Source: AQA.

Periglacial processes and associated landforms

Periglacial areas are those which, although not actually glaciated, are exposed to very cold conditions with intense frost action and the development of permanently frozen ground or permafrost. At present, areas such as the tundra of northern Russia, Alaska and northern Canada, together with high mountainous areas such as the Alps, experience a periglacial climate. In the past, however, as ice sheets and glaciers spread, many areas which are now temperate were subject to such conditions.

The climate of periglacial regions is marked by persistently low temperatures. Summers are short but temperatures can sometimes reach above 15°C. In winter, the temperature remains well below zero and in some areas may fall below –50°C at times.

Permafrost

Where subsoil temperatures remain below zero for at least 2 consecutive years, permafrost will occur. Today, it is estimated that permafrost covers around a quarter of the Earth's surface. When summer temperatures rise above freezing, the surface layer thaws from the surface downwards to form an **active layer**. The thickness of this layer depends upon local conditions, but may extend to 4 m. As the ice in this layer melts, large volumes of water are released. This water is unable to drain through the permafrost layer and, as low temperatures do not encourage much evaporation, the surface becomes very wet. On slopes as gentle as 2°, saturation of this upper layer encourages soil movement downslope, a periglacial process known as **solifluction** (see below).

There are three kinds of permafrost (Figure 4.22):

- **Continuous permafrost** is found in the coldest regions, reaching deep into the surface layers. In Siberia today, it is estimated that the permafrost can reach down over 1,500 m. In the very coldest areas, there is hardly any melting of the uppermost layer.

Figure 4.22
Variations in the depth of permafrost

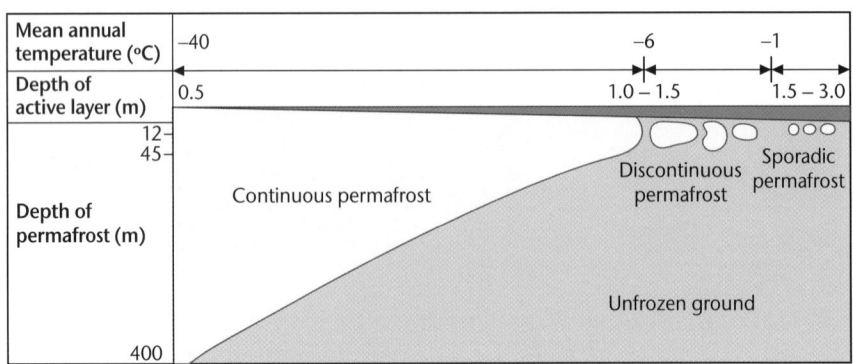

Mean annual temperature (°C)	–40	–6	–1
Depth of active layer (m)	0.5	1.0 – 1.5	1.5 – 3.0

Continuous permafrost — Discontinuous permafrost — Sporadic permafrost

Depth of permafrost (m): 12–45, 400

Unfrozen ground

- **Discontinuous permafrost** occurs in regions that are slightly warmer, where the ground is not frozen to such great depths. On average the frozen area will extend 20–30 metres below the ground surface, although it can reach 45 m. There are also gaps in the permafrost under rivers, lakes and near the sea.
- **Sporadic permafrost** is found where mean annual temperatures are around or just below freezing point. In these places, permafrost occurs only in isolated spots where the local climate is cold enough to prevent complete thawing of the soil during the summer.

Periglacial processes

There is a clear link between the landforms of periglacial regions and the processes that form them. The main processes are described below.

Freeze–thaw action (frost shattering)

This process has already been described above because it provides a great deal of the erosive material in glaciers. In periglacial areas, **screes** develop at the foot of slopes as a result of frost shattering. On relatively flat areas, extensive spreads of angular boulders are left, which are known as **blockfield** or **felsenmeer** (sea of rocks) (Figure 4.23).

Nivation

Nivation takes place beneath patches of snow in hollows, particularly on north- and east-facing slopes. Freeze–thaw action and possibly chemical weathering, operating under the snow, cause the underlying rock to disintegrate. As some of the snow melts in spring, the weathered particles are moved downslope by the meltwater and also by solifluction (see below). Over some period of time, this leads to the formation of **nivation hollows** which, when enlarged, can be the beginnings of a corrie (cirque).

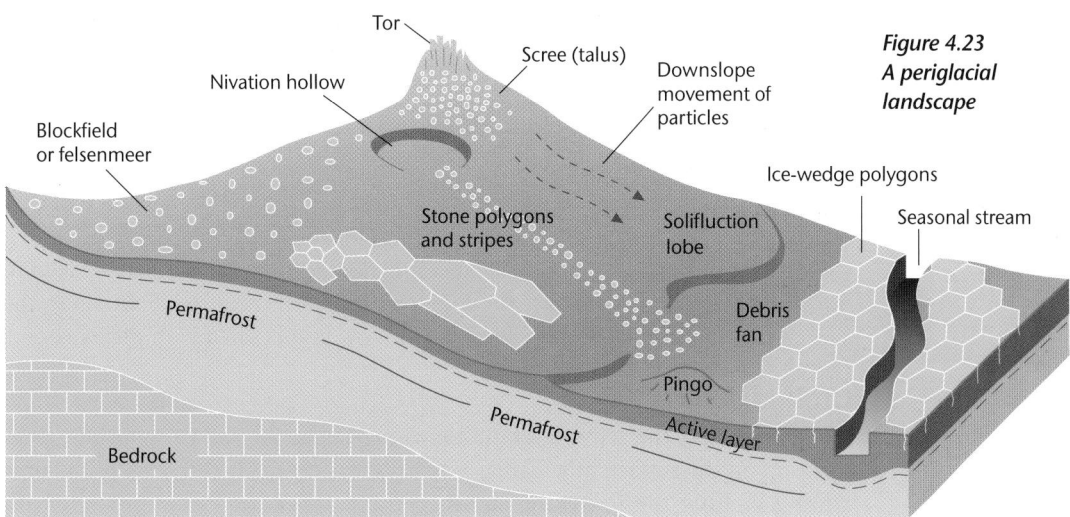

*Figure 4.23
A periglacial
landscape*

Solifluction

When the active layer thaws in summer, excessive lubrication reduces the friction between soil particles. Even on slopes as shallow as 2°, parts of the active layer then begin to move downslope. This leads to **solifluction sheets** or **lobes** — rounded, tongue-like features often forming terraces on the sides of valleys. Solifluction was widespread in southern Britain during the Quaternary ice age, and such deposits are often known here as **head.**

Frost heave

As the active layer starts to refreeze, ice crystals begin to develop. They increase the volume of the soil and cause an upward expansion of the soil surface. Frost heave is most significant in fine-grained material and, as it is uneven, it forms small domes on the surface.

Within the fine-grained material there are stones which, because of their lower specific heat capacity, heat up and cool faster than the surrounding finer material. Cold penetrating from the surface passes through the stones faster than through the surrounding material. This means that the soil immediately beneath a stone is likely to freeze and expand before the other material, pushing the stone upwards until it reaches the surface. On small domes, the larger stones move outwards, effectively sorting the material which, when viewed from above, takes on a pattern. This **patterned ground** on gentler slopes takes the form of **stone polygons**, but where the ground is steeper (slopes exceeding 6°), the stones move downhill to form **stone stripes** (Figure 4.24).

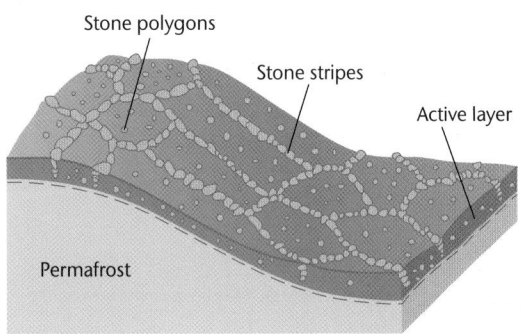

Stone polygons

Stone stripes

Active layer

Permafrost

Figure 4.24
Stone polygons and stripes

Groundwater freezing

Where the permafrost is thin or discontinuous, water is able to seep into the upper layers of the ground and then freeze. The expansion of this ice causes the overlying sediments to heave upwards into a dome-shaped feature known as a **pingo** which may rise as high as 50 m. This type of pingo is referred to as an open-system or East Greenland type.

In low-lying areas with continuous permafrost on the site of small lakes, groundwater can be trapped by freezing from above and the permafrost beneath. As this water freezes, it will expand, pushing up the overlying sediments into a closed-system pingo or Mackenzie type. It is named after the Mackenzie delta in northern Canada where over 1,000 pingos have been recorded. Sometimes the surface of a pingo will collapse, leaving a hollow that is filled with meltwater.

Ground contraction

The refreezing of the active layer during winter causes the soil to contract. Cracks open up on the surface in a similar way to cracks on the beds of dried-up lakes. During melting the following summer, the cracks open again and fill with meltwater. As the meltwater contains fine sediment, this also begins to fill the

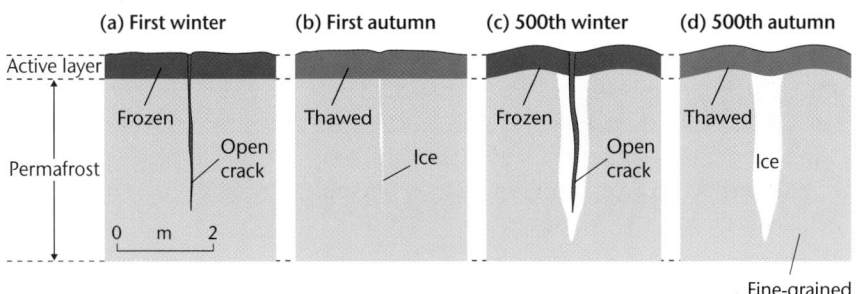

Figure 4.25
The formation
of ice wedges

crack. The process occurs repeatedly through the cycle of winter and summer, widening and deepening the crack to form an **ice wedge** which eventually, over a period of hundreds of years, can become at least 1 metre wide and 2–3 m deep (Figure 4.25). The cracking produces a pattern on the surface which, when viewed from above, is similar to the polygons produced by frost heaving. These are therefore known as **ice-wedge polygons**.

Water and wind action

Owing to the open and sparsely vegetated periglacial landscape, rates of erosion caused by water and wind can be high.

Water erosion is highly seasonal, occurring mainly in spring and summer when the active layer melts. This can cause short periods of very high discharge in rivers, bringing about far more fluvial erosion than would be expected given the relatively low mean precipitation. Drainage is typically **braided** because of the high amount of debris being carried by meltwater streams.

Unobstructed **winds** blowing across periglacial landscapes can reach high velocities. They cause erosion through abrasion and dislodge the fine unconsolidated materials that cover the area. The effects of erosion can be seen in grooved and polished rock surfaces and in stones shaped by the wind, known as **ventifacts**. The fine material of the outwash plain is picked up by the wind and carried long distances to be deposited elsewhere as extensive areas of **loess**. Loess is found in many parts of North America and Eurasia, south of the boundary of the Pleistocene ice sheets. In England, loess deposits, rarely more than 2 m in depth, cover parts of East Anglia and the London Basin where they are known as brick-earth deposits. In China, loess deposits are widespread and in places reach depths of over 300 m.

Chapter 5

Option Q
Coastal environments

Processes affecting coastal landforms

There are a number of factors that determine the shape, form and appearance of a coastline:

- **wave** size, frequency, type, energy produced and direction
- **local sea currents**
- **longshore drift**
- **tides**
- **depth of water offshore**
- type and amount of **sediments offshore**
- **rock type and structure**
- **sub-aerial processes**: runoff, weathering and mass movement
- **land-based agents of erosion**: rivers and glaciers
- **climate and weather**
- **fetch**: the distance over open sea that a wind blows to generate waves; the longer the fetch, the greater the potential for large waves
- **long-term sea-level change**: eustatic (worldwide) and isostatic (local)
- **coastal ecosystems**: sand dunes, salt marshes and mangroves
- the presence of **coral**
- **human activity**

Marine processes

Waves

Waves (Figure 5.1) are created by transfer of energy from the wind blowing over the sea surface. The energy acquired by waves depends upon the strength of the wind, the length of time it is blowing and the distance over which it blows (**fetch**). As waves approach shallow water, friction with the sea bed increases and the base of the wave begins to slow down. This has the effect of increasing the height and steepness of the wave

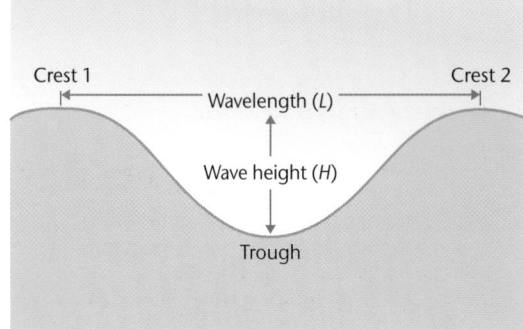

Figure 5.1 Wave terminology

Figure 5.2
Wave movement

Crest of wave rises
as it moves forward:
velocity and wave
length decrease

Wave steepens until it
reaches a wave height:
wavelength ratio of 1:7,
when it will break

Water rushes up the
beach as swash

Friction with sea bed
slows down base of wave

Increasingly
elliptical orbit

Water from previous
wave returns as
backwash

Sea bed

Source: Waugh, D. *An Integrated Approach.*

until the upper part plunges forward and the wave 'breaks' onto the beach (Figure 5.2). The rush of water up the beach is known as **swash** and any water running back down the beach into the sea is the **backwash**. Waves can be described as constructive or destructive.

Constructive waves

Constructive waves (Figure 5.3) are usually associated with long fetch. They tend to be low waves, but with a long wavelength, often up to 100 m. They have a low frequency of around six to eight per minute. As they approach the beach, the wave front steepens only slowly, giving a gentle spill onto the beach surface. Swash rapidly loses volume and energy as water percolates through the beach material. This tends to give a very weak backwash which has insufficient force to pull sediment off the beach or to impede swash from the next wave. As a consequence, material is slowly, but constantly, moved up the beach, leading to the formation of ridges (or **berms**).

Destructive waves

Destructive waves (Figure 5.4) tend to occur when the fetch is shorter. They are high waves with a steep form and a high frequency (10–14 per minute). As they approach the beach, they rapidly steepen and, when breaking, they plunge down. This creates a powerful backwash as there is little forward movement of water. It also inhibits the swash from the next wave. Very little material is moved up the beach, leaving the backwash to pull material away. Destructive waves are commonly associated with steeper beach profiles. The force of each wave may project some shingle well towards the rear of the beach where it forms a large ridge known as the **storm beach**.

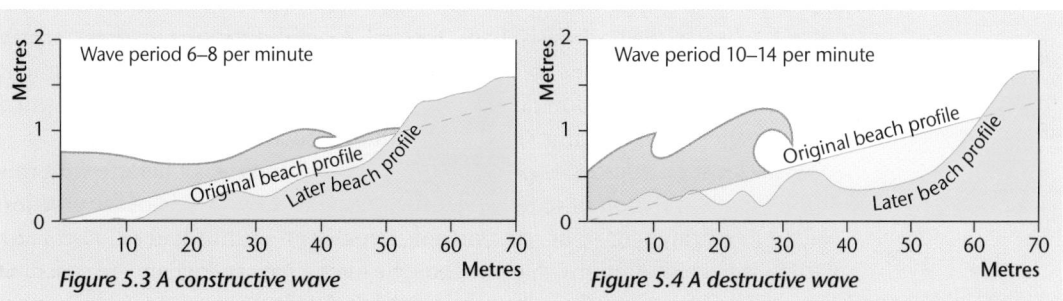

Figure 5.3 A constructive wave

Figure 5.4 A destructive wave

Figure 5.5
Wave refraction

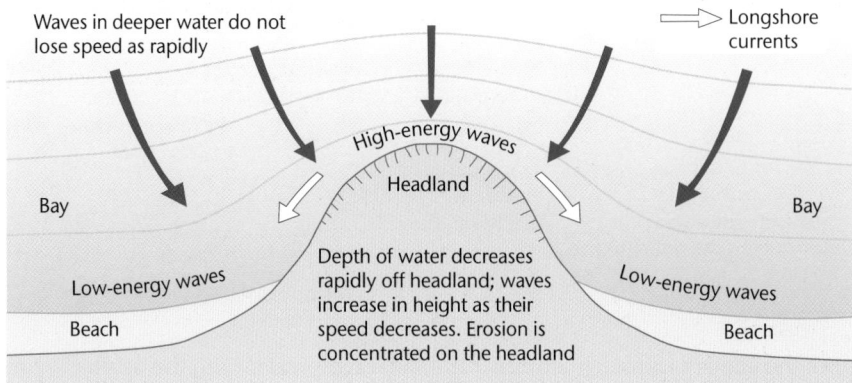

Effects of waves

Most beaches are subject to the alternating action of constructive and destructive waves. Constructive waves build up the beach and result in a steeper beach profile. This encourages waves to become more destructive (as destructive waves are associated with steeper profiles). With time, though, destructive waves move material back towards the sea, reducing the beach angle and encouraging more constructive waves. So the pattern repeats itself. This type of **negative feedback** should encourage a state of **equilibrium**, but this is impossible as other factors, such as wind strength and direction, are not constant.

When waves approach a coastline that is not of a regular shape, they are **refracted** and become increasingly parallel to the coastline. Figure 5.5 shows a headland separated by two bays. As each wave nears the coast, it tends to drag in the shallow water which meets the headland. This causes the wave to become higher and steeper with a shorter wavelength. That part of the wave in deeper water moves forward faster, causing the wave to bend. The overall effect is that wave energy becomes concentrated on the headland, causing greater erosion. The low-energy waves spill into the bay, resulting in beach deposition. As the waves pile against the headland, there may be a slight local rise in sea level that results in a longshore current from the headland, moving some of the eroded material towards the bays and contributing to the build up of the beaches.

Longshore drift

When waves approach the shore at an angle, material is pushed up the beach by the swash in the same direction as the wave approach. As the water runs back down the beach, the backwash drags material down the steepest gradient, which is generally at right angles to the beach line. Over a period of time, sediment moves in this zig-zag fashion down the coast (Figure 5.6). If the material is carried some distance it will become smaller, more rounded and better sorted.

Obstacles such as groynes (wooden breakwaters) and piers interfere with this drift, and accumulation of sediment occurs on the windward side of the groynes, leading to entrapment of beach material. Deposition of this material also takes place in sheltered locations, such as at the head of a bay, and where the coastline changes direction abruptly — here spits tend to develop.

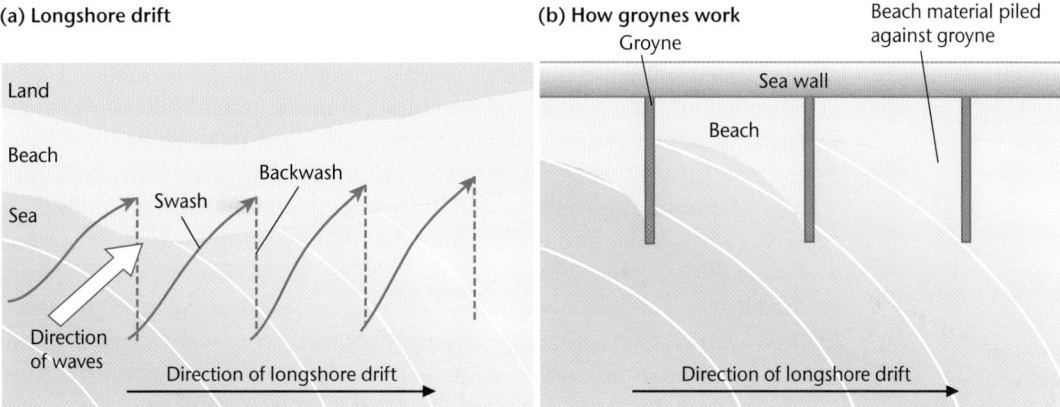

Figure 5.6
Longshore drift

Tides

Tides are the periodic rise and fall in the level of the sea. They are caused by the gravitational pull of the sun and moon, although the moon has much the greatest influence because it is nearer. The moon pulls water towards it, creating a high tide, and there is a compensatory bulge on the opposite side of the Earth (Figure 5.7). In the areas of the world between the two bulges, the tide is at its lowest.

As the moon orbits the Earth, the high tides follow it. Twice in a lunar month, when the Moon, Sun and Earth are in a straight line, the tide-raising force is strongest. This produces the highest monthly tidal range or **spring tide**. Also twice a month, the Moon and Sun are positioned at 90° to each other in relation to the Earth. This alignment gives the lowest monthly tidal range, or **neap tide**.

Figure 5.7
The causes of tides

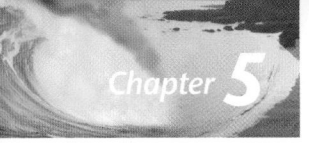

Marine erosion

When waves break on a coastline they often do so with considerable energy. It has been estimated that waves breaking against the foot of a cliff can generate energy of 25–30 tonnes m^{-2}. There are several ways in which waves are able to erode coastlines:

- **Hydraulic action (wave quarrying)** A breaking wave traps air as it hits a cliff face. The force of water compresses this air into any gap in the rock face, creating enormous pressure within the fissure or joint. As the water pulls back, there is an explosive effect of the air under pressure being released. The overall effect of this over time is to weaken the cliff face. Storms may then remove large chunks of it. This process can also lead to extensive damage to sea defences. Some coastal experts also point out that the sheer force of water itself (without debris) can exert an enormous pressure upon a rock surface, causing it to weaken. Such an activity is sometimes referred to as **pounding**.
- **Abrasion/corrasion** The material the sea has picked up also wears away rock faces. Sand, shingle and boulders hurled against a cliff line will do enormous damage. This is also apparent on intertidal rock platforms, where sediment is drawn back and forth, grinding away at the platform.
- **Attrition** The rocks in the sea which carry out abrasion are slowly worn down into smaller and more rounded pieces.
- **Solution (corrosion)** Although this is a form of weathering rather than erosion, it is included here as it contributes to coastal erosion. It includes the dissolving of calcium-based rocks (e.g. limestone) by the chemicals in sea water and the evaporation of salts from water in the rocks to produce crystals. These expand when they form and put stress upon rocks. Salt from sea-water spray is capable of corroding several types of rock.

There are many factors that affect the rate of erosion:

- **Wave steepness and breaking point** Steeper waves are higher-energy waves and have a greater erosive power than low-energy waves. The point at which waves break is also important — waves that break at the foot of a cliff release more energy and cause more damage than those that break some distance away.
- **Fetch** How far a wave has travelled determines the amount of energy it has collected.
- **Sea depth** A steeply-shelving sea bed at the coast will create higher and steeper waves.
- **Coastal configuration** Headlands attract wave energy through refraction.
- **Beach presence** Beaches absorb wave energy and can therefore provide protection against marine erosion. Steep, narrow beaches easily dissipate the energy from flatter waves, while flattish, wide beaches spread out the incoming wave energy and are best at dissipating high and rapid energy inputs. Shingle beaches also deal with steep waves as energy is rapidly dissipated through friction and percolation.
- **Human activity** People may remove protective materials from beaches (sand), which may lead to more erosion, or they may reduce erosion by the construction of sea defences (discussed later). Sea defences in one place, however, may lead to increased rates of erosion elsewhere on the same coastline.

Geology

Several geological aspects affect the rate of erosion.

Lithology refers to the characteristics of rocks, especially resistance to erosion and permeability. Very resistant rocks such as granite, and to a lesser extent chalk, tend to be eroded less than weaker materials such as clay. Some rocks are well-jointed (e.g. limestone), which means that the sea can penetrate along lines of weakness, making them more vulnerable to erosion. Variation in the rates at which rocks wear away is known as **differential erosion**.

The **structure** and **arrangement** of the rocks also affects erosion. When rocks lie parallel to the coast, they produce a very different type of coastline than when they lie at right angles. Figure 5.8 shows two very different types of coastline that can be found close to one another in Purbeck (southern England). The southern part of the coast has the rocks running parallel to it — known as a **concordant** coastline. Here the resistant Portland limestone forms cliffs, and these have protected the coast from erosion, only allowing the sea to break through in a few places (the large area of Worbarrow Bay and the small area of Lulworth Cove) to the clay behind.

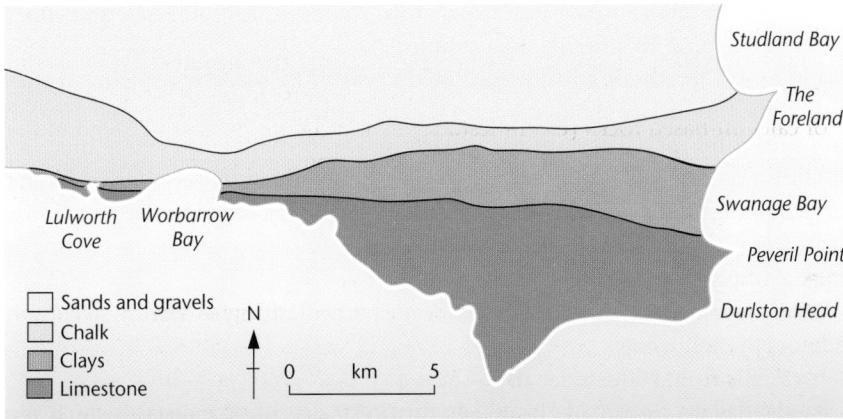

Figure 5.8
The Purbeck coast

To the east, the rocks run at right angles to the coast (known as **discordant**), allowing the sea to penetrate along the weaker clays and gravels and produce large bays (e.g. Swanage Bay) flanked by outstanding headlands (The Foreland and Peveril Point).

The **dip** of the rocks is also a major factor. The steepest cliffs tend to form in rocks that have horizontal strata or which dip gently inland, whereas rocks that dip towards the coast tend to produce much more gently sloping features (Figure 5.9).

Sub-aerial processes

In addition to marine processes, there are also sub-aerial (land-based) processes which shape the coastline. These processes come under the general headings of **weathering** and **mass movement**. Solution was included in the marine processes listed above because it is a major process that combines with erosion to produce

Figure 5.9
The influence of
rock strata on
coastlines

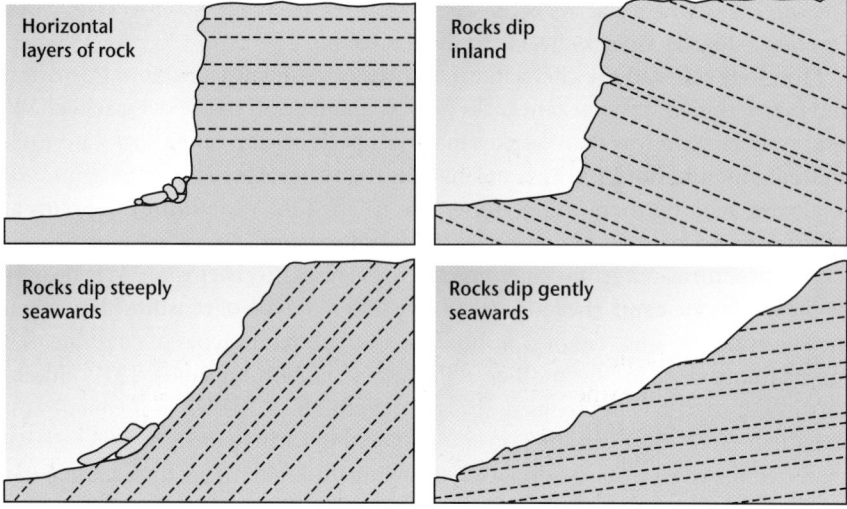

coastlines. Other weathering processes that can be effective include frost shattering, exfoliation (thermal expansion), biological weathering and other forms of chemical weathering than solution.

Biological weathering is quite active on coastlines. Some marine organisms, such as the piddock (a shellfish), have specially adapted shells to enable them to drill into solid rock. They are particularly active in chalk areas where they can produce a sponge-like rock pitted with holes. Seaweed attaches itself to rocks and the action of the sea can be enough to cause swaying seaweed to prise away loose rocks from the sea floor. Some organisms, algae for example, secrete chemicals capable of promoting solution.

Mass movement is common on coastlines, particularly those that are steep, and includes the following:

- **rock falls** from cliffs undercut by the sea
- **landslides** on cliffs made from softer rocks or deposited material, which slip down when lubricated
- **mudflows** — heavy rain can cause fine material to flow downhill where it sometimes takes on the appearance, and movement, of a glacier
- where softer material overlies much more resistant materials, cliffs are often subject to **slumping**. With excessive lubrication, whole sections of the cliff face may move downwards with a slide plane that is concave, producing a rotational movement. Slumps are a very common feature of the British coast,

Figure 5.10
Rotational
slumping

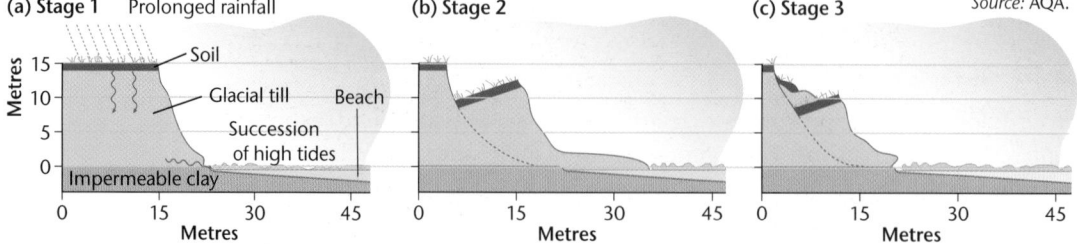

particularly where glacial deposits form the coastal areas, e.g. east Yorkshire and north Norfolk. Figure 5.10 shows a typical rotational slump in an area where glacial deposits form cliffs on top of an impermeable clay layer.

Features produced by coastal erosion

Headlands and bays

Figure 5.8 shows the impact of geology on a coastline. There are many similar parts of the British coastline where there are areas of alternating resistant and less resistant rocks. The less resistant rocks experience most erosion and develop into bays, while the more resistant rocks become headlands. Because of refraction, the headlands receive the highest-energy waves and are more vulnerable to the forces of erosion than are the bays. The bays experience low-energy waves that allow sediment to accumulate and form beaches. These act to protect that part of the coastline.

Where the rocks run parallel with the coast, as in Purbeck (Figure 5.8), it is possible for continued erosion to break through the more resistant rocks on the coast and begin to attack the weaker strata behind. If that happens, a cove will

Kitchenham

Photograph 5.1 Aerial view of Lulworth Cove, Dorset. Worbarrow Bay is visible beyond

Figure 5.11
The geology of
south Purbeck

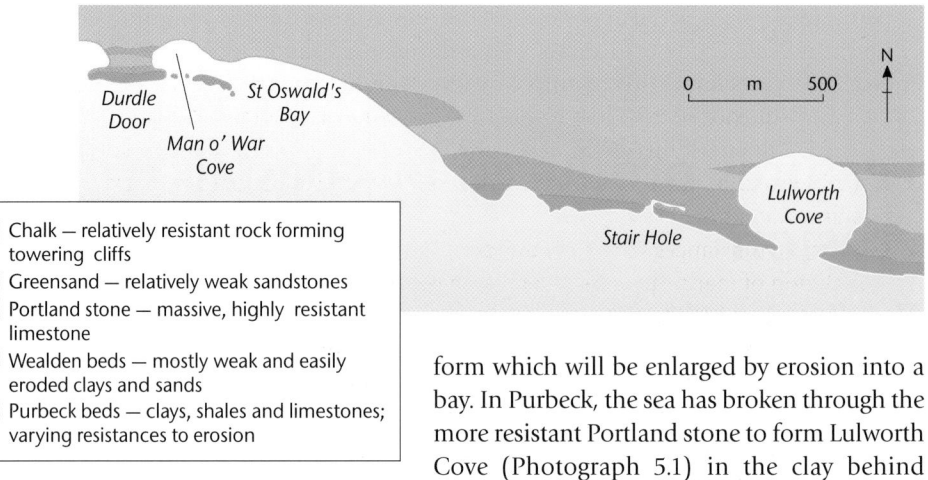

Durdle
Door

St Oswald's
Bay

Man o' War
Cove

0   m   500

N

Lulworth
Cove

Stair Hole

Chalk — relatively resistant rock forming
towering cliffs

Greensand — relatively weak sandstones

Portland stone — massive, highly resistant
limestone

Wealden beds — mostly weak and easily
eroded clays and sands

Purbeck beds — clays, shales and limestones;
varying resistances to erosion

form which will be enlarged by erosion into a bay. In Purbeck, the sea has broken through the more resistant Portland stone to form Lulworth Cove (Photograph 5.1) in the clay behind (although there is some evidence that this could have been a former river mouth). Just along from Lulworth Cove is Stair Hole. Here the sea enters through two arches and has begun to work its way along the weaker clays (Figure 5.11).

Wave-cut platforms

When high and steep waves break at the foot of a cliff they concentrate their erosive capabilities into only a small area of the rock face. This concentration eventually leads to the cliff being undercut, forming a feature known as a **wave-cut notch**. Continued activity at this point increases the stress on the cliff and in time it collapses.

The cliff begins to retreat, leaving at its base a gently sloping (less than 5°) wave-cut platform (Figure 5.12). When viewed from a distance, the platform looks remarkably even as it cuts across the rocks, regardless of their hardness. On closer inspection, the platform is often deeply cut into by abrasion from the huge amount of material that is daily carried across it, and by the effects of chemical action.

The platform continues to grow and, as it does, the waves break further out to sea and have to travel across more platform before reaching the cliff line. This leads to a greater dissipation of wave energy, reducing the rate of erosion on the headland and slowing down the growth rate of the platform. There tends, therefore, to be a limit to how big the feature can grow and some experts have suggested that growth beyond 0.5 km is unusual.

Figure 5.12
Formation of a
wave-cut platform

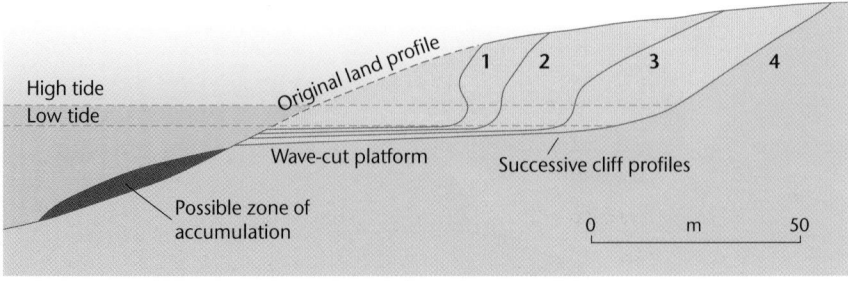

High tide

Low tide

Original land profile

1   2   3   4

Wave-cut platform

Successive cliff profiles

Possible zone of
accumulation

0   m   50

Geos, caves, blowholes, arches, stacks and stumps

These features are all independently observable on British coastlines, but they also represent a sequence of events in the erosion of a cliff or headland. On any cliff line the sea will attack the weakest parts such as cracks, joints or along bedding planes. Along a joint, the sea will cut inland, widening the crack to form a narrow, steep-sided inlet known as a **geo**.

In other circumstances the cliff is undercut and a **cave** is formed, usually from a combination of marine processes. If erosion continues upwards, it is possible for the cave to be extended to the top of the cliff, where a **blowhole** will form. Much more likely is that the cave will extend backwards to meet another, eventually creating a hole all the way through the headland, known as an **arch**.

As the cliff recedes and the wave-cut platform develops, the arch will eventually collapse, leaving its isolated portion as a **sea stack** standing above the platform. With time, the sea will exploit the wave-cut notch at the base of the stack, leading eventually to its collapse. A small raised portion of the wave-cut platform may be left marking the former position of the stack. This is known as a **stump**.

There are several well-known areas of Britain where these features stand out. Flamborough Head in Yorkshire has a well-developed wave-cut platform in chalk, along with sea caves, arches and a large blowhole. In the old red sandstone rocks of the Orkney Islands there is a well-known stack, the Old Man of Hoy, and the Needles on the Isle of Wight are another example of the same feature, although they look different because they are formed from chalk.

On Purbeck, in the Portland stone (a highly resistant limestone), the sea has cut the well-known arch of Durdle Door. Also on Purbeck (Figure 5.8) the chalk escarpment culminates in The Foreland and its detached pieces that are known as Old Harry Rocks. Figure 5.13 is a sketch of this area, where the sequence from headland to stack can be clearly seen. Marine erosion and sub-aerial processes will eventually reduce the upstanding parts of this area to a wave-cut platform.

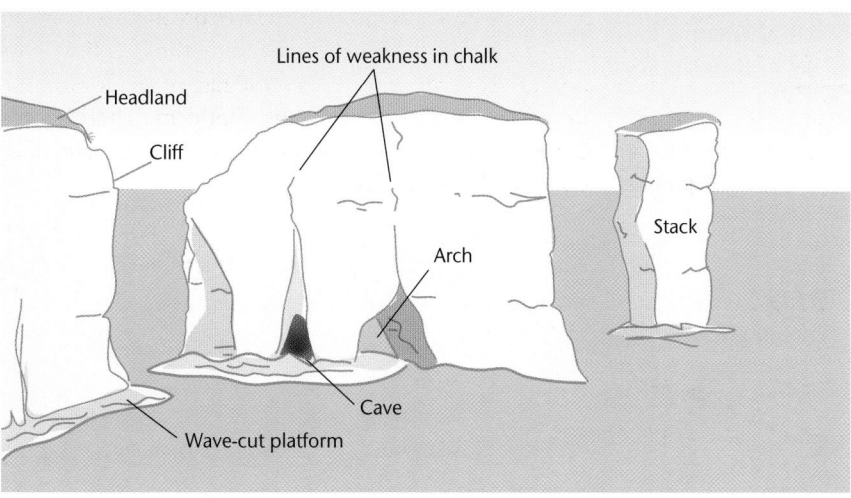

Figure 5.13 Coastal erosion features seen at Old Harry Rocks, Purbeck

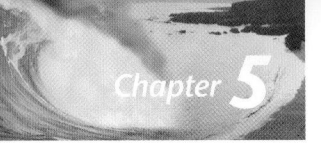

Features of coastal deposition

Deposition occurs on coastlines where sand and shingle accumulate faster than they are removed. It often takes place where the waves are low energy or where rapid coastal erosion provides an abundant supply of material.

Beaches

Beaches represent the accumulation of material deposited between low spring tides and the highest point reached by storm waves. They are mainly constructed from sand and shingle. Sand produces beaches with a gentle gradient (usually under 5°) because its small particle size means the sand becomes compact when wet, and allows very little percolation. Most of the swash therefore returns as backwash, little energy is lost to friction, and material is carried down the beach. This leads to the development of **ridges** and **runnels** in the sand at the low-water mark. These run parallel to the shoreline and are broken by channels that drain the water off the beach (Figure 5.14).

Shingle may make up the whole, or just the upper parts of the beach. The larger the size of the material, generally the steeper is the gradient of the beach (usually 10–20°). This is because water rapidly percolates through shingle, so the backwash is somewhat limited. This, together with the uneven surface, means that very little material is moved down the beach.

At the back, strong swash at spring high-tide level will create a **storm beach**, a ridge composed of the biggest boulders thrown by the largest waves. Below this will be a series of ridges marking the successively lower high tides as the cycle goes from spring to neap. These beach ridges are known as **berms** and are built up by constructive waves. **Cusps** are semicircular-shaped depressions which form when waves break directly on to the beach and swash and backwash are strong. They usually occur at the junction of the shingle and sand beaches. The sides of the cusps channel incoming swash into the centre of the embayment and this produces a stronger backwash in the central area which drags material down the beach, deepening the cusp.

Below this, **ripples** are developed on the sand by wave action or tidal currents.

Figure 5.14
Beach features

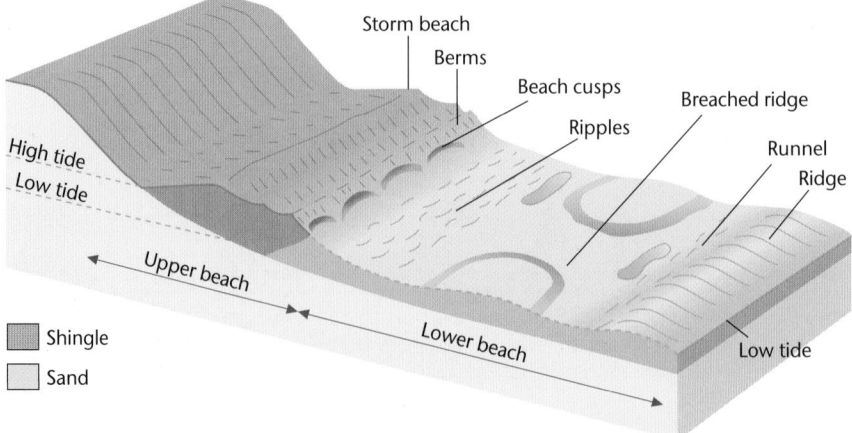

Spits and bars

A spit is a long, narrow piece of land that has one end joined to the mainland and projects out into the sea or across an estuary. Like other depositional features, it is composed of sand and/or shingle and the mixture very much depends upon the availability of material and the wave energy required to move it.

Figure 5.15 shows the formation of a spit. On the diagram, the prevailing winds and maximum fetch are from the southwest, so material will be carried from west to east along the coast by the process of longshore drift. Where the coastline changes to a more north–south orientation, there is a build-up of sand and shingle in the more sheltered water in the lee of the headland. As this material begins to project eastwards, storms build up more material above the high-water mark, giving a greater degree of permanence to the feature. Finer material is then carried further eastward, extending the feature into the deeper water of the estuary.

Increasingly, though, the end of the spit begins to curve round as wave refraction carries material round into the more sheltered water. The second most dominant wind direction and fetch may contribute to this, pushing the spit material back towards the mainland. Several curved ends may develop during a period of southeast weather until the dominant southwest movement reasserts itself. The spit cannot grow all the way across the estuary as the material will be carried seaward by the river and the deeper water at the centre inhibits growth.

Spits are often associated with two other features:

■ **Sand dunes** form as dried out sand is blown to the back of the spit where it increasingly accumulates. Stability is achieved if vegetation such as marram grass begins to colonise the area and hold the dunes together.

■ **Salt marshes** form as low-energy, gentle waves enter the sheltered area behind the spit and deposit the finer material such as silt and mud. This builds up to form a feature which is then colonised by vegetation.

Around the British coasts, well-known spits are found at Borth (west Wales), Dawlish Warren (Devon), Hurst Castle (Solent), Orford Ness (East Anglia, Figure 5.16 and Photograph 5.2) and Spurn Head (Humber estuary). A spit that joins an island to the mainland is known as a **tombolo**. The best example in Britain is Chesil Beach on the south coast of England. This links the Isle of Portland to the mainland and is about 30 km long.

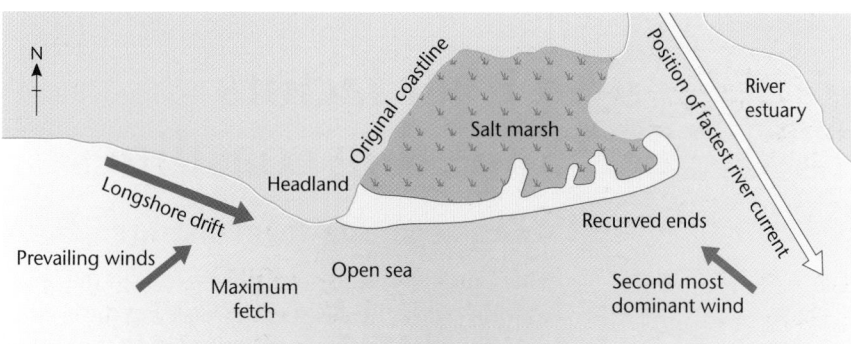

Figure 5.15
The formation
of a spit

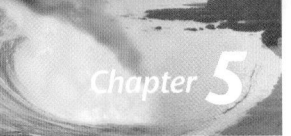

*Photograph 5.2
Aerial view of
Orford Ness*

Aerofilms

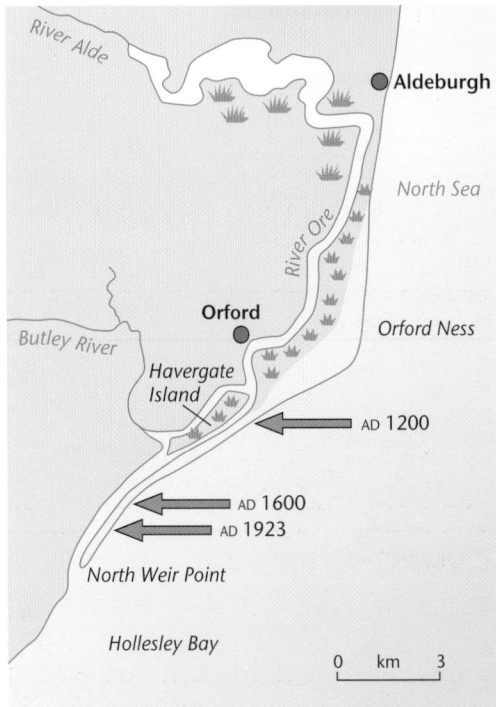

Figure 5.16 The spit at Orford Ness

If a spit develops across a bay where there is no strong flow of water from the landward side, it is possible for the sediment to reach across to the other side. In this case, the feature is known as a **bar**. Some bars, however, may simply be the result of the onshore migration of material from offshore as sea levels rose following the last ice age. Slapton Ley, a bar formed in Devon, is believed to have come about in this way. Recent work on Chesil Beach has suggested a similar cause, although the spit was probably formed by a combination of onshore migration of sea bed materials and longshore drift.

Other factors affecting coastlines

Long-term sea-level change

Tides are responsible for daily changes in the levels at which waves break on to the land, but the average position of sea level relative to the land has changed through time. Although sea levels have been fairly static since the last ice age, many changes took place during the Quaternary glaciation that reflected both the advance and retreat of the ice. A typical sequence would have run as follows:

- **Stage 1** As the climate begins to get colder, marking the onset of a new glacial period, an increasing amount of precipitation falls as snow. Eventually, this snow turns into glacier ice. Snow and ice act as a store for water, so the hydrological cycle slows down — water cycled from the sea to the land (evaporation, condensation, then precipitation) does not return to the sea. As a consequence, sea levels fall and this affects the whole planet. Such a worldwide phenomenon is known as a **eustatic** fall.
- **Stage 2** The weight of ice causes the land surface to sink. This affects only some coastlines and then to a varying degree. Such a movement is said to be **isostatic** and it moderates the eustatic sea-level fall in some areas.
- **Stage 3** The climate begins to get warmer. Eventually the ice masses on the land begin to melt. This starts to replenish the main store and sea levels rise worldwide (eustatic). In many areas this floods the lower parts of the land to produce **submergent** features such as flooded river valleys (**rias**) and flooded glacial valleys (**fjords**).
- **Stage 4** As the ice is removed from some land areas they begin to move back up to their previous levels (isostatic readjustment). If the isostatic movement is faster than the eustatic, then **emergent** features are produced such as **raised beaches**. Isostatic recovery is complicated as it affects different places in different ways. In some parts of the world it is still taking place as the land continues to adjust to having masses of ice removed. Today, the southeast of the British Isles is sinking while the northwest is rising.

Submergent features

Rias are created by rising sea levels drowning river valleys. The floodplain of a river will vanish beneath the rising waters, but on the edges of uplands only the middle and upper course valleys will be filled with sea water, leaving the higher land dry and producing this feature. In Devon and Cornwall, for example, sea level rose

Figure 5.17
A ria

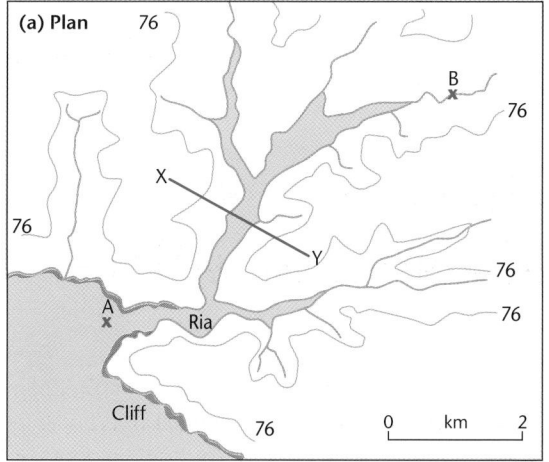

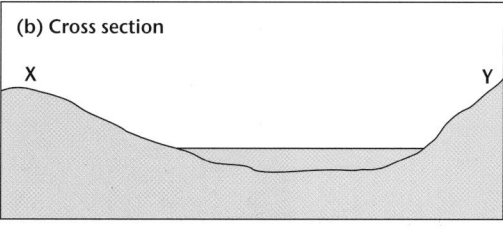

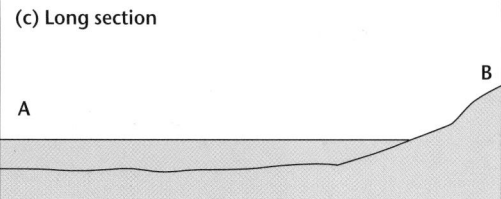

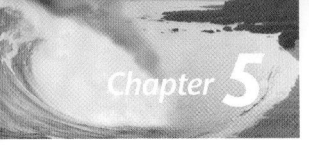
and drowned the valleys of the rivers flowing off Dartmoor and the uplands of Cornwall. Good examples are the Fowey estuary in Cornwall and the Kingsbridge estuary in south Devon. Rias have a long section and cross profile typical of a river valley, and usually a dendritic system of drainage (Figure 5.17).

Fjords are drowned glacial valleys typically found on the coasts of Norway, southwestern New Zealand, British Columbia in Canada, southern Chile and

Figure 5.18
A fjord

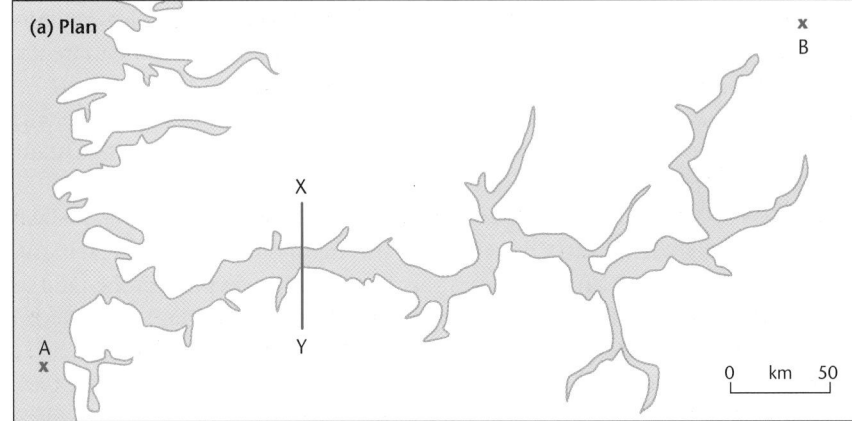

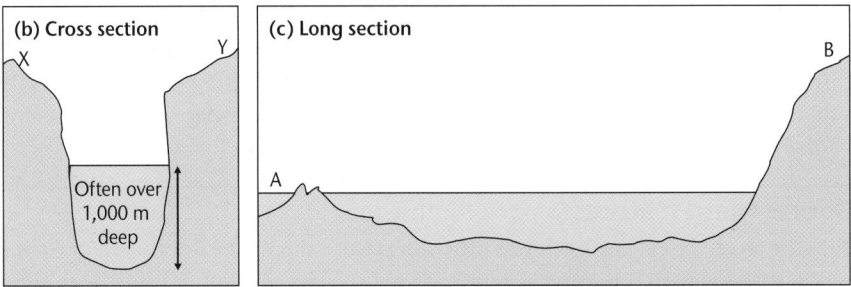

Photograph 5.3
The entrance to
Milford Sound fjord,
New Zealand

Greenland. The coast of western Scotland contains fjords which are not as well developed as those in the areas above because the ice was not as thick and did not last for the same length of time.

Fjords have steep valley sides (cliff-like in places) and are fairly straight and narrow (Figure 5.18). Like glacial valleys, they have a typical U-shaped cross section with hanging valleys on either side. Unlike rias, they are not deepest at the mouth, but generally consist of a glacial rock basin with a shallower section at the end, known as the **threshold**. They were formed when the sea drowned the lower part of glacial valleys that were cut to a much lower sea level. Good examples include Sogne Fjord in Norway, which is nearly 200 km long, and Milford Sound in New Zealand (Photograph 5.3).

Emergent features

Raised beaches are areas of former wave-cut platforms and their beaches which are at a level higher than the present sea level. Behind the beach it is not unusual to find old cliff lines with wave-cut notches, sea caves, arches and stacks. Raised beaches are common around the coasts of western Scotland where three levels have been recognised, at 8 m, 15 m and 30 m. Because of differential uplift these are only approximate heights.

On the west of the Isle of Arran there is a well-developed raised beach north of Drumadoon. This has a relict cliff, arches, stacks and caves, including the well-known King's Cave. This beach is around 4 or 5 m above present sea level and is probably the equivalent of the 8 m beach. It was clearly produced when the sea was at that level, which initially suggests that the sea has fallen to its present level. However, we know that sea levels have *risen* considerably (eustatic) since the end of the last ice age, so the beach must have reached its raised position by isostatic rising of the land. The land locally must have risen faster than the eustatic rise in sea level to create this emergence.

Vegetation, sedimentation and stabilisation

Coastal sand dunes

Coastal sand dunes are accumulations of sand shaped into mounds by the wind. They represent a dynamic landform. Beaches are the source of the sand which, when dried out, is blown inland to form dunes. Sand is moved inland by a process known as **saltation** (a bouncing action which is also seen in particles moved along by running water).

During the day, the wind on the coastal fringe is generally from the sea. Air moves in response to the small pressure differentials set up by the warmer land and the colder sea. When there is a large tidal range, large amounts of sand may be exposed at low tide, and this further contributes to dune formation. The sequence of sand-dune development (Figure 5.19) is as follows:

- Sand may become trapped by obstacles (seaweed, rock, driftwood) at the back of the beach, possibly on the highest berm or storm beach. Sand is not a very hospitable environment for plant growth so only very hardy plants will begin to colonise here. Such plants are called **pioneers** — examples of sand-dune pioneers are sea rocket and prickly saltwort.

- The first dunes to develop are known as **embryo dunes**. They are suitable for colonisation by grasses such as sea couch, lyme and marram. These are able to grow upwards through accumulating wind-blown sand, stabilising the surface. As a result low, hummocky dunes are formed. Marram is a robust plant which spreads vigorously by underground shoots (rhizomes). This is still a difficult environment and plants need certain features to survive. Sea couch has succulent leaves to store water, prickly saltwort has thorn-like leaves which reduce transpiration and conserve water, and marram possesses long tap roots to draw moisture from the water table. Plant growth of this kind adds organic matter to the dunes which aids water retention.

- Upward growth of embryo dunes raises the height to create dunes that are beyond the reach of all but the highest storm tides. These **foredunes (mobile dunes)** are initially yellow, because they contain little organic matter, but as colonisation increases, plants like marram begin to add humus to the sand. As a result, the dunes look more grey in colour and may reach heights in excess of 20 metres.

- The dunes inland gradually become **fixed.** An organic layer develops which improves nutrient supply and water retention, allowing more plant colonisation. Lichens, mosses and flowering plants begin to appear and marram is slowly replaced by red fescue grass. Other plants include creeping willow and dewberry.

- In places **dune slacks** develop. These are depressions within the dunes where the water table is on or near the surface and conditions are often damp. Rushes, reeds, mosses and willows can be found, but the plants present will very much depend upon the amount of moisture.

- Behind the yellow and grey dunes, the supply of beach sand is gradually cut off, giving smaller dune features. This area is often referred to as **dune heath,** and heather, gorse, broom and buckthorn are the main plants. Towards the rear of the dune system, woodland may occur with trees such as pine, birch and the occasional oak. This is beginning to lead into the climatic climax vegetation for the British Isles, but in many areas the dunes may be planted with conifers to stabilise the area. If this is the case, then the vegetation is said to be a plagio-climax (short of the climatic climax because of human interference). Within this system, it is possible to find **blowouts** where wind has been funnelled through areas and removed the sand.

Figure 5.19
A typical sand-dune transect

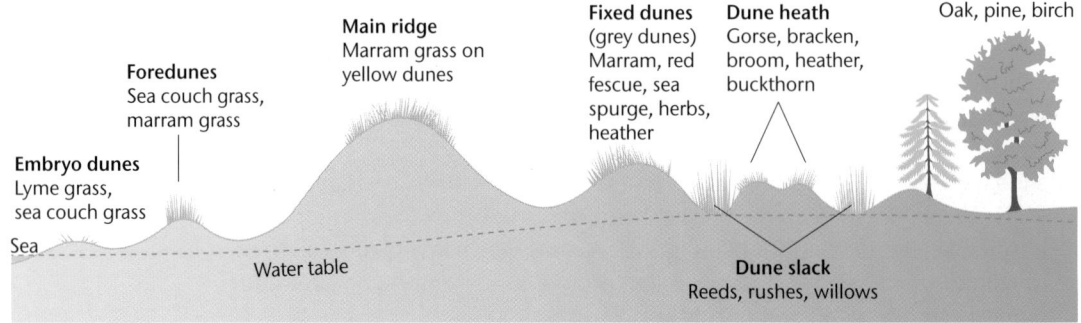

Sand dunes are an example of a **succession**, a plant community where the structure develops over time. At each stage, certain species have evolved to exploit the particular conditions present. Initially only a small number of species will be capable of thriving in a harsh environment. These hardy pioneering plants gradually modify the conditions by altering such things as the mineral and moisture content of the soil and the amount of shade. As each new plant species takes hold, the process is repeated. Changes made by the plants present allow other species, better suited to this modified habitat, to succeed the old species. When the succession has reached a point where it is in balance with the climatic conditions, a climax is said to have been reached. A succession that develops on sand is called a **psammosere**.

Salt marshes

In sheltered river estuaries or behind spits, silt and mud are deposited by rivers or gentle tides to form intertidal mudflats. Upon these, vegetation will develop which, like that of the sand-dune environment, changes through time. The succession that develops (Figure 5.20) is known in this case as a **halosere** (tolerant of salty conditions) and follows these stages:

- Mudflats are formed by deposition of fine material. This may be aided by the growth of eelgrass that slows currents and leads to further, uneven, deposition.
- **Pioneers** begin to colonise the area. These are plants able to tolerate salt and periodic submergence by the sea. They are known as **halophytes** and examples include glasswort, sea blite and *Spartina*. *Spartina* has two root systems — a fine mat of surface roots to bind the mud, and long, thick, deep roots that can secure it in up to 2 metres of deposited material. This enables the plant to trap more mud than other pioneers, and thus it has become the dominant vegetation on tidal flats in the British Isles.
- The pioneers gradually develop a close vegetation over the mud and this allows colonisation by other plants such as sea aster, marsh grass and sea lavender. These form a dense mat of vegetation up to 15 cm high. The growth of vegetation has the effect of slowing the tidal currents even further and this, together with the vegetation's ability to trap particles, leads to more mud and silt accumulation. Dead organic matter also helps to build up the surface, which grows in height at anywhere between 1 and 25 mm per year.

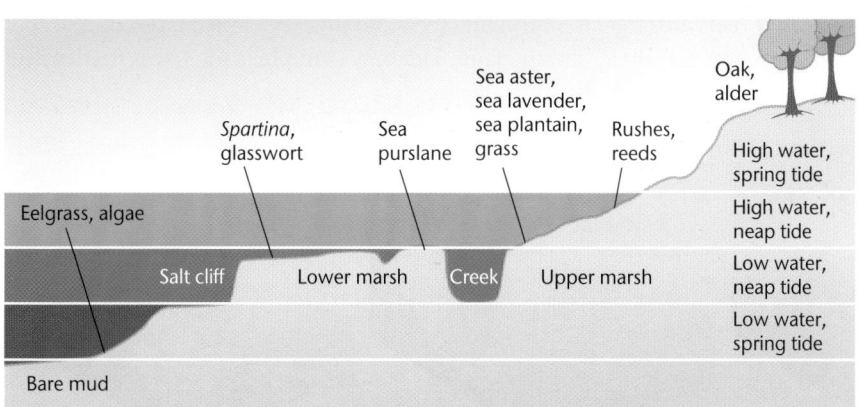

*Figure 5.20
The structure of
a salt marsh*

119

■ As mud levels rise, complex creek systems develop that channel the tides and these deepen as the marsh becomes higher. Hollows may form where seawater becomes trapped and evaporates, leaving salt-pans in which the salinity is too great for plants to survive. As the land rises above sea level, rushes (such as *Juncus* species) and reeds become established, eventually leading to the growth of trees, such as alder, ash and then oak to complete the succession. This land is now rarely covered by the sea.

Photograph 5.4
Coral reef

Coral reef formation

Coral reefs are unique among marine ecosystems in that they are built up entirely by biological activity. Reefs are produced by coral polyps — very simple organisms with small cylindrical bodies and a ring of tentacles at one end surrounding a mouth. Many live in colonies in shallow tropical seas, where they build reefs (Photograph 5.4). Coral polyps can only live within a relatively confined set of conditions:

■ water temperature of at least 18°C, and preferably 23–25°C
■ water depth of 25 m or less
■ enough light for photosynthesis by the tiny algae (zooxanthellae) which live symbiotically with corals and provide them with 98% of their food requirement
■ minimum salinity levels of 30–32 psu
■ relatively sediment-free water — too much sediment will clog up their feeding mechanisms
■ wave action to ensure water is well-oxygenated, but not so strong that the polyps are damaged

There are three types of coral reef (Figure 5.21):
■ **Fringing reefs** surround islands, such as many of those in the Caribbean. They are separated from the island by a narrow lagoon.
■ **Barrier reefs** are larger than fringing reefs and there is a much larger and deeper lagoon between them and the land. The best example is the Great Barrier Reef off northeastern Australia.

Figure 5.21
Types of coral reef

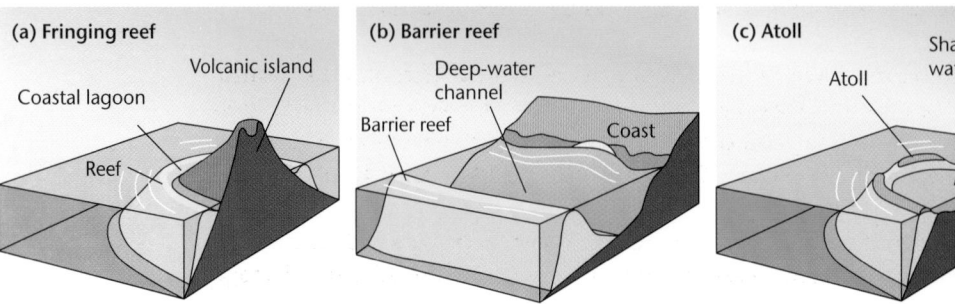

- **Coral atolls** consist of a ring-shaped reef that encloses a lagoon, but there is no island in the centre. Atolls usually rise from very deep water — the Maldives chain of islands is a typical example.

It is difficult to reconcile the fact that atolls rise from very deep water with the conditions listed above, particularly the requirement for shallow water. The theory of reef and atoll formation was first developed by Charles Darwin, who proposed the following stages (Figure 5.22):

- A volcanic island appears in a tropical region as the result of an undersea eruption. Fringing reefs grow around the island in shallow water.
- The island subsides, or sea level rises relative to the land. If this process is slow enough, the coral will grow upwards to keep pace with the changing level and become a barrier reef. This means the reef becomes separated from the island by a larger lagoon.
- As the process continues, the island eventually vanishes, leaving the atoll. Coral continues to grow on the outside of the reef, where the waters are well oxygenated. In the central lagoon, the water is quieter and there is increased sedimentation.

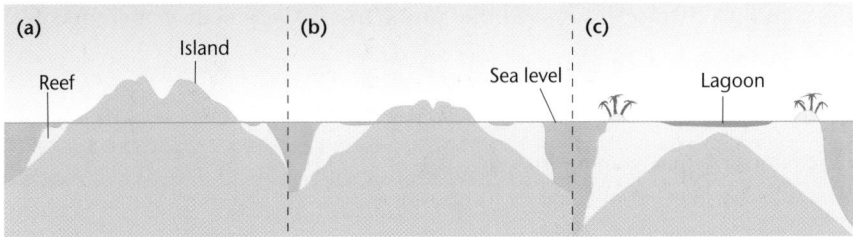

*Figure 5.22
The Darwinian
theory of reef
formation*

Coral reefs are under enormous environmental pressure and it is believed that at least 10% of them are dead or degraded beyond recovery. Oceanographers have estimated that, at the present rate of decay, coral reefs in some areas will be wiped out by the middle of the twenty-first century. Reefs are being damaged at a greater rate than they can self-repair by new coral growth. A number of initiatives were developed in the 1990s to address the problem. These included the International Coral Reef Initiative (1995) and the International Year of the Coral Reef (1997). The main threats to coral reefs come from the following:

- **Global warming** As oceans warm, corals become stressed and expel their zoo-xanthellae. It is the algae that provide colour and food, so the corals turn white or 'bleach' and die. Change in sea temperature can also be caused by the local effects of El Niño. Global warming causes sea levels to rise and corals cannot keep pace.
- **Increased storm levels** Hurricanes have a devastating effect on reefs, particularly if the coral is diseased and under stress. Storm waves often pound reefs into rubble. They also produce vast amounts of rainfall that can wash quantities of sediment into the reef area. Global warming could increase the number and magnitude of storms.
- **The crown of thorns starfish** This creature occurs in plagues and attacks coral reefs. A rise in numbers in several areas has resulted in excessive reef damage.

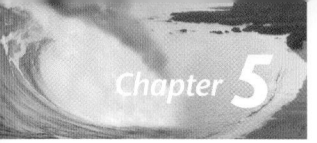

Biologists believe the rise in starfish numbers is caused by nutrient enrichment from the land and overfishing which has killed off a number of their predators.

- **Rapid population growth in coastal areas** This has a number of effects including deforestation of coastal areas, leading to greater sedimentation of reef areas; construction of harbours; pollution from agricultural chemicals, from the growth of industry and from sewage.
- **Increased tourism** This leads to damage by boat anchors and divers; pollution from boats; dredging for harbours and marinas.
- **Commercial exploitation** Coral mining for building materials, lime and for making jewellery.
- **Overfishing** Direct damage is caused by certain fishing activities (e.g. blast fishing), and overfishing upsets the balance of the food chain.

Coastal management

Coastal areas contain a variety of landforms which are increasingly coming under pressure from both natural processes and human activities. In response to this, a range of protection and management strategies has been put into place in many coastal areas. These solutions are often relatively successful but, in some cases, the solutions themselves cause other problems. Coastal management has two main aims:

- to provide defence against flooding
- to provide protection against coastal erosion

Other aims of management include:

- stabilising beaches affected by longshore drift
- stabilising sand-dune areas
- protecting salt marshes

There are a number of approaches to defence of the coast.

Figure 5.23
Examples of hard engineering solutions

The hard engineering approach

Hard engineering involves the building of some type of sea defence, with a specific purpose (Figure 5.23):

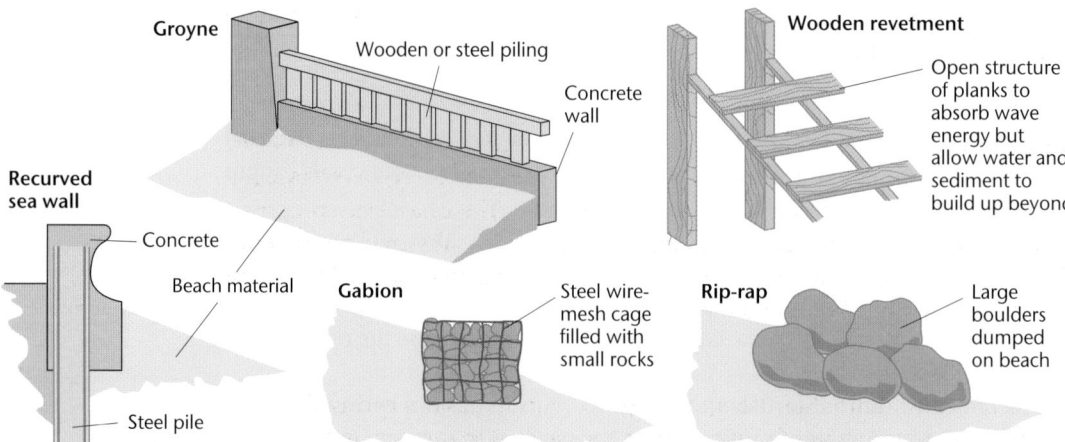

- **Sea walls (sometimes recurved)** aim to absorb wave energy. The recurved structure throws waves backwards. Sea walls must have a continuous facing because any slight gap will be exploited by hydraulic action. They also need drain outlets so that water does not accumulate behind them.
- **Rock armour (rip-rap)** consists of large boulders dumped in front of a cliff or sea wall to take the full force of the waves.
- **Gabions** operate on the same principle as rip-rap, but the boulders are contained within a steel wire-mesh cage.
- **Revetments** are concrete or wooden structures placed across the beach to take the full force of wave energy.
- **Groynes** are wooden or steel breakwaters built nearly at right angles to waves (usually 5–10° to the perpendicular to prevent scouring on the downdrift side of the groyne). They are built to control longshore drift but will also break up the waves as they hit the coast. Halting the bulk of longshore drift in an area may have serious effects down the coast where it will cut off the supply of beach material and could leave the coast exposed to erosion.
- **Cliff fixing** is often done by driving iron bars into the cliff face, both to stabilise it and to absorb some wave power.
- **Offshore reefs** get the waves to break offshore which reduces their impact on the base of cliffs.

Hard engineering has several disadvantages:
- structures can be expensive to build and to maintain (to repair a sea wall can cost up to £5,000 per metre)
- defence in one place can have serious consequences for another
- defence structures may not keep pace with rising sea levels
- structures are sometimes an eyesore, spoiling the landscape

The soft engineering approach
Soft engineering uses natural systems for coastal defence, such as beaches, dunes and salt marshes which can absorb and adjust to wave and tide energy. It involves manipulating and maintaining these systems, without changing their fundamental structures. **Beach nourishment** is the attempt to replace material that has been lost through longshore drift. It is not unknown for local councils to move material from one end of a beach to the other before the start of the tourist season.

Managed retreat
This involves abandoning the current line of sea defences and then developing the exposed land in some way, perhaps with salt marshes, to reduce wave power. In this way the scale of hard sea defences can be reduced. There have been proposals in some areas to ban new developments on the coast. In California, for example, there are already requirements on some stretches of coastline that building must be a certain distance from the shore.

Do nothing!
In recent years a school of thought has grown up that asks whether the coast *should* be protected. Tens of millions of pounds are spent annually in the UK on

coastal protection and it might be cheaper to let nature take its course and pay compensation to those affected. Figure 5.24 is a newspaper report on the findings of the House of Commons select committee on agriculture in August 1998. The committee suggested that large tracts of land should be 'surrendered to the sea' as trying to protect them was a waste of money.

Figure 5.24

We must surrender our land to the sea, say MPs

Huge tracts of Britain's coastal land, especially along the east and south coast, should be surrendered to the sea as part of a 'peaceful accommodation' with nature, MPs said yesterday.

They set out a stark vision of a dramatically different coastal and riverside landscape complete with floodplains and regularly water-logged farmlands as mankind showed more 'humility' in the face of the sea.

Describing the millions spent on flood prevention and coastal defences as an unsustainable and 'deluded' waste of money, the agricultural select committee said it was time to give up the fight along much of the East Anglian and southeast coasts.

Daily Telegraph, 6 August 1998

Case study *Coastal management on the Isle of Wight*

The Isle of Wight Council, like all others on the UK coast, has four coastal defence options (Figure 5.25):

- **Hold the line** Retain the existing coastline by maintaining current defences or building new ones where existing structures no longer provide sufficient protection.
- **Do nothing but monitor** On some stretches of coastline it is not technically, economically or environmentally viable to undertake defence works. The value of the built environment here does not exceed the cost of installing coastal defences.
- **Retreat the line** Actively manage the rate and process by which the coast retreats.
- **Advance the line** Build new defences seaward of the existing line.

The examples of different approaches highlighted on Figure 5.25 are described below.

Monk's Bay (1)

Cliff failure resulting from a combination of high-energy destructive waves and high rainfall during

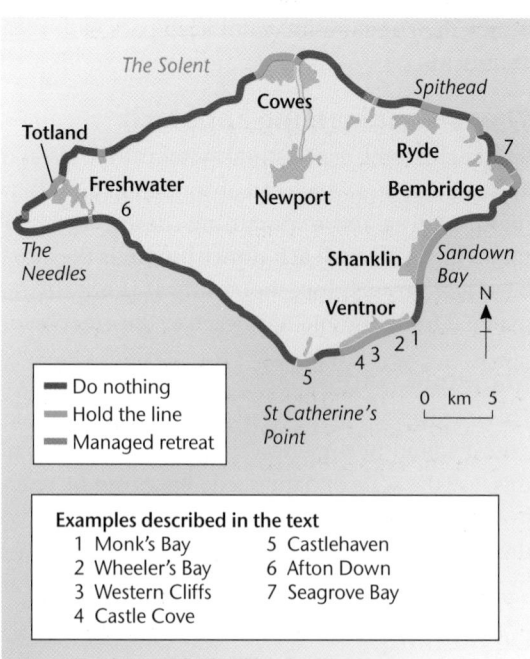

Figure 5.25 Isle of Wight coastal defences

the severe storms of the winter of 1990/91 gave additional impetus for upgrading coastal defence here. The scheme involved constructing an **offshore breakwater**, six **rock groynes** and a **rock revetment** to reinforce the existing sea wall using 25,000 tonnes of Swedish granite. It also required **beach nourishment** using 17,000 m^3 of sand and gravel, **reprofiling** the slope and installing **land drainage** to check the active mass movement of the cliffs on the western side of the bay.

The collective value of the property far exceeded the £1.4 million cost of the scheme. It was completed in 1992 but sedimentation of the rock groynes has since been a problem.

Wheeler's Bay (2)

The ageing sea walls were in danger of collapse which would have reactivated ancient landslides. Property on the cliff behind was becoming unsaleable. Over 15,000 tonnes of Norwegian granite was placed seaward of the existing defences to form a **rock revetment** and the coastal slopes were **regraded to make a shallower profile** before installing **land drainage**. The scheme was completed in 2000 at a cost of £1.6 million and has led to a recovery in property values.

Western Cliffs (Ventnor) (3)

High-energy waves were removing chalk blocks which protected an ancient landslide complex upon which houses had been built. There was a danger that the landslides would be reactivated if sufficient chalk was removed. £1.2 million of Carboniferous limestone was brought in from the Mendip Hills in Somerset to construct a 700 m long **rock revetment** along the base of the cliffs using blocks weighing 6–8 tonnes. A series of limestone **rock groynes** was also constructed at 100 m intervals at the base of the cliffs.

Castle Cove (4)

The existing wooden revetments were being left progressively ineffective as the clay cliffs retreated. Property valued in excess of £10 million was increasingly at risk as coastal processes activated ancient landslides. The scheme stabilised the environmentally-sensitive cliffs by removing the topsoil, **stabilising the slopes** with thousands of tonnes of chalk and installing **land drainage** before replacing the topsoil. The cliff was protected by a **rock revetment** of Somerset limestone, a concrete walkway and a **gabion wall**. The defences cost £2.3 million and were completed in 1996.

Castlehaven (5)

Landslides have removed part of a road here and are threatening property worth over £20 million. A scheme to stabilise the cliffs and protect them from active coastal processes has been proposed, but the council is unable to obtain funding from the Department for Environment, Food and Rural Affairs (DEFRA) because of objections from nature conservationists. They are concerned about the impact of land drainage on soft cliff habitats that support many rare insects and plants. A public enquiry will be held to see if the scheme is environmentally acceptable.

East of Freshwater (6)

Where the A3055 passes over a chalk ridge at Afton Down, it is now within 11 metres of the cliff edge. Sea defences to prevent further cliff erosion would be economically unjustifiable and environmentally unacceptable. The council has therefore devised a scheme that **stabilises the cliff top** by anchoring the top of the cliff-face chalk on the landward side of the road. The cost is £750,000.

Seagrove Bay (7)

The crumbling sea walls (maintained by residents) and unstable slopes forced the council to intervene and protect property as part of its 'hold the line' policy. A scheme costing just under £1 million, and completed in 2000, included a new concrete **sea wall** with 200 metres of **rock revetment** placed in front of the wall to dissipate the energy of the waves. **Rock groynes** were constructed as a further layer of protection. Mass movement on the soft clay cliffs has been significantly reduced by installing **land drainage**.

Chapter 6

Option R Urban physical environments

Weather and climate in urban areas

Element of climate	Effect of urban area (compared to nearby rural areas)
Temperature	
Annual mean	0.5–0.8°C increase
Winter minimum	1.0–1.5°C increase
Precipitation	
Quantity	5–10% increase
Days with less than 5 mm	10% increase
Relative humidity	
Annual mean	6% decrease
Winter	2% decrease
Summer	8% decrease
Visibility	
Fog in winter	100% increase
Fog in summer	30% increase
Wind speed	
Annual mean	20–30% decrease
Calms	5–20% increase
Extreme gusts	10–20% decrease
Radiation	
Ultraviolet in winter	30% lower
Ultraviolet in summer	5% lower
Total on horizontal surface	15–20% lower
Pollution	
Dust particles	1,000% increase

Table 6.1
The effects of urban areas on local climate

Urban climate is a good example of the impact of human activity on the atmosphere. Cities create their own climate and weather. Some geographers refer to this as the 'climatic dome' within which the weather is different from that of surrounding rural areas in terms of temperature, relative humidity, precipitation, visibility and wind speed (Table 6.1). For a large city, the dome may extend upwards to 250–300 metres and its influence may well continue for tens of kilometres downwind (Figure 6.1).

Within the urban dome, two levels can be recognised. Below roof level there is an urban canopy where processes act in the spaces between buildings (sometimes referred to as 'canyons'). Above this is the urban boundary layer, whose characteristics are governed by the nature of the urban surface. The dome extends downwind as a plume into the surrounding rural areas. Figure 6.1 shows that this phenomenon occurs at height. The effects of the plume are absent at ground level.

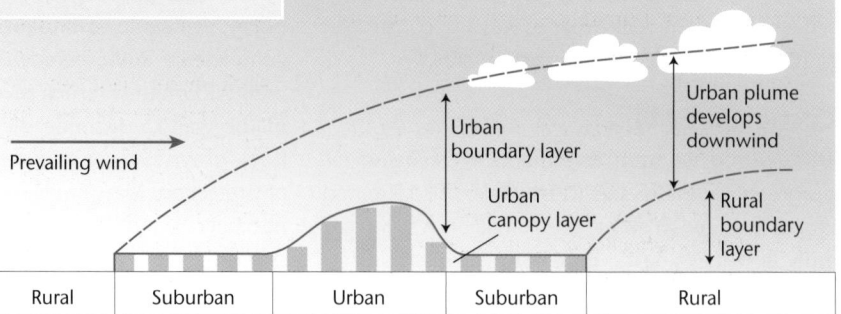

Figure 6.1
The urban climate dome

Temperatures within urban areas

The urban heat island effect

An urban area can be identified as a warm spot in the 'sea' of surrounding cooler rural air. This is the **urban heat island**. Cities tend to be warmer than the surrounding rural areas for the following reasons:

- Building materials such as concrete, bricks and tarmac act like bare rock surfaces in that they absorb large quantities of heat which are slowly released at night. Some of these surfaces also have a high reflective capacity and multistorey buildings tend to concentrate the heating effect in the surrounding streets by reflecting energy downwards.
- Heat comes from industries, buildings and vehicles which all burn fuel (anthropogenic heat, i.e. caused by human activities) as well as from the large numbers of people present.
- Air pollution from industries and vehicles increases cloud cover and creates a 'pollution dome' which allows in the short-wave radiation but absorbs a large amount of the outgoing radiation as well as reflecting it back to the surface.
- In urban areas, water falling on to the surface is disposed of as quickly as possible. This changes the urban moisture and heat budget — reduced evapo-transpiration means that more energy goes into heating the atmosphere.

Key terms

Albedo The amount of solar radiation (insolation) that is reflected by the Earth's surface and the atmosphere. Lighter-coloured surfaces reflect larger amounts.

Conservation The protection, and possible enhancement, of natural and man-made landscapes for future use.

Ecology The study of the relationships among organisms and the environment in which they live, including all living and non-living components. The chief environmental factors governing the distribution of plants and animals are temperature, humidity, soil, light intensity, length of day, food supply and interaction with other organisms.

Niche A specialised area within a habitat, which has the conditions necessary for the existence of a living organism or species. Some niches are referred to as 'microhabitats' — small, distinctly specialised and effectively isolated habitats. Niche can also refer to the ecological role or status of a

plant or animal in a community, especially with regard to the organism's activities, food consumption, and relationships with other organisms.

Particulates The name given to constituents of the atmosphere that are solid rather than gaseous. In urban areas they derive mainly from the burning of fuels, particularly in power stations and vehicles. Particulates can cause health problems such as asthma and eye irritation.

Succession The series of changes which takes place in a plant community through time.

Temperature inversion An atmospheric condition in which temperature increases with height rather than the more usual decrease. As inversions are extremely stable conditions and do not allow convection, they trap pollution in the lower layer of the atmosphere.

Urban heat island The zone around and above an urban area which has higher temperatures than the surrounding rural areas.

*Figure 6.2
Summer
temperatures of
different surfaces
in a European city*

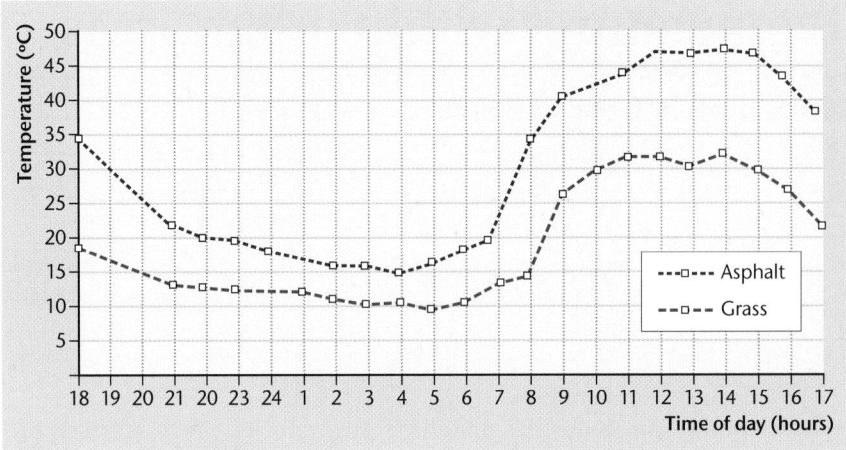

Changes over time

The heat island effect develops best under certain meteorological conditions. The contrast between urban and rural areas is greatest under calm, high-pressure conditions, particularly with a temperature inversion in the boundary layer. Heat islands are also better developed in winter when there is a bigger impact from city heating systems. Urban–rural contrasts are much more distinct at night when the impact of insolation is absent and surfaces which absorbed heat by day slowly release it back into the atmosphere. Heat islands are not constant — they vary both seasonally and diurnally.

*Figure 6.3
Various urban
albedos*

Changes over space

The heat island effect is also spatially variable. The edge of the island is usually well defined and temperatures change abruptly at the rural–urban boundary. Some climatologists have likened the effect to a 'cliff' in temperatures. From this point temperature rises steadily to a peak in the city centre where building densities are highest. The rise tends to be gentle, at an average of 2–4°C per kilometre.

After the cliff of the city edge, this steady rise has often been referred to as a 'plateau'. Within the plateau, though, there are variations that reflect the distribution of industries, power stations, water areas and open spaces. Figure 6.2 shows the temperatures of different surfaces in a city over a 24-hour period and indicates that at the middle of the day there can be a 15°C difference or more between the temperatures of roads and parks.

Figure 6.3 shows the albedo for various surfaces in the city. Highly reflective surfaces absorb very little insolation.

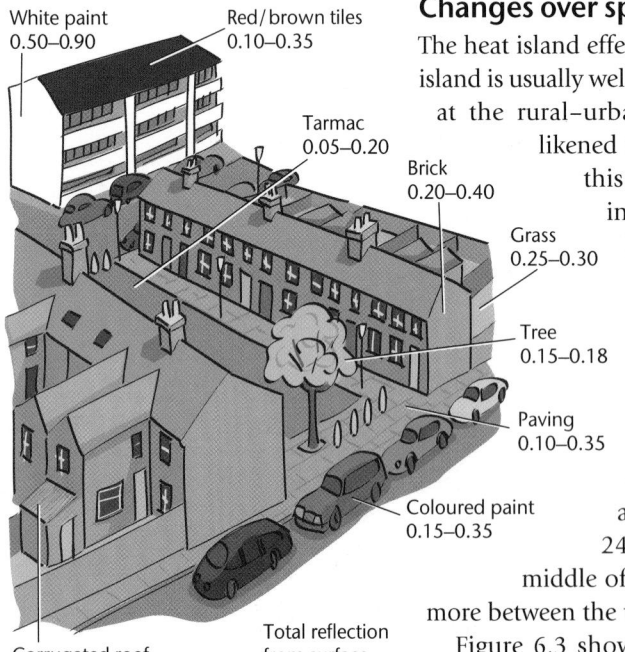

White paint
0.50–0.90

Red/brown tiles
0.10–0.35

Tarmac
0.05–0.20

Brick
0.20–0.40

Grass
0.25–0.30

Tree
0.15–0.18

Paving
0.10–0.35

Coloured paint
0.15–0.35

Total reflection
from surface
= 1.00

Corrugated roof
0.10–0.15

They can reflect it back out into the atmosphere and keep urban areas cool, or reflect it so that it focuses into a small area which heats up. Darker surfaces tend to absorb insolation much better and then re-radiate it as long-wave energy that heats up the urban area. Surfaces in the city tend to be much less reflective than those in rural areas — lots of tarmac but not a great deal of grass. In winter, rural areas keep snow for a much longer period and therefore have a greater albedo. Rural surface albedos include snow (0.86–0.95), sand (0.37), deciduous forest (0.17) and pine forest (0.14).

Figure 6.4 shows the temperature distribution over Montreal in Canada. The following can be clearly seen:

■ the lower temperatures on the side of the prevailing wind
■ the plateau-like temperature zone running from the city edge towards the city centre which has the highest temperatures
■ the correlation between the highest density of building (city centre) and the highest temperatures
■ the fall in temperature over the park area which is an open, vegetated space

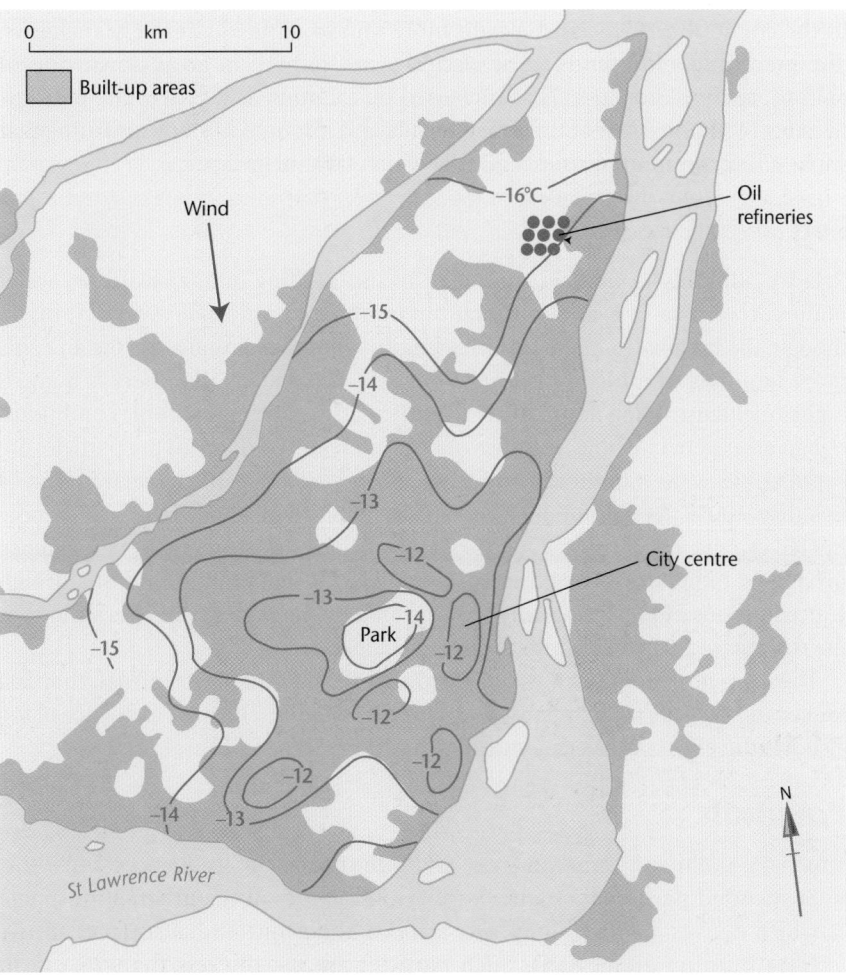

Figure 6.4
Temperature distribution over Montreal, 7 March, 07.00 hours, with winds from the north at 0.5 m s⁻¹

Precipitation, fog and pollution

Precipitation

There is some evidence that rainfall, particularly that from summer thunder-storms, can be higher over urban than over rural areas. Convective storms tend to be heavier and more frequent, as does the incidence of thunder. There are several possible reasons for this:

- the urban heat island generates convection
- the presence of high-rise buildings and a mixture of building heights induces air turbulence and promotes increased vertical motion
- cities may produce large amounts of water vapour from industrial sources and power stations, as well as various pollutants that act as hygroscopic (water attracting) nuclei and assist in raindrop formation

There is also some evidence that cities increase precipitation downwind. Recent research by the University of Salford has shown that the building of tower blocks in the city in the 1970s has brought more rain to parts of Greater Manchester. The prevailing wind tends to be westerly, and rainfall in areas downwind of Salford, such as Stockport, has increased by as much as 7% over the past few decades. With the effect of the urban heat island, though, it is not surprising that snow is less common in cities, and that which falls melts faster.

Fog

Relative humidity is lower in cities than in surrounding rural areas (fewer water bodies, lower rate of evapotranspiration, more rapid runoff of water). At night, though, the urban area maintains its humidity, whereas in rural areas the air cools more rapidly and moisture is lost to dewfall. This means that there is a much higher incidence of fog in urban areas, particularly under anticyclonic conditions. Urban areas generate huge volumes of pollution, and as many of the particles are hygroscopic, water vapour can condense around them when relative humidity is less than 100%. As the temperature falls at night, fog forms. These hygroscopic nuclei can be concentrated in the lower urban atmosphere by subsidence inversions, rather than being dispersed, and it is this that produces unusually high concentrations of fog. This is often exaggerated by the presence of river valleys, as the river provides more vapour for the formation of fog.

There are more fogs in cities (100% more in winter than rural areas), they last longer and they are more intense, with a lower visibility. The high incidence of fog also has the effect of lowering sunshine levels in cities.

Pollution

The major pollutants of urban areas, and some of their consequences, are:

- **Suspended particulate matter** — the solid matter in the urban atmosphere which derives mainly from power stations and vehicle exhausts (particularly from burning diesel fuel). Such particles are usually less than 25 µm in diameter and are responsible for fog/smog, respiratory problems, soiling of buildings and may contain carcinogens. Other particulates in the atmosphere include cement dust, tobacco smoke, ash, coal dust and pollen. Coastal cities also have a vast number of sea salt particulates. Particulates are sometimes referred to as **PM10s**, as the bulk of particles have a diameter of less than 10 μm.

- **Sulphur dioxide** which produces haze, acid rain, respiratory problems (including asthma), damage to lichens and plants and corrosion of buildings.
- **Oxides of nitrogen** which cause accelerated weathering of buildings, photochemical reactions, respiratory problems, acid rain and haze.
- **Carbon monoxide** which is associated with heart problems, headaches and fatigue.
- **Photochemical oxidants** (ozone and peroxyacetyl nitrate [PAN]) which are associated with smog, damage to plants and a range of discomforts to people including headaches, eye irritation, coughs and chest pains.

Air pollution varies with the time of year and with the air pressure conditions. Concentration of pollutants may increase five to six times in winter because of temperature inversions trapping them over the city (Figure 6.5).

The mixture of fog and smoke particulates produces **smog**. This was a common occurrence in European cities through the nineteenth and first half of the twentieth centuries, because of the high incidence of coal burning, particularly on domestic fires. Britain suffered particularly badly, many of the smogs being so thick that they were known as 'pea-soupers'. In December 1952, a smog in London lasted for several days and was claimed to be responsible for over 4,000 deaths (Photograph 6.1). This event persuaded the British government of the time that legislation needed to be introduced to control coal burning.

More recently, there has been an increase in **photochemical smog**. The action of sunlight on the nitrogen oxides and hydrocarbons in vehicle exhaust gases leads to the production of ozone. (Do not confuse this low-level ozone with the high-level ozone in the atmosphere which protects the Earth from damaging ultraviolet radiation.) Los Angeles has had a serious problem with photochemical smog because of its high density of vehicles, frequent sunshine and the favourable topography that traps the high concentration of photooxidant gases at low levels.

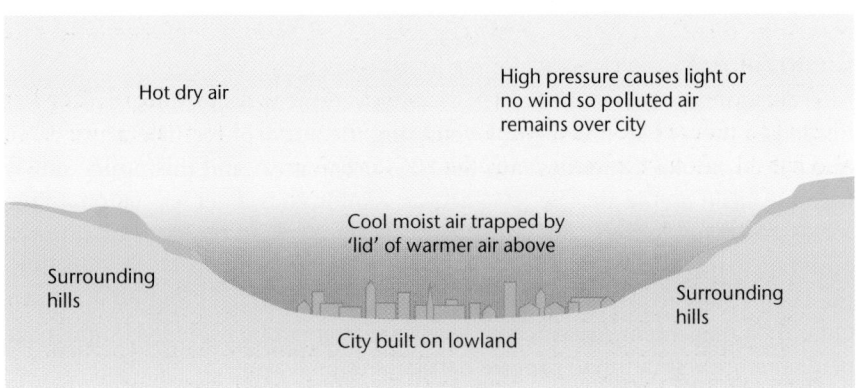

Hot dry air

High pressure causes light or no wind so polluted air remains over city

Cool moist air trapped by 'lid' of warmer air above

Surrounding hills

Surrounding hills

City built on lowland

Figure 6.5
An urban temperature inversion

*Photograph 6.1
Smog in Trafalgar
Square, London,
December 1952*

High levels of ozone can have serious consequences, causing people to suffer from eye irritation, respiratory complaints and headaches. Ozone is also potentially toxic to many species of trees and plants. European cities have seen a growth in photochemical smog in the last 50 years, particularly when anticyclonic conditions prevail and trap the pollutants at low levels. Athens is often quoted as the worst sufferer in Europe, but conditions can be equally bad in British cities, particularly London. London also suffers from high levels of nitrogen dioxide at certain times. These come from vehicle exhaust emissions. Figure 6.6, a newspaper article, gives an account of smog in the city in December 1991.

Pollution reduction policies

There are a number of ways in which governments and other organisations have tried to reduce atmospheric pollution in cities.

Clean Air Acts

After the London pea-souper of 1952, the government decided legislation was needed to prevent so much smoke entering the atmosphere. The act of 1956 introduced smoke-free zones into the UK's urban areas and this policy slowly began to clean up the air. The 1956 act was reinforced by later legislation. In the 1990s, for example, very tough regulations were imposed on levels of airborne pollution, particularly on the level of PM10s in the atmosphere. Local councils in the UK are now required to monitor pollution in their areas and establish **Air Quality Management Areas** where levels are likely to be exceeded. Some have planted more vegetation to **capture particulates** on leaves.

Figure 6.6

COLD WEATHER AND CARS REVIVE THE DREADED PEA-SOUPER

Health warning as smog covers London

London returned to the days of the pea-soup fog this weekend as pollution from car exhausts combined with freezing temperatures to produce smog that exceeded levels in many of the world's dirtiest cities.

Britain's leading lung experts called for tighter controls on city-centre traffic, as doctors linked air pollution to a rise in hospital admissions for acute asthma attacks.

Specialists advised that babies, the elderly and anyone with respiratory problems should not go into busy car-filled streets. Even the healthiest were urged to avoid vigorous exercise such as jogging and cycling in cities, because of the danger of pushing pollutants more deeply into their lungs.

High levels of nitrogen dioxide from car exhaust fumes were recorded in cities through the country yesterday as motorists ignored warnings of fog and black ice and took to their cars for Christmas shopping.

Sunday Times, 15 December 1991

Vehicle control in inner urban areas

A number of cities have looked at ways of controlling pollution by trying to reduce the number of vehicles that come into central urban areas. In Athens, for example, the city declared an area of about 2.5 km² in the centre **traffic free**. Many British towns and cities have **pedestrianised** their CBDs. In London, attempts to control vehicle numbers have included introducing a **congestion charge** (in effect, a road toll) which means vehicle owners have to pay if they wish to drive into the centre. In Mexico City, the city council passed **driving restriction** legislation known as the *Hoy no Circula* (don't drive today). This bans all vehicles from being driven in the city on one weekday per week, the vehicle's registration number determining the day. If conditions worsen, the legislation allows for a ban of two weekdays and one weekend day.

More public transport

Attempts have been made to persuade people to use public transport instead of cars. Such schemes have included Manchester's development of a **tram system** (Metrolink), the development of **bus-only lanes** into city centres, the growth of **park-and-ride** schemes in many British cities and the encouragement of **car-sharing** schemes.

Zoning of industry

Industry has been placed downwind in cities if at all possible and planning legislation has forced companies to build higher factory chimneys to emit pollutants above the inversion layer.

Vehicle emissions legislation

Motor vehicle manufacturers have been made to develop more efficient fuel-burning engines and to introduce **catalytic converters** which remove some of the polluting gases from exhaust fumes. The switch to **lead-free petrol** has also reduced pollution.

There is no doubt that all this legislation has worked, although it has not eliminated the problem. Reduction in coal burning has certainly produced cleaner air over British cities. In his book *The Skeptical Environmentalist*, Bjørn Lomborg states that 'London air has not been as clean as it is today since the Middle Ages. Almost all the modern period has been more polluted with smoke than it is today.'

There is still a great problem with pollution from vehicle exhausts but in many cities those levels are falling too. In Los Angeles, for example, despite large increases in both population and number of motor vehicles since 1970, peak ozone levels have declined markedly and the area subject to high ozone concentrations has shrunk in size.

Urban winds

There are three main types of effects that urban areas have on winds:

- The surface area of cities is very uneven due to the varying height of the buildings. Buildings, in general, exert a powerful frictional drag on air moving over and around them (Figure 6.7). This creates turbulence, giving rapid and abrupt changes in both wind direction and speed. Average wind speeds are lower in cities than the surrounding areas and they are also lower in city centres than in suburbs.

Figure 6.7
The effect of terrain on wind speeds

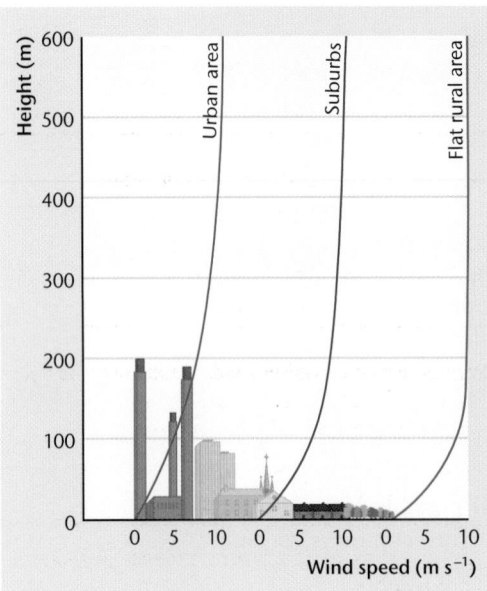

- High-rise buildings may slow down air movement but they also channel air into the 'canyons' between them. Winds in such places can be so powerful that they make buildings sway and knock pedestrians off their feet.
- On calm and clear nights when the urban heat island effect is at its greatest, there is a surface inflow from the cooler areas outside the city to the warmer areas in the city centre. Such breezes transport pollution from the outer parts of an urban area to the city centre, accentuating the pollution problem during periods of photo-chemical smogs.

Winds are therefore affected by the size and shape of buildings. Figure 6.9 shows how a single building can modify an airflow passing over it. Air is displaced upwards and around the sides of the building and is also pushed downwards in the lee of the structure.

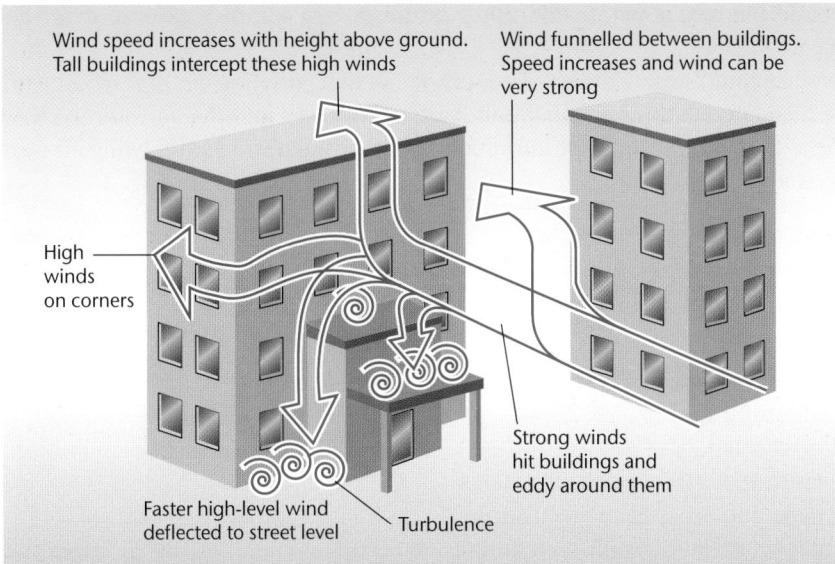

Figure 6.8
The effects of large urban buildings on winds

Wind speed increases with height above ground. Tall buildings intercept these high winds

Wind funnelled between buildings. Speed increases and wind can be very strong

High winds on corners

Faster high-level wind deflected to street level

Turbulence

Strong winds hit buildings and eddy around them

Figure 6.8 shows the windward side of a building in more detail. The air will push against the wall on this side with relatively high pressures. As the air flows around the sides of the building it becomes separated from the walls and roof and sets up suction in these areas. On the windward side the overpressure, which increases with height, causes a descending flow which forms a vortex when it reaches the ground and sweeps around the windward corners. This vortex is considerably increased if there is a small building to windward.

In the lee of the building there is a zone of lower pressure, causing vortices behind it. If two separate buildings allow airflow between them, then the movement may be subject to the **Venturi effect** in which the pressure within the gap causes the wind to pick up speed and reach high velocities. Some buildings have gaps in them, or are built on stilts, to avoid this problem, but a reasonable flow of air at street level is essential to remove harmful pollution.

Usually buildings are part of a group and the disturbance to the airflow depends upon the height of the buildings and the spacing between them. If they are widely spaced, each building will act as an isolated block, but if they are closer, the wake of each building interferes with the airflow

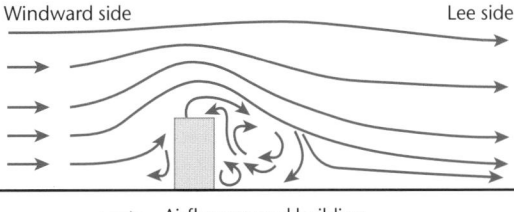

Windward side   Lee side

→ Airflow around building

Figure 6.9 Airflow modified by a single building

(a) Widely-spaced bulidings act like single buildings

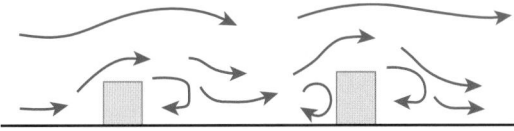

(b) Narrower-spaced buildings – flows interfere

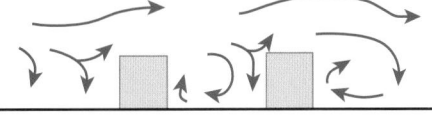

(c) Very close spacing causes winds to skim over the top

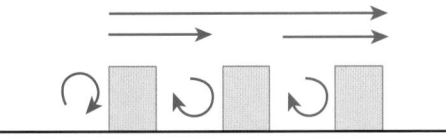

Figure 6.10 Airflow in urban areas modified by more than one building

135

around the next structure and this produces a very complex pattern of airflow (Figure 6.10).

When buildings are designed it is important that pollution emitters (chimneys) are high enough to ensure that pollutants are released into the undisturbed flow above the building and not into the lee eddy or the downward-flowing air near the walls.

The ecology of urban areas

Urban areas contain a wide variety of habitats including:

- industrial sites
- derelict land
- residential gardens and allotments
- parks and other open green areas
- transport routes (both used and disused) such as canals, roadside verges and railway embankments
- waste disposal areas
- urban forests
- water bodies

*Figure 6.11
Origins of introduced species on a small urban common in Sheffield*

This variety means that it is difficult to make generalisations about urban ecology — all these habitats contain different mixes of flora and fauna. Human impact also makes urban habitats somewhat unstable. Many of the plant and animal species found in urban areas are recently introduced, and there are relatively few indigenous (native) species. Cities are centres for the establishment and spread

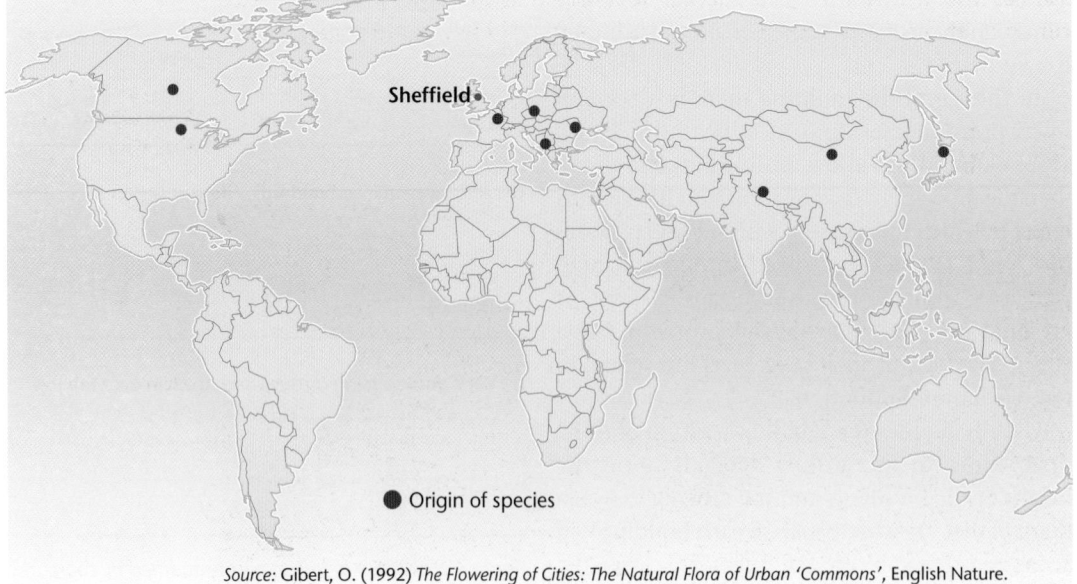

● Origin of species

Source: Gibert, O. (1992) *The Flowering of Cities: The Natural Flora of Urban 'Commons'*, English Nature.

of foreign species — their vegetation contains a far higher proportion of exotic plants than that of rural areas. Figure 6.11, for example, shows the results of a survey of a small urban common in Sheffield. Species that were recorded included:

- from **North America**: Canadian goldenrod, Michaelmas daisy
- from **Europe**: sycamore, laburnum, wormwood, goat's-rue
- from **China and Japan**: buddleia, Japanese knotweed

Such species could have been introduced to the area by escapes from gardens, plants brought in by collectors or amateur gardeners, wind-blown seed, or seed carried by animals and forms of transport.

Urban areas are attractive for immigrant species because of the variety of habitats, the constant creation of new habitats and the reduced level of competition. Many of these habitats are very specialised and within one site a number of niches or microhabitats might be available for plants and animals to colonise. Photograph 6.2 shows an abandoned and neglected urban site. On this small part of the site, the following available niches for plants would include:

- horizontal bare tarmac
- vertical stone walls
- vertical brick walls
- tops of walls
- rubble-strewn ground (different types of rubble create their own microhabitats)

Plant succession in urban areas

Plant succession, the change in a community of species over time, is brought about by changes in the microenvironment occurring because of the supply of new species, the competition between species and changes in habitat. A number of different successions occur within urban areas, depending upon the type of habitat initially colonised, but one of the most studied is that which takes place on an abandoned or neglected area (such as that shown in Photograph 6.2).

Phil Banks

Photograph 6.2
Urban niches on a
derelict site

On such an area a lithosere-type succession (bare rock succession) develops. The types of plants which can initially colonise such a site are influenced by the following:

- **slope**: on horizontal surfaces and gentle slopes, debris will accumulate that will eventually develop into soil
- **moisture availability**: on horizontal surfaces and gentle slopes, rainwater will accumulate or drain away slowly; on steep slopes, faster runoff will create dry areas
- **aspect**: south-facing slopes are warmer and drier
- **porosity** (the ability of the surface to hold water): surfaces that can hold water are colonised more quickly
- **surface roughness**: the ability of the plants to get a hold on the surface — glass and metal are too smooth for most plants
- **pollution levels**: these depend very much on the previous use of the site. There are many places where substances such as lead, which are toxic to a lot of plants, contaminate the ground

The succession on an abandoned industrial site, for example, would occur as follows:

- **Stage 1 The pioneers** Mosses and lichens are the first plants to develop on the bare surfaces. They are able to exist in areas where there is little water, obtaining nutrients by photosynthesis and from the bare concrete beneath them. The concrete is slowly weathered by the production of acids. When the plants die, they provide a thin mat of organic matter which, mixed with the weathered mineral matter, produces a protosoil that other plant species can root into.
- **Stage 2 Oxford ragwort** Cracks in the surface provide sheltered places for seeds to germinate as well as retaining moisture and dust which help plants to root. The most common invaders are plants with windblown seeds such as Oxford ragwort. This has a long flowering season (180–190 days May–November) which enables it to produce millions of seeds. Other common plants at this stage include American willowherb, annual meadow-grass, buddleia, groundsel, knotgrass, dandelion, mugwort, wormwood, white clover and perennial rye-grass. Many of these plants are known as ruderal species, because they are able to tolerate waste ground, rubbish and debris. At this stage, plant succession is usually very rapid.
- **Stage 3 Tall herbs** As these higher plants die off, they produce a thicker and more nutrient-rich soil. Taller plants that are more demanding of good growth conditions can then become established. One of the most common is rosebay willowherb, which spreads initially by seeds and then by rhizomes which can extend up to 1 m a year (rhizomes are elongated horizontal underground plant stems producing shoots above and below ground). Other common plants at this stage include Michaelmas daisy, goldenrod, fennel, goat's-rue, garden lupin, tansy, Jacob's ladder and columbine. These plants gradually shade out the smaller plants, stopping them photosynthesising.
- **Stage 4 Grassland** As soil enrichment continues, the amount of grass in the vegetation increases. The smaller meadow grasses and bents of earlier stages are

replaced by taller species. At this stage the area takes on the appearance of grassland containing scattered clumps of tall herbs. One of the invaders is Japanese knotweed. Thickets of this plant can grow up to 3 m in height and their dense canopies shade out most of the species beneath them.

- **Stage 5 Scrub woodland** As the processes of soil enrichment and competition continue, the taller herbaceous plants are replaced by shrubs and, eventually, trees. The early woody plant colonists (e.g. grey willow, birch) all possess light, windborne seeds, but when the herbaceous vegetation thickens it becomes increasingly difficult for these small-seeded plants to establish unless there has been some disturbance to the succession, such as fire. The later trees have larger seeds that can enter closed vegetation. These include sycamore, laburnum, rowan and hawthorn. Dense thickets of bramble and other such scrubby plants develop. These are able to compete because they can grow roots into deeper crevices in the rock.

As the plant succession develops, there are also changes in the fauna. The soil fauna, such as earthworms, increase in number as soil improves. As the vegetation develops there is an increase in the number and diversity of the insect population. These in turn provide food for small mammals which then allow the presence of predators such as kestrels and the urban fox. The arrival of trees may see the introduction of squirrels, feeding on the available nuts and seeds.

On any site such as this, there will be variations caused by differences in the nature of the surface being colonised. These are known as **sub-stratum** variations and can lead to several parallel successions developing. Surfaces can be acid or alkaline, and can include wetland, concrete, tarmac and rubble. On one site in Sheffield it was discovered that, even on rubble-strewn areas, there were different successions on different types of rubble. Investigation revealed that there were three sub-strata present:

- crushed brick and mortar rubble
- several metres of whole and half-bricks on a slight slope
- a granular layer of ash and slag

Each of these had distinct successions taking place at different rates, with different species involved.

Ecologies of other habitats

Transport routes

Routeways are distinctive habitats because exotic species of plants and insects may be brought in by traffic or train. They also represent wildlife corridors comparable to rural hedgerows.

Railway lines enable animals to move around the city with little or no interference from traffic. During the days of steam, there were frequent fires on the lines which burnt off tall species and allowed light through, therefore encouraging light-demanding species to establish (e.g. primrose and foxglove). Windborne seeds can be sucked along by trains, allowing the establishment of plants such as Oxford ragwort. A lack of human interference, because the track is fenced off, encourages wildlife such as badgers and urban foxes to live here and there are many areas filled with brambles which provide nesting sites for a variety of birds.

Roads act in a similar way, providing homes on verges and embankments for kestrels and scavenging birds. The nitrogen-rich exhaust fumes boost the growth of some wild flowers and these in turn increase the presence of insects and animals further up the food chain. The number of wild flowers, however, can be reduced by mowing, depending on when it is carried out. Some roadsides are managed: trees and shrubs are planted to act as noise screens and flowers are put in to brighten the landscape, as part of the Britain in Bloom competition or to provide advertising for local businesses.

Canals act like long ponds, providing a habitat for a variety of aquatic plants (e.g. yellow flag iris), waterfowl (e.g. moorhens, ducks, kingfishers) and water-loving insects (e.g. dragonflies, damselflies).

Gardens and parks

Gardens (private and public), parks, cemeteries, playing fields and school property are all areas where the vegetation is managed. Species are introduced, many from overseas, and others are removed or controlled by mowing, weeding or the use of pesticides and herbicides. Sports fields, for example, reduce the diversity of plant species by maintaining grass pitches where once there were meadows with a variety of species. There are many other reasons for management of such areas, including the following:

- **altruistic motives** — giving a dull urban landscape more colour and developing green areas to provide amenity space within the urban area. The amenity value for residents might also be increased by improvements in the spiritual and aesthetic value of the area
- **improving the visual outlook** will hide eyesores (**screening** of factories, for example) and might encourage either businesses or residents to move in
- schools may produce a diverse environment for **study purposes** and therefore increase the area's educational value
- local businesses (shops and factories) may want a pleasant site to **attract customers**
- local authorities may provide the public with an **arboretum**
- some groups, such as birdwatchers, may wish for a diverse environment for study and to **attract new species**
- to act as **noise and pollution inhibitors**
- to **provide shade** in hot urban environments
- to **reduce soil erosion** on embankments

In recent years a lot of attention has been directed towards **urban parks**. In 1998 a report suggested that of 5,000 urban parks in the UK, only 12% were in good condition and at least a quarter were derelict. Many local authorities in charge of parks have abandoned traditional planted beds and reduced many areas to little more than grass and tarmac for cheaper maintenance. A committee of MPs (the Environmental, Transport and Regional Affairs Committee) stated that public parks should be a key feature in 'Government plans for the renaissance of urban areas.' In 2000 English Heritage announced a £4 million campaign to restore many of London's 600 garden squares.

Conservation areas

Conservation areas are developed for a variety of reasons, some of which are included in the list above. Other reasons include:

- encouraging wildlife back into cities
- making cheap use of an otherwise derelict area that would be more expensive to set up as a park
- reducing maintenance costs in an area
- maintaining a diverse species base and reintroducing locally extinct species

A great variety of work is done in such areas including planting of trees and other species, planting of native species, dredging of ponds and other water bodies, and soil improvements. Groups and organisations behind such conservation include local authorities, national government, English Nature (Joint Nature Conservation Council), conservation volunteers, the Groundwork Trust, the National Urban Forestry Unit, the National Trust, English Heritage, potential users of the site and local inhabitants.

There is a range of **attitudes** to conservation of vegetated areas within urban environments. Different groups have different priorities, and these affect their view of conservation. For example, local authorities have planning needs, and have to balance the desire to make use of derelict land against the potential cost to local taxpayers. Conservation groups want to create environments where traditional species can re-establish. Local people often want a safe environment for leisure pursuits and may, through the National Playing Fields Association, wish to see sports fields. Urban wildlife groups prefer areas that provide cover for wildlife. Issues of conservation include the eventual management plan for an area, the resolution of ownership, cost, and the satisfaction of the needs of various user groups.

Case study *Dulwich Upper Wood conservation area*

Dulwich Upper Wood is in southeast London, close to the site of the old Crystal Palace. It is a 2 hectare remnant of a much larger wooded area which once stretched some distance across south London. The park is open at all times and has a network of trails, some of which are suitable for wheelchairs.

Species in the wood

The wood developed from the abandoned gardens of old Victorian houses (now demolished) and a small core of ancient woodland. Trees on the site include sycamores, oak, ash, yew and chestnut, and there is also a magnificent line of lime trees. As most of the ground in the wood is deeply shaded, few of the garden plants remain, apart from shrubs, such as rhododendron and laurel. Plants from the ancient woodland have survived, including wood anemone, lords and ladies, bluebell and yellow pimpernel.

There are over 250 different types of fungi (mushrooms and toadstools) which are best seen in autumn. These live on dead wood or leaf litter, helping to break down these materials and return nutrients to the soil.

Many mammals, such as foxes, bats, mice and hedgehogs, live in the wood and over 40 species of birds nest here including woodpeckers and owls.

There are also butterflies, moths and a great variety of other insects.

What makes the site interesting?

Figure 6.12 and the description above give some suggestions as to why the area is so interesting to ecologists and conservationists. Reasons include the following:

- conservation of both abandoned Victorian gardens and ancient woodland
- a number of both preserved and recreated habitats including coppiced areas, wet areas and a pond, herb garden and foxglove area

- the site is both managed and allowed to grow wild in different areas
- there is a range of different habitats
- there is plenty of wildlife on the site including mammals, over 40 species of birds and a wide variety of insects
- 'original' habitats have been preserved, enabling native species of plants and animals to survive
- it is a good example of how habitats can be preserved and created and yet still allow the public access through a network of trails
- the site has an educational value with a posted nature trail

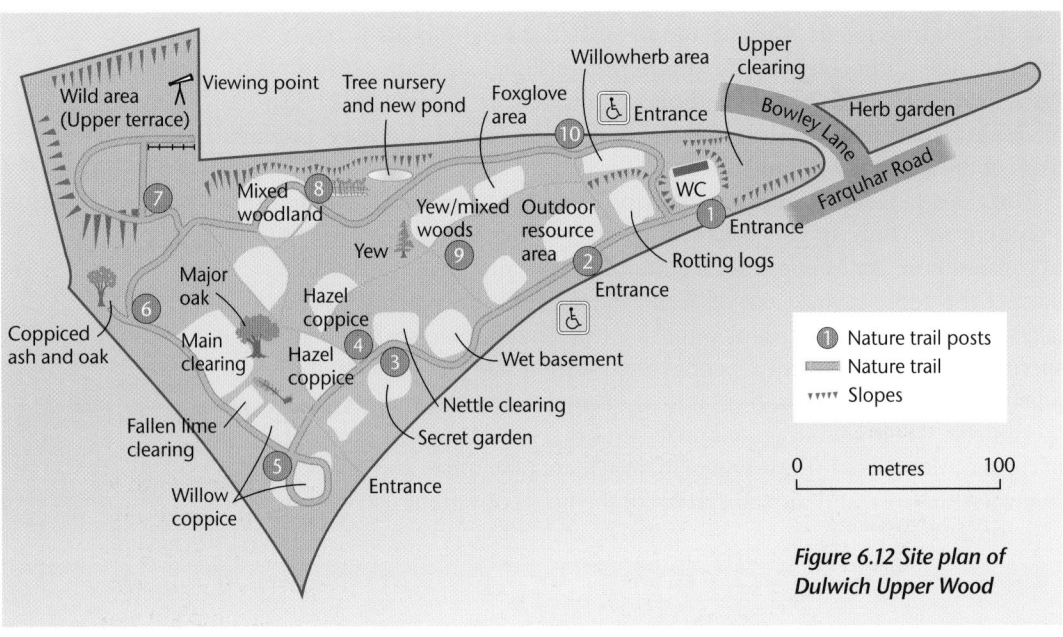

Figure 6.12 Site plan of Dulwich Upper Wood

Assessment exercises

Option P Glacial environments

1 a Look at Figure A. In the context of the budget of a glacier, label the diagram using the list of terms below:

crevasse   snout   tributary glacier   ablation   accumulation   (4 marks)

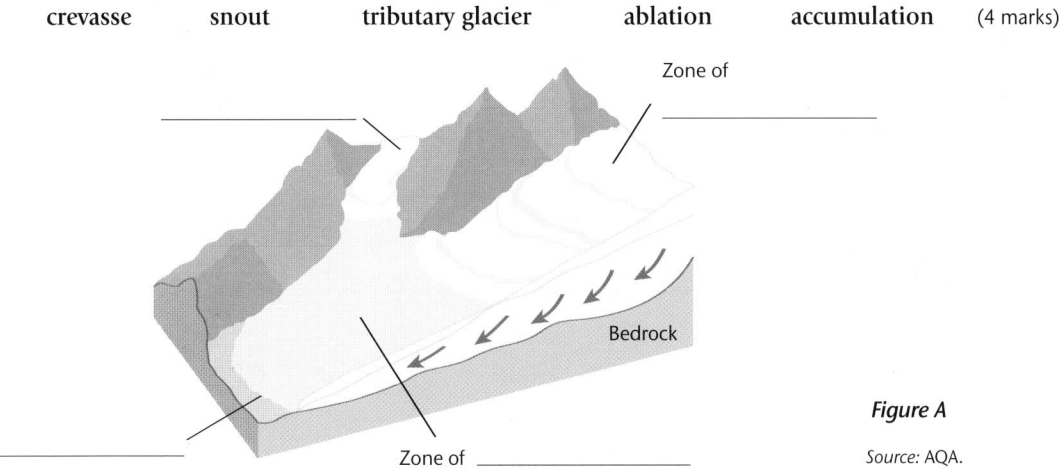

Zone of

Bedrock

Zone of

Figure A

Source: AQA.

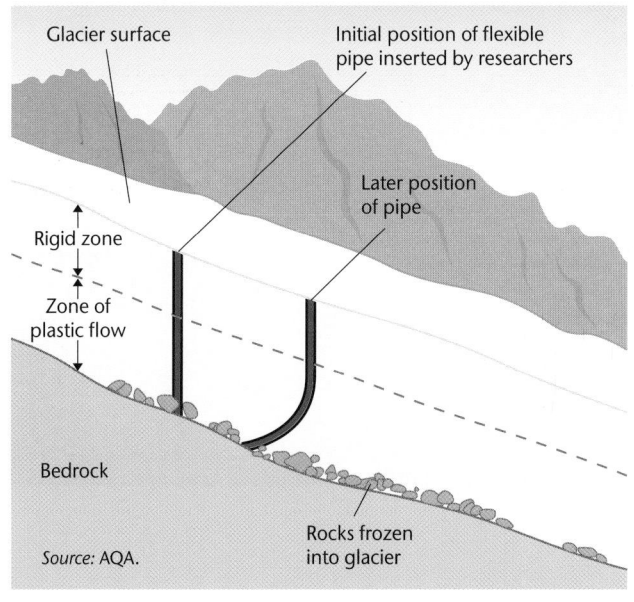

Glacier surface

Initial position of flexible pipe inserted by researchers

Later position of pipe

Rigid zone

Zone of plastic flow

Bedrock

Source: AQA.

Rocks frozen into glacier

Figure B

b Look at Figure B, which shows the effect of ice movement on a flexible pipe inserted vertically into a glacier. Using the diagram and your own knowledge, explain why the pipe has deformed in the way it has.   (6 marks)

2 a In the uplands of the British Isles, corrie basins tend to have a dominant orientation (direction faced).
   (i) What is that direction?
   (ii) How may their orientation have helped their formation? (6 marks)
   b Photograph A shows an upland area in the British Isles.
   (i) Using the grid provided, identify and locate *three* landforms in the photograph that have been produced mainly by glacial erosion. (6 marks)
   (ii) Choose one of the landforms you have identified from the photograph and explain how glacial processes produced that landform. (8 marks)

Photograph A

3 Figure C shows the typical features left behind on a lowland area after the retreat of a glacier or ice sheet.
  a In which direction did the ice advance?
  b Which major feature of glacial/fluvioglacial deposition is not included in Figure C?
  c Explain the presence of the lake deposits.
  d For any other feature shown on the sketch, explain its formation.   (10 marks)

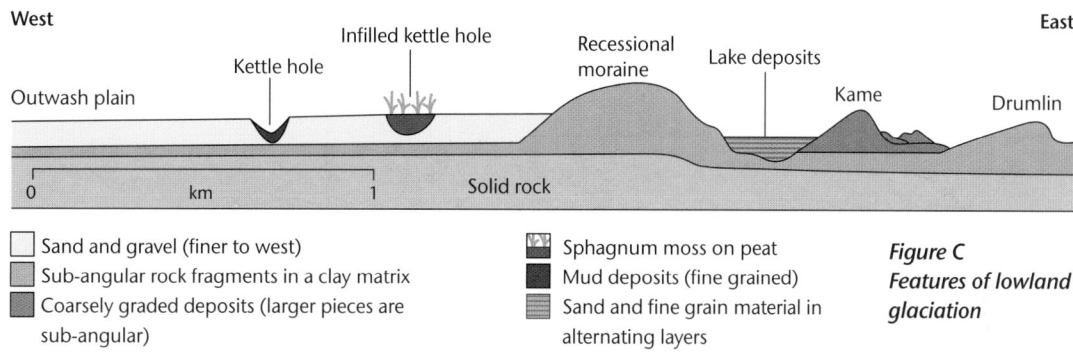

West   East

Infilled kettle hole

Kettle hole

Recessional moraine   Lake deposits

Outwash plain   Kame

Drumlin

0   km   1   Solid rock

☐ Sand and gravel (finer to west)
■ Sub-angular rock fragments in a clay matrix
■ Coarsely graded deposits (larger pieces are sub-angular)

▨ Sphagnum moss on peat
■ Mud deposits (fine grained)
▤ Sand and fine grain material in alternating layers

Figure C
Features of lowland glaciation

4 Look at Figure D, a sketch showing typical landforms found in periglacial regions.
  a Name each of the features of the periglacial area labelled A, B and C.   (3 marks)
  b Choose one of the features that you have named and explain how it was formed.   (7 marks)

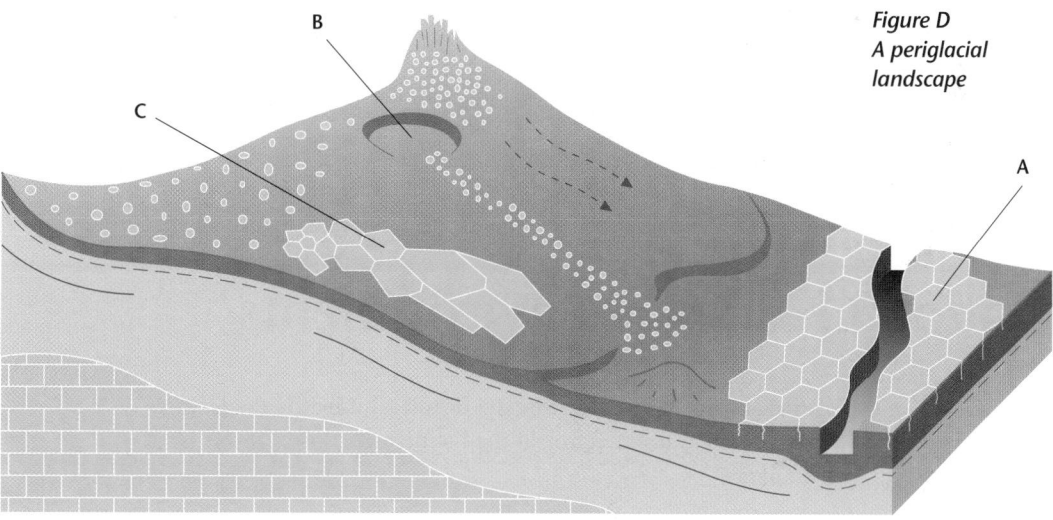

Figure D
A periglacial landscape

Assessment exercises

5 Figure E is a map showing Lake Harrison, a proglacial lake that developed in the English Midlands during the ice ages.

a Explain how this lake developed. (5 marks)

b Why were the overflow channels found to the south and east? (3 marks)

Figure E

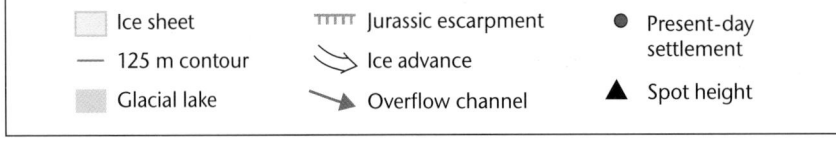

Ice sheet	ᴛᴛᴛᴛ Jurassic escarpment	● Present-day settlement
— 125 m contour	↘ Ice advance	
Glacial lake	↘ Overflow channel	▲ Spot height

Option Q Coastal environments

1 a Look at Figure F, which shows wave refraction at a headland. Identify the features labelled 1–6 by matching them to the terms in the list below:

- direction of wave energy
- waves bend to adopt the shape of the coastline
- wave energy concentrated here
- wave energy dissipated here
- headland
- wave crests (6 marks)

146

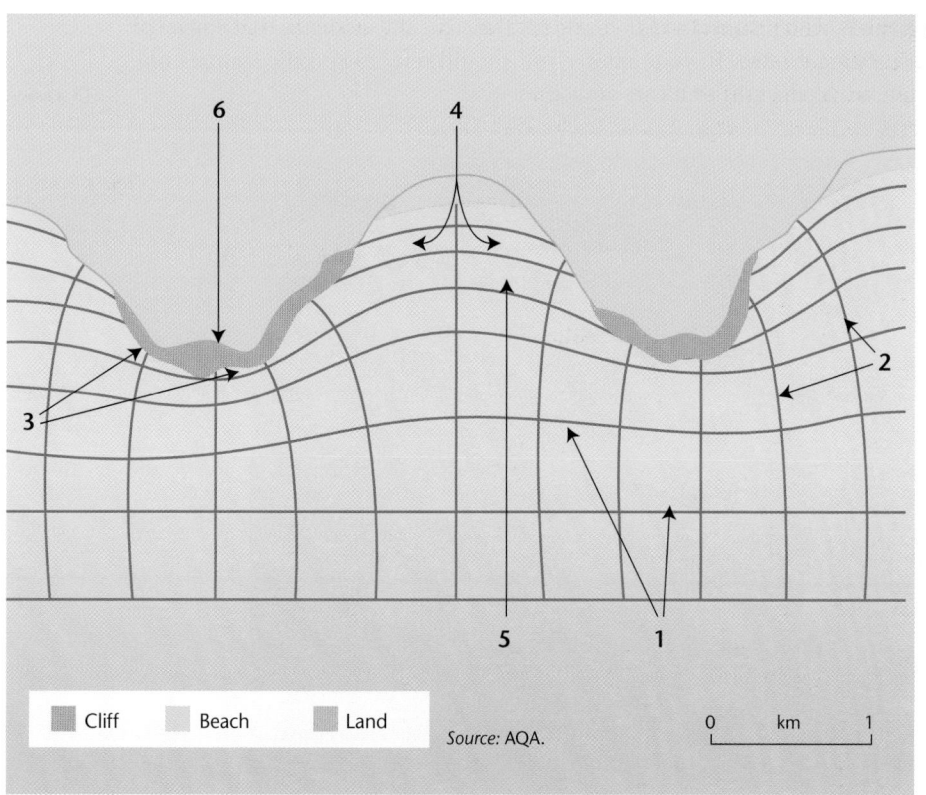

Figure F

Cliff   Beach   Land

Source: AQA.

0   km   1

b Look at Figure G, which shows the features of a beach.

(i) Name the features marked A and B. (2 marks)

(ii) The storm beach labelled on Figure G has been formed by destructive waves.
Draw a labelled diagram to show the main features of destructive waves. (4 marks)

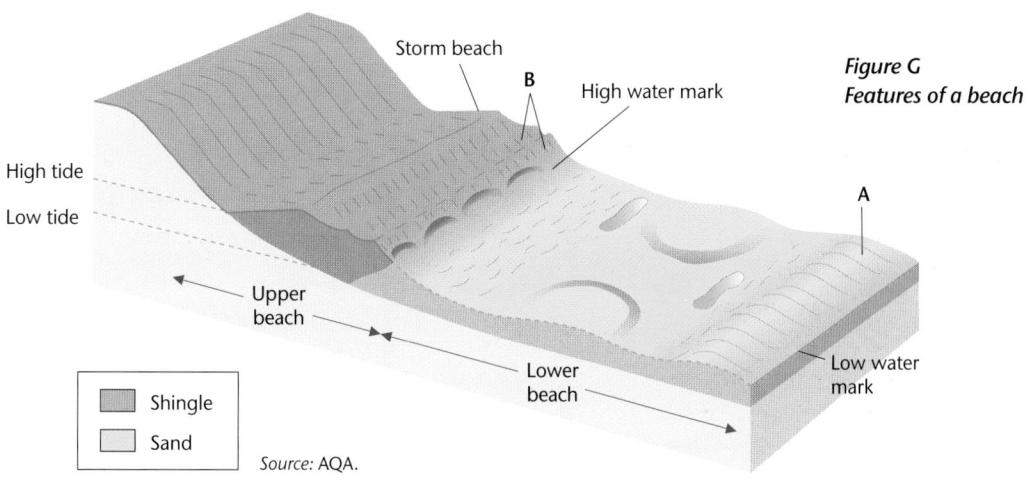

Storm beach

B

High water mark

Figure G
Features of a beach

High tide

Low tide

Upper
beach

A

Lower
beach

Low water
mark

Shingle

Sand

Source: AQA.

147

2 Study Photograph B, which shows a cliff coastline. Describe the evidence that suggests this coastline is being eroded. In your answer you should refer to specific features on the photograph, using the grid to locate your evidence. (8 marks)

Photograph B
Old Harry Rocks
and The Foreland

Kitchenham

3 Look at Figure H, which shows coastal features in Norfolk.
  a Name the landform at A. (1 mark)
  b Define the term 'longshore drift'. (2 marks)
  c Explain how wind and longshore drift could have caused the formation of the landform at A. (5 marks)

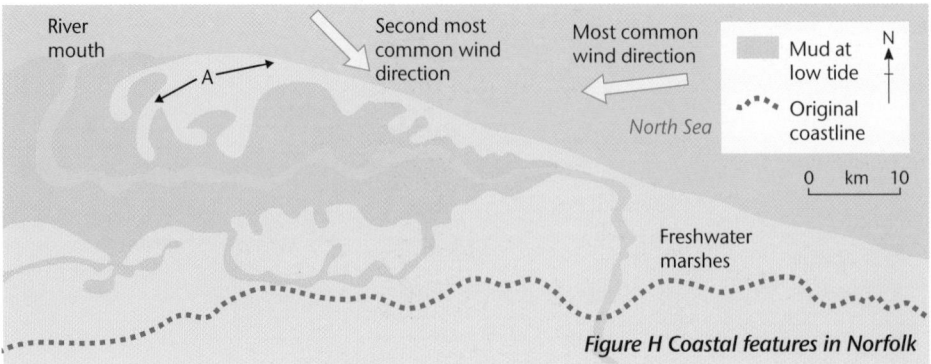

Figure H Coastal features in Norfolk

4 a Draw a labelled diagram to show the main features of either a salt marsh or a belt of coastal sand dunes. How has vegetation helped to create the feature you have chosen?

(10 marks)

b Look at Figure I, which is a tourist advertisement. How might the activities on offer damage this part of the Great Barrier Reef?

(5 marks)

Why Agincourt Reef?

Agincourt Reef is the only part of the very outer edge of the Great Barrier Reef to be visited by day cruisers.

It offers

- excellent year-round visibility
- ideal swimming and snorkelling areas
- spectacular scuba diving on one of the reef's most acclaimed sites
- over 1,000 species of multi-coloured tropical reef fish
- over 200 varieties of spectacular and colourful corals to see from your semi-submersible
- an abundance of giant clams, anemones and other marine life

Why Wavepiercer Boats?

Quicksilver is the first marine tourism operator to use these large, fast, modern 37-metre Wavepiercers. They feature spacious, air-conditioned passenger-cabin areas and a fully licensed bar. The revolutionary Wavepiercer's torpedo-shaped hulls actually pierce the waves, providing the smoothest possible ride in all wave conditions.

Fare $85 Departs 10.00 a.m. daily

Quicksilver

is the Great Barrier Reef

Figure I

Source: AQA.

5 Suggest the purpose of the sea defences shown in Figure J.

(5 marks)

Figure J

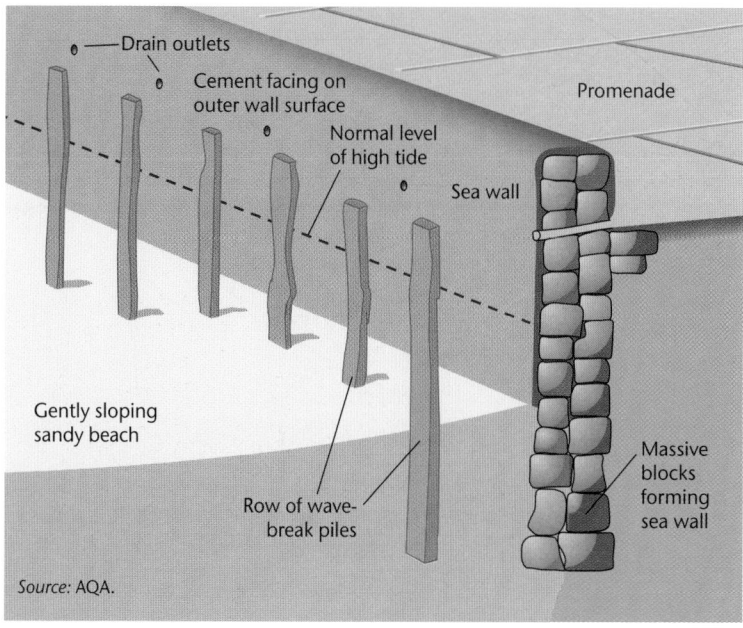

- Drain outlets
- Cement facing on outer wall surface
- Normal level of high tide
- Promenade
- Sea wall
- Gently sloping sandy beach
- Row of wave-break piles
- Massive blocks forming sea wall

Source: AQA.

Option R Urban physical environments

1 Look at Figure K which shows temperature variations in central and east Oxford
at 21.00 hours on 26 February 1987.

 a Describe the variations in temperature from Oxford city centre to the surrounding
 rural area as shown in Figure K. (5 marks)

 b Explain the variations in temperature that you have described. (8 marks)

*Figure K
Temperature
variations in
Oxford, 21.00 hours,
26 February 1987*

2 Figure L shows the distribution of rainfall during a
storm over the Washington,
DC area on 9 July 1970.

 a Describe the distribution
 of the rainfall in the
 area shown on Figure L.

 (4 marks)

 b Suggest reasons why there
 is both more rainfall and
 more intense rainfall in
 urban areas than in the
 surrounding rural areas.

 (5 marks)

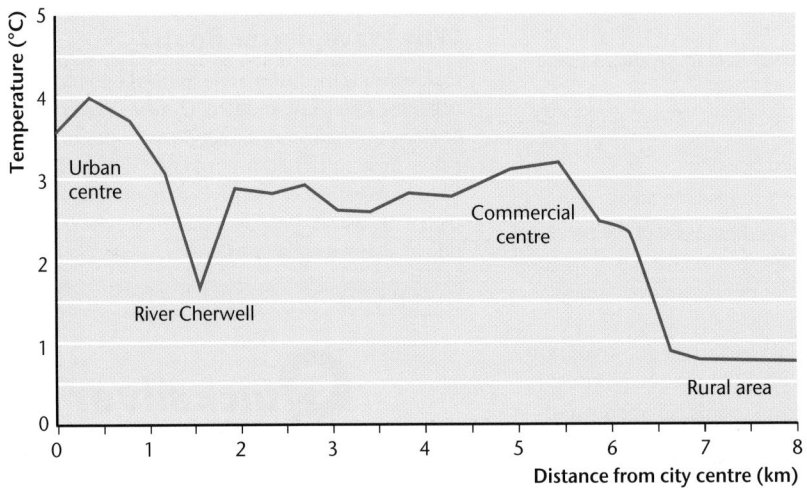

*Figure L
Rainfall over Washington, DC,
9 July 1970*

— 50 — Isohyet (mm) ◆ Central urban area

3 Look at Figure M, which shows data about particulate air pollutants released over cities.

    a Compare the size range of vehicular exhaust particulates with that of cement dust.   (4 marks)

    b Explain the roles of particulates and other factors in the formation and distribution of fog in urban areas.   (8 marks)

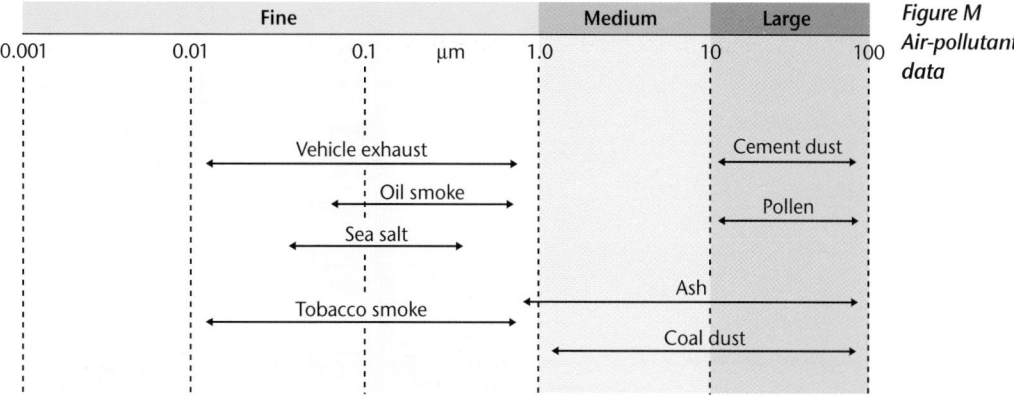

Figure M
Air-pollutant data

Note: 1 μm = 0.001 mm.

4 Study Figure N and then describe the effects that certain urban structures have on wind direction and turbulence.   (5 marks)

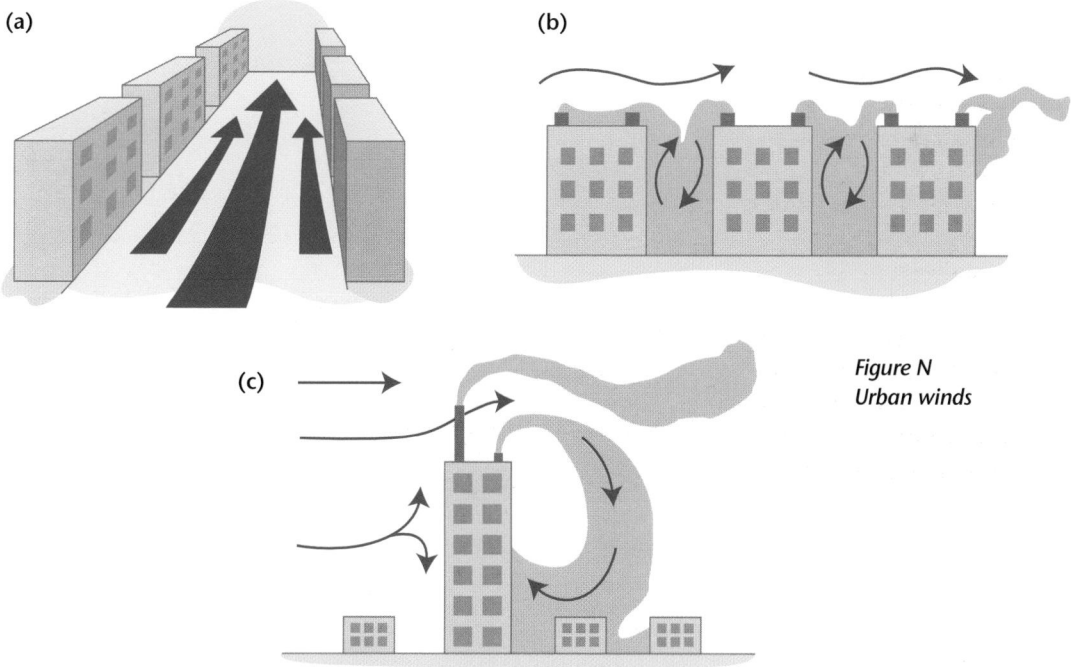

Figure N
Urban winds

5 Look at Photographs C and D, which show
two views of the same ecological conservation area.

a Give three pieces of evidence from
the photographs that indicate the
development caters for the various
needs of the local community.   (3 marks)

b Using only evidence from the
photographs, describe the attempts
made to establish an ecological
conservation area.   (4 marks)

c Describe the various attitudes to
ecological conservation within
urban areas. Using examples, show
how these attitudes have led to the
planned introduction of some
species and the destruction of
others.   (10 marks)

Photograph C

Malcolm Skinner

Photograph D

Malcolm Skinner

AS

The human options

Chapter 7

Option S Urban change in the last 30 years

The general characteristics of urbanisation

Urbanisation is the increase in the proportion of a country's population which lives in towns and cities. There are two main causes of urbanisation — natural population growth and migration into urban areas from rural areas. In more economically developed countries (MEDCs) urbanisation took place during the nineteenth and early twentieth centuries. Urbanisation has been much more rapid in the less economically developed countries (LEDCs) during the last 50 years. It is estimated that by 2020 over half the world's population will live in urban areas, and this proportion may be as high as 60%. Urbanisation is clearly a process that will continue for some time to come.

Figure 7.1
Average increase in urban population per year, 1991–2001

Figure 7.1 shows that the highest rates of urbanisation are in Africa and the Middle East, and the lowest rates in North America and Europe. There are moderate rates of urban growth in Central and South America, and in Indonesia.

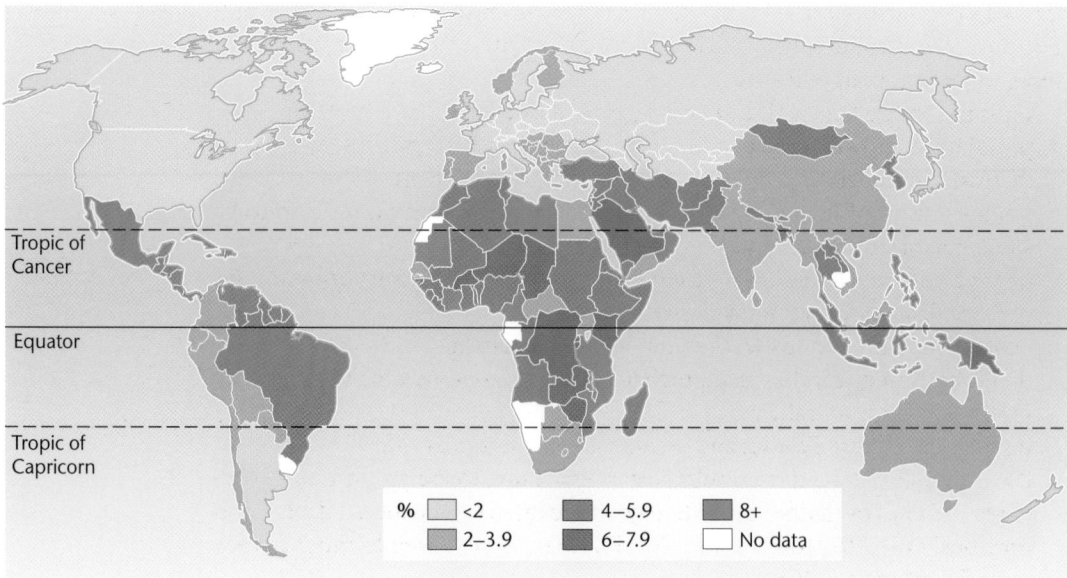

Tropic of Cancer

Equator

Tropic of Capricorn

% | <2 | 4–5.9 | 8+
2–3.9 | 6–7.9 | No data

There are, however, some interesting anomalies in these generalised distributions. For example, within Africa there are low rates of urbanisation in Egypt and the Central African Republic. Similarly, Argentina has much lower rates within South America. Within Europe there are relatively high rates of urbanisation in the southeast, for example in the Balkan countries.

Table 7.1 shows the growth in millionaire cities — cities with over 1 million inhabitants. Much of this growth has been in LEDCs (Table 7.2). In 1900 there were only three such cities in South America and one in Africa. Today there are 16 in South America and 12 in Africa.

Year	Number
1900	11
1930	50
1960	99
1980	180
2000	250

Table 7.1 Growth of millionaire cities

City	1950	1980	2000
Mexico City	3	15	28
New York	12	16	16
São Paulo	3	13	24
Rio de Janeiro	3	9	14
Mumbai	3	9	16
Delhi	1	6	14
Calcutta	5	10	17
Shanghai	6	12	14
Seoul	1	9	14
Tokyo	7	17	17

Table 7.2 The world's largest cities, 1950–2000 (million inhabitants)

Why is urbanisation taking place in LEDCs?

Natural increase

Urban areas tend to have relatively low age profiles. Young adults (15–40 years) have traditionally migrated from rural areas, for reasons outlined in the next section. These are the fertile years, during which people have children, and because of this rates of natural increase are higher in cities than in the surrounding rural areas, and so the population increases.

In LEDCs falling death rates combined with relatively high birth rates occur in both rural and urban areas, but are exaggerated in cities. The factors causing these changes in birth and death rates are covered in chapter 2.

Rural–urban migration

The reasons for rural–urban migration are often divided into 'push' and 'pull' factors. **Push factors** cause people to move away from rural areas, whereas **pull factors** attract them to urban areas (Figure 7.2). In LEDCs push factors tend to be more important than pull factors.

Push factors are largely due to poverty caused by:

- population growth which means the same area of land has to support increasing numbers of people, causing over-farming, soil erosion and low yields
- fragmentation of land due to a system of inheritance that causes land to be subdivided into smaller and smaller plots
- systems of tenure that do not allow tenants to have a long-term perspective for their land, so they do not invest in it
- debt on high-interest loans taken out to support agricultural change
- desertification due to low and unreliable rainfall amounts which has resulted in very low agricultural yields
- high levels of local diseases and inadequate medical provision
- the conversion of land from subsistence agriculture to the production of cash crops. This has been done to try to pay off the interest on national debts. Land previously used to grow food for local farmers is now used to produce cash crops for sale to MEDCs

*Figure 7.2
Push and pull
factors which
contribute to
urbanisation*

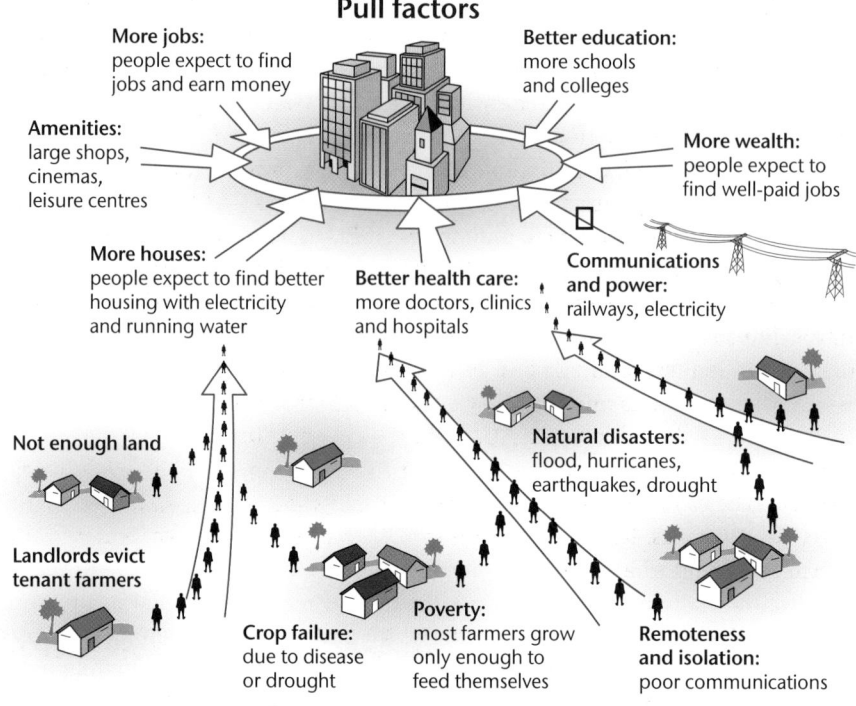

Pull factors

More jobs: people expect to find jobs and earn money

Better education: more schools and colleges

Amenities: large shops, cinemas, leisure centres

More wealth: people expect to find well-paid jobs

More houses: people expect to find better housing with electricity and running water

Better health care: more doctors, clinics and hospitals

Communications and power: railways, electricity

Not enough land

Natural disasters: flood, hurricanes, earthquakes, drought

Landlords evict tenant farmers

Crop failure: due to disease or drought

Poverty: most farmers grow only enough to feed themselves

Remoteness and isolation: poor communications

Push factors

- natural disasters such as floods (Mozambique), tropical storms (Bangladesh) and earthquakes (Gujarat, India) have all caused people to flee previously fertile rural areas and not to return
- wars and civil strife in countries including Afghanistan, Sudan, Kashmir and Rwanda have caused people to flee their land. Fear and unexploded bombs may prevent them returning

Pull factors include the prospect of:

- employment in factories and service industries (e.g. hotels), earning better wages than those in rural areas
- earning money from the informal sector, e.g. selling goods on the street, providing transport (taxi/rickshaw driver), prostitution
- better-quality social provisions, from basic needs such as education and health care to entertainment and tourism
- a perceived better quality of life in the city, fed by images in the media

Attitudes to urbanisation in LEDCs

Those in favour of urbanisation believe it creates a welcome reduction in the population density of rural areas. Those left behind, such as women and the elderly, welcome the prospect of money being sent home by family members who have migrated to the city. Commercial farmers also welcome the process of urbanisation as it creates a potential increase in the market for agricultural produce in the city.

Those against urbanisation believe it creates a demographic and social imbalance in rural communities, which are left with an excess of the elderly and the very young. This means there are fewer people available to work the agricultural land. Combined with the movement of those with entrepreneurial skills to the city, this could cause rural areas to face economic decline.

There are also varying attitudes to urbanisation within the urban areas themselves. Until recently, city authorities regularly evicted squatters and destroyed squatter settlements. Nowadays, they are seen in a more positive manner.

For example, in **Delhi (India)**, 2.4 million people live in squatter settlements known as **jhuggis**. The city authorities have introduced **self-help schemes** which aim to improve living conditions by helping the residents to help themselves. They grant squatters legal title to their land, provide essential infrastructure such as mains water, roads and electricity, and give loans to residents to purchase building materials and hire builders to improve their homes.

Site and service schemes are another popular low-cost solution to the housing problem. The city authorities provide serviced plots with mains water, electricity, sewerage and roads. Residents build their own houses on the plots, either from scratch or around a basic shell.

Case study Urbanisation in Manila, the Philippines

Manila is the capital of the Philippines. Its population has grown in recent years to 8.6 million.

History

Manila traces its origins to a small seaport established in the twelfth century at the mouth of the Pasig River. Captured by Spain in 1570, the wealthy Muslim kingdom of Maynila was proclaimed capital of the Philippines. Manila remained under Spanish rule for nearly four centuries, exporting agricultural products to Spain. In the aftermath of the Spanish–American War the Philippines was ceded to the USA. From the outset, the goal of American policy was to integrate the new colony into the American market. Following independence in 1947, there was a major shift in Philippine economic policy from export promotion to import substitution. The main beneficiary of the import substitution policies was the capital region of Manila.

The industrial base of the city has broadened in recent decades to include textile production, publishing and printing, food processing, and the manufacture of tobacco, paints, drugs, oils, soap and wood products such as plywood and veneers.

Problems in the city

Metro Manila has serious infrastructure and environmental problems with land, water, air, sewerage, drainage, waste and traffic.

- Land use in the city has largely been shaped by the activities of the private sector. Due to speculation, land prices have risen by 100–200%, reducing access to affordable housing in the city.
- The water supply system is unsatisfactory, and is unable to reach much of the population on the periphery of the city. Water pumps are being used, which is increasing pollution.
- The sewerage system is extremely inadequate. Only about 11% of the population has piped sewerage. The majority of sewage is conveyed untreated through open ditches (*esteros*) and canals into Manila Bay.
- Air pollution is another major environmental problem, caused by motor vehicles and industry.
- The vast amounts of solid waste produced by the city each day find their way to huge dumps such as Smokey Mountain, a 22 ha site on the city limits. Waste often clogs the poor drainage pipelines and causes flooding.

- The high organic content (sewage and agricultural pesticides) of water in Manila Bay has produced 'red tides' in the bay, caused by the uncontrolled growth of algae.
- Like most other conurbations, Metro Manila suffers from serious traffic congestion.

Informal settlements

Metro Manila lacks direct planning of development, and systems have evolved in the city largely as the result of market forces. The government has attempted to motivate leaders of the smallest political units, the *barangay*, to organise environmental improvement projects that encourage community self-reliance and active participation of citizens. These self-help projects (Figure 7.3) are concentrated in the existing illegal squatter settlements in the city.

All these settlements are located on wasteland outside the centre of the urban area, often on steep-sided land and marshland on the banks of the river. These are poor-quality areas that other land-users do not want.

Squatter settlements are also built alongside the road and railway networks and near the harbour — industrial areas which provide employment. Roads offer the opportunity too for people to work as street sellers. A large area of settlements exists at the junction of arterial and ring roads in Quezon City, and there is a line of settlements alongside the Pasig River south of San Juan (Figure 7.4).

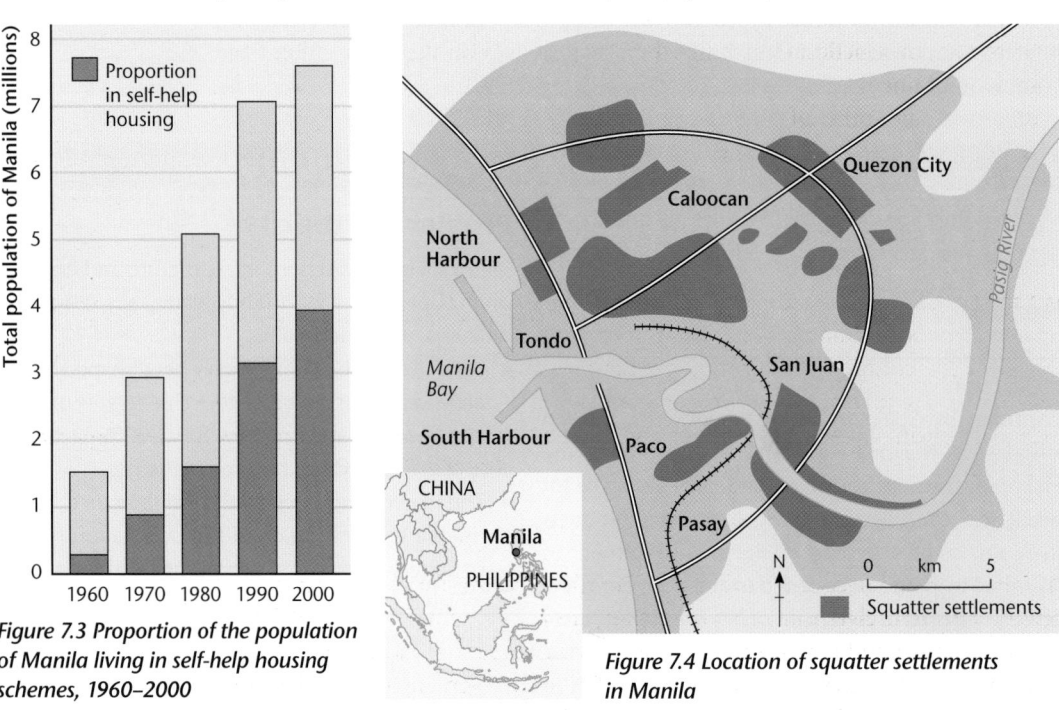

Figure 7.3 Proportion of the population of Manila living in self-help housing schemes, 1960–2000

Figure 7.4 Location of squatter settlements in Manila

Suburbanisation and counter-urbanisation in MEDCs

Suburbanisation is the outward growth of urban development to engulf surrounding villages and rural areas. It has been possible because the growth of public and private transport systems has allowed the inhabitants of the new urban areas to commute to the main town/city. The transport systems that have enabled this over time have been railways, trams, buses and private cars.

Counter-urbanisation is the process of migration of population from major urban areas to more rural settlements, largely due to dissatisfaction with urban living. The rise of new communication technologies, especially electronic systems, which allow home-working, may have encouraged this movement.

The following are evidence of counter-urbanisation taking place in an area:

- increased use of a commuter railway station near the new or expanded settlement
- higher levels of traffic movement in the area
- increased value of houses in the settlement
- increased construction of executive housing in the settlement
- an increased number of conversions of farm buildings to exclusive housing

Figure 7.5 provides some evidence of counter-urbanisation taking place in the UK. Note the large growth of new homes in rural areas such as East Anglia, Kent, Devon and Hampshire. However, the same map can also support the case for little counter-urbanisation, as many new homes are planned in large urban areas such as London, Greater Manchester and West Yorkshire.

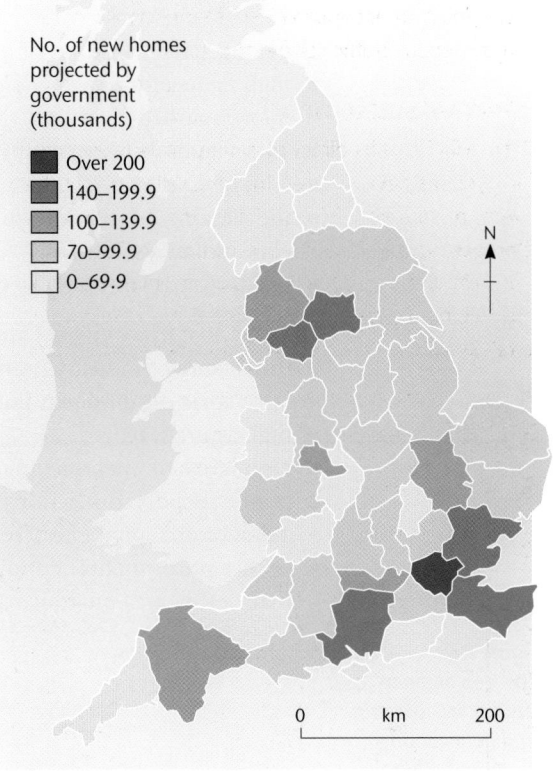

Figure 7.5 Projected number of new homes in England by county, 1991–2016

Attitudes to counter-urbanisation in MEDCs

Those in favour of counter-urbanisation will have the following views. The newcomers themselves think that the settlement is a delightful place in which to live — it is quiet and has clean air. There is a pleasant lack of haste in the village, and there are attractive countryside walks. They may even encourage friends to move as well. Local shopkeepers see a potential for more custom, but they need to modify the produce sold, for example, an increase in video/DVD sales and rentals, alcohol and frozen food.

Those against counter-urbanisation are local residents who dislike the newcomers or 'weekenders'. The incomers do not contribute to the stability of the village, frequently disappearing during the day and/or week. They have different social norms from the locals — more wine drinking, barbeques and fast cars. The rising demand increases the price of housing and local people may be unable to afford it. The newcomers do not use public services, such as local buses, post offices and even schools, which speeds up their decline.

Inner-city decline and regeneration in the UK

The characteristic features of inner-city decline are:

- high figures for out-migration of population
- many boarded-up shops
- 'to let' signs in large numbers
- a lot of empty and derelict property
- the closing of schools, particularly primary schools, and low levels of educational attainment
- census wards with high levels of unemployment

The causes of inner-city decline are:

- the closure of industry (particularly old manufacturing industry)
- large areas of run-down housing, caused by the out-migration of people with high aspirations
- high levels of unemployment, resulting in a lack of spending power
- poor transport connections to the area
- a downward spiral effect, resulting from social and economic malaise
- poor environmental conditions, with derelict land, graffiti and vandalism making the area unattractive to new investors (Figure 7.6)

Figure 7.6
The web of inner-city decline, despair and deprivation

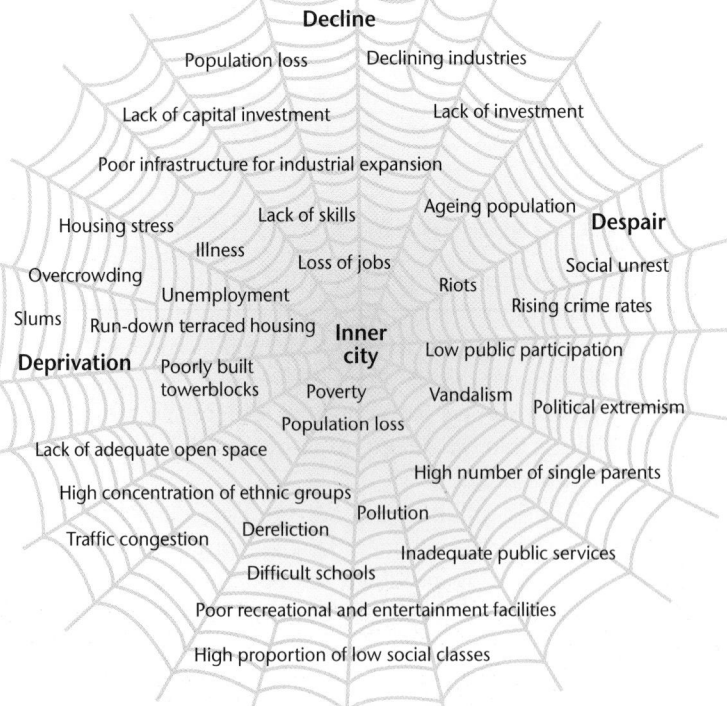

Case study *Inner-city decline: London Docklands*

The London Docklands is now an area of urban regeneration stretching downriver from London Bridge through Wapping, Limehouse and the Isle of Dogs to the Surrey Docks and the Royal Docks. In the 1960s this area, which had been an international port, suffered major decline. The shipping of goods around the world switched to use larger vessels and the container-based system, and these ships were too large for the London docks. New and expanded ports were established around the country, at places such as Felixstowe and Dover. London declined as a port, and the docks suffered huge dereliction. Associated industries of food-processing, engineering, ship-building and ship repairing also suffered, causing high levels of unemployment and social deprivation within the area.

The reasons for and effects of gentrification

Gentrification is a process of housing improvement. It is associated with a change in neighbourhood composition in which low-income groups are displaced by more affluent people, usually in professional or managerial occupations. Gentrification is a process by which regeneration of inner cities can take place, but it is different from the schemes described in the next section in that it is carried out by individuals or groups of individuals, and not by supported bodies. Gentrification involves the rehabilitation of old houses and streets on an individual basis, but is openly encouraged by groups such as estate agents, building societies and local authorities.

One of the clear positive outcomes of gentrification is that the social mix of the area is changed and becomes more affluent. The purchasing power of the residents is greater, which leads to a rise in the general level of prosperity. The area becomes

Alan Young

*Photograph 7.1
A gentrified street
in Notting Dale,
North Kensington,
London. One house
is still swathed in
scaffolding*

dominated by 'yuppies', with a subsequent increase in the number of bars, restaurants and other higher-status services. The very nature of the refurbishment that takes place in each house leads to the creation of local employment in areas such as design, building work, furnishings and decoration.

There are, however, clear disadvantages of gentrification. Local people on low incomes find it increasingly difficult to purchase houses, as the price of refurbished property rises markedly. Indeed, the size of the privately rented sector diminishes as more properties are sold off. Friction and conflict arise between the 'newcomers' and the original residents.

Gentrification is taking place in the central parts of many towns and cities in the UK (Photograph 7.1). Well-documented examples include Notting Hill and Islington in London.

Inner-city regeneration policies

Property-led regeneration

Urban Development Corporations (UDCs) were set up in the 1980s and 1990s to take responsibility for the physical, economic and social regeneration of selected inner-city areas with large amounts of derelict and vacant land. They are an example of what is known as property-led regeneration. They were given planning approval powers over and above those of the local authority, and were encouraged to spend public money on the purchase of land, the building of infrastructure and on marketing to attract private investment. The intention was that private investment would be four to five times greater than the public money initially invested.

The appointed boards of UDCs, mostly made up of people from the local business community, had the power to acquire, re-claim and service land prior to private-sector involvement and to provide financial incent-ives to attract private investors. In 1981, two UDCs were established — the London Docklands Development Corporation (LDDC) and a Merseyside UDC. Eleven others followed, in areas such as the Lower Don Valley in Sheffield, Birmingham Heartlands, Trafford Park in Manchester and Cardiff Bay (Figure 7.7).

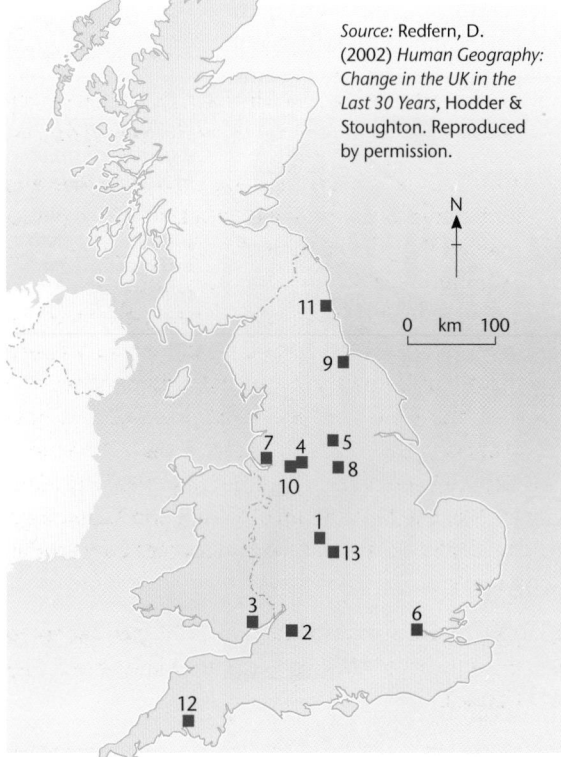

Source: Redfern, D. (2002) *Human Geography: Change in the UK in the Last 30 Years*, Hodder & Stoughton. Reproduced by permission.

1 Black Country	8 Sheffield
2 Bristol	9 Teesside
3 Cardiff Bay	10 Trafford Park
4 Central Manchester	11 Tyne and Wear
5 Leeds	12 Plymouth
6 London Docklands	13 Birmingham
7 Merseyside	Heartlands

Figure 7.7 Urban Development Corporations

By 1993, UDCs accounted for nearly 40% of all urban regeneration policy expenditure. Over £12 billion of private-sector investment had been attracted, along with £4 billion from the public sector. They had built or refurbished 35,000 housing units, and created 190,000 jobs (Table 7.3).

*Table 7.3
Expenditure and
targets of UDCs*

Location	Date started	Expenditure (£million) 1992/93	Expenditure (£million) 1995/96	Lifetime targets Land reclaimed	Lifetime targets Housing (units)	Lifetime targets Jobs
London Docklands	1981	293.9	88	846.5	24,036	75,458
Merseyside	1981	42.1	34	384.0	3,544	23,357
Trafford Park	1987	61.3	29.7	400.6	3,774	21,440
Black Country	1987	68.0	36.6	525.3	1,403	10,212
Teesside	1987	34.5	47.5	210.8	311	25,618
Tyne and Wear	1987	50.2	43.5	517.7	4,842	34,043
Central Manchester	1988	20.5	13.7	60.0	661	4,590
Cardiff Bay	1988	–	–	250.0	950	2,200
Leeds	1988	9.6	–	35.3	2,581	5,074
Sheffield	1988	15.9	11.6	68.0	561	8,369
Bristol	1989	20.4	8.7	259.6	0	17,616
Birmingham Heartlands	1992	5.0	11.7	129.1	878	5,983
Plymouth	1993	n/a	10.6	12.7	93	491

Case study *Central Manchester Development Corporation (CMDC)*

CMDC is an example of a Development Corporation established after the LDDC, in 1988.

A partnership between the local authority and private developers was created. Its aim was to regenerate 200 ha of land and buildings in the southern sector of Manchester city centre. The area contained decaying warehouses, offices, former mills and contaminated land, unsightly railway viaducts and neglected waterways. It had been declared a conservation area in 1979.

Some of the buildings were refurbished into a range of uses including housing. For example, in the Whitworth Street district, warehouses were converted and redeveloped to create a village-like community of more than 1,000 household units, pubs, bars, restaurants and shops. The canals in the area were cleaned, and their banks were improved by the addition of lighting, seats and plants, all in an effort to improve

the aesthetics of the area. This has now become a very popular entertainment-based area for young people.

The CMDC engaged in widespread consultation and formulated a development strategy that complemented the plans of Manchester City Council. For example, Castlefield, which was once an area of disused canals, wharves and warehouses, became a mixture of housing (including some luxury apartments), office developments and leisure facilities. The area also developed its tourist potential and now attracts over 2 million visitors a year. Attractions include the world famous tour of Granada Studios, the Manchester Museum of Science and Technology, the GMEX Centre and the Bridgewater Concert Hall complex.

The CMDC was disbanded in 1996, and planning powers have now reverted to Manchester City Council.

Criticisms of UDCs

Some people argued that this amount of new employment was inadequate. There were two more significant criticisms. First, the UDCs were too dependent on property speculation and they lost huge sums of money through the compulsory purchase of land which subsequently fell in value. Second, because they had greater powers than local authorities, democratic accountability was removed. Local people often complained that they had no involvement in the developments taking place. Indeed, there were some examples, particularly in the London Docklands, where the local people felt physically and socially excluded by prestigious new housing and high-technology office developments.

Partnerships between government and the private sector

City Challenge Partnerships represented a major switch of funding mechanisms towards competitive bidding. To gain funding a local authority had to come up with an imaginative project and form a partnership in its local inner-city area with the private sector and the local communities. The partnership then submitted a 5-year plan to central government in competition with other inner-city areas. The most successful schemes combined social aims with economic and environmental outcomes. By 1993, over 30 City Challenge Partnerships had been established and another 20 or more bids had been unsuccessful. By the end of that year partnerships accounted for over 20% of expenditure on inner-city regeneration.

How City Challenge worked

The City Challenge initiative was designed to address some of the weaknesses of the earlier regeneration schemes. The participating organisations — the partners — were better coordinated and more involved. This particularly applied to the residents of the area and the local authority. Separate schemes and initiatives operating in the same area, as had happened before, were not allowed — the various strands of the projects had to work together. Many earlier initiatives had concentrated on improving buildings, whereas City Challenge gave equal importance to buildings, people and values. Cooperation between local authorities and private and public groups, some of which were voluntary, was prioritised.

All the City Challenge areas suffered from high youth and long-term unemployment, a low skills base, poor levels of educational attainment, environmental deterioration, increasing areas of derelict land and growing commercial property vacancy. Public-sector housing was deteriorating in almost all the City Challenge areas due to a combination of poor initial design and inadequate maintenance. The population of these areas usually had a higher than national average incidence of health-care problems, high levels of personal crime and fear of crime, a high proportion of single-parent families and households dependent on Social Security.

The priorities of the different City Challenge areas varied. In Liverpool, priority was given to environmental improvement, while in Wolverhampton a science park formed the centrepiece of the project. In Hulme, Manchester, housing improvement was the main focus (see case study below).

Was the initiative successful?

Overall, the competition between areas for funding was believed to be successful — improving the overall quality of proposals and encouraging new and more imaginative ideas. The private sector, in particular, found the competitive principle attractive and argued that competition had encouraged local authorities to suggest solutions as well as identifying problems. However, the competitive nature of the scheme was criticised by others on the grounds that large sums of money should have been allocated according to need, not competitive advantage.

In some cases neighbouring authorities competed against each other when they could have worked together. It is rare for the limits of disadvantaged areas to coincide with an administrative boundary. The policy that all successful bidders should receive exactly the same sum of money, irrespective of need, was also criticised. Finally, competing authorities were not given clear information about the criteria on which their application was to be judged — for some it was a stab in the dark.

By 1997 the Conservative government was able to publish statistics pointing towards the success of City Challenge. Over 40,000 houses had been improved, 53,000 jobs had been created, nearly 2,000 ha of derelict land had been reclaimed, and over 3,000 new businesses had been established.

Case study *Hulme City Challenge Partnership*

The Hulme area of Manchester was redeveloped as part of a slum clearance programme in the 1960s and a number of high-rise flats were built. Of the 5,500 dwellings, 98% were council owned. Over half of the dwellings were part of a deck access system, with many of the bad design features of prefabricated construction. The area had a low level of families with children, and a disproportionate number of single-person households. There was also a high number of single parents, and other people with social difficulties. There was some evidence that the local authority had used the area to 'dump' some of its more unfortunate residents.

Redevelopment

In 1992, under the Hulme City Challenge Partnership, plans were drawn up to build 3,000 new homes, with new shopping areas, new roads and community facilities. A more traditional pattern of housing development was designed, with streets, squares, two-storey houses and low-rise flats. By 1995, 50 ha of land had been reclaimed, the majority of the former deck access flats had been demolished, 600 new homes for rent had been built, and over 400 homes had been improved and refurbished. The main shopping area was totally refurbished, including the addition of an ASDA supermarket. A new community centre, the Zion Centre, was also constructed. Crime in the area has been greatly reduced, and there is more of a social mix of people living in the area. The appearance of Hulme has altered radically.

The partners

A number of agencies and organisations were responsible for this transformation, including the Guinness Trust and Bellway Homes. These worked in close collaboration with each other and with Manchester City Council. The company responsible for Manchester airport also invested capital in the project. Hulme is a good example of how the public and private sectors can work together to improve a previously declining and socially challenging area.

Case study *A flagship project: the St Stephen's Development, Kingston upon Hull*

The site of this development lies west of Ferensway and comprises 17 hectares. It includes major work on the existing Grade II listed railway station.

Site

The brownfield site currently comprises a mixture of ownerships, buildings and vacant lots. The general ambience of the area is one of neglect and decay. Various piecemeal redevelopment schemes over the years have come to nought. The development sponsors, Kingston upon Hull City Council and Yorkshire Forward, now own or have acquired a significant portion of the site; the remainder is subject to a confirmed Compulsory Purchase Order or acquisitions by agreement.

Scheme

The scheme will provide:

- a flagship development that is intended to complement and reinforce city-centre activities by having a mixture of retail outlets, a foodstore, a leisure complex, a hotel, a new home for the Hull Truck Theatre Company, a music centre, 1,600 car parking spaces and over 200 new residential units

- a new integrated transport interchange

The basic design concept is built around a diagonal pedestrian route across the site which curves gently at either end to form an 'S'. At the east end of this route there is a mix of retail, leisure, hotel and cultural uses, with associated parking and servicing. The route will be roofed over where it passes through the retail and leisure accommodation and this will form a covered street in the tradition of nineteenth-century arcade developments. At the west end of the pedestrian route there is an area of open space. New-build residential accommodation is proposed to the north and south of the open space.

Transport interchange

The aim of the £10 million transport interchange scheme is to transform the existing Paragon Rail Station and add a new bus station to provide an integrated transport interchange and new gateway to the city.

The interchange is promoted by a partnership between Kingston upon Hull City Council, Yorkshire Forward, Railtrack, Arriva Trains Northern Ltd, East Yorkshire Motor Services and Stagecoach Ltd.

Schemes and strategies of the twenty-first century

In the early years of the twenty-first century, the Labour government moved in two main directions in its attempts to regenerate and redevelop urban environments in the UK.

- It created **prestige project developments** (also known as **flagship projects**), such as the waterfront developments in Cardiff Bay, the Convention Centre area of Birmingham and the St Stephen's Development in Kingston upon Hull (see case study above).
- It began to develop **sustainable communities** in a variety of UK towns and cities. In theoretical terms, urban **economic sustainability** should allow the individuals and communities who live in cities to have access to a home, a job and a reliable income. Urban **social sustainability** should provide a reasonable quality of life and opportunities to maximise personal potential through education and health provision, and through participation in local democracies. (See London case study below.)

Case study *Sustainable communities in London*

The Labour government has stated that the sustainable communities initiative has the following general aims:

> We will work closely with our key regional partners to identify practical steps to ensure that we have communities that:
> - are prosperous
> - have decent homes for sale or rent at a price people can afford
> - safeguard green and open space
> - enjoy a well-designed, accessible and pleasant living and working environment
> - are effectively and fairly governed with a strong sense of community

There is an urgent need for more affordable homes all over London to accommodate its growing population and to reduce homelessness. It is also essential that workers who are key to the delivery of the capital's public services are able to afford to live and work within its communities. The Government Office for London (GOL) is working with the Greater London Authority (GLA), local authorities and relevant agencies (the Housing Corporation, English Partnerships, the Commission for Architecture and the Built Environment [CABE] and English Heritage) to achieve these aims.

Examples

The Holly Street estate redevelopment in Hackney has transformed a whole community. In addition to the newly created neighbourhood of small streets, small blocks of flats and brick-built houses with pitched roofs and gardens, a sports and community centre, an elderly persons' day centre and a health centre have also been provided. Alongside the aim of redeveloping the housing on the estate, the project sought to remove the fear of crime, improve security and improve the mental and physical health of residents, thus reducing the pressure on health services.

Greenwich Millennium Village is being developed on a brownfield site and is part of the larger Greenwich Peninsula development. Over a period of 5 years, 1,377 homes are being built, including homes for social rent or shared ownership. Sustainability, energy efficiency, waste management and quality in design and construction are key features of this project. The Millennium Primary School and Health Centre in the village is providing education, training, health care, crèche and other community facilities on one site.

Coin Street Community Builders (CSCB) is a social enterprise that has built social housing and commercial developments, including Oxo Tower Wharf, on London's South Bank. CSCB does not distribute profits from its commercial activities but uses them to cross-subsidise activities which otherwise would not be viable, including an arts and leisure programme. Its four housing developments are run by 'fully-mutual' cooperatives and provide 220 affordable homes for people in housing need.

Photograph 7.2 Oxo Tower Wharf

The role of Housing Associations

Housing Associations are non-profit-making organisations set up to provide rented accommodation. Initially they were the third type of housing provider after the private sector and local authorities, but during the last 20 years their influence has increased. They use a system in which they borrow private capital either to

Case study The Stockbridge Village Trust Housing Association scheme

The Cantril Farm estate, on the eastern edge of Liverpool, is part of Knowsley Metropolitan Borough Council. The estate was built in the 1960s to rehouse people displaced by slum clearance in the city centre. In many ways the estate was a typical housing redevelopment of the time, including two-storey houses, maisonettes and high-rise flats.

Aims of the trust

In 1983 the estate was purchased by the Stockbridge Village Trust with loans from banks, building societies and Knowsley Council itself. The estate consisted of 3,000 homes in various states of deterioration, and there were signs of the population moving out. It was predicted that half the estate would be empty by the mid-1990s.

The trust was established as a non-profit-making association with four main aims:

- to demolish the most unpopular and unsafe buildings
- to refurbish the remaining housing stock
- to develop new private housing for sale and rent
- to redevelop the service provision of the estate, in particular shopping and leisure facilities

The trust was supported by the government of the time. The Merseyside Task Force had been established to coordinate the variety of initiatives in the Merseyside area and an allocation of urban programme funds was made available.

Problems

By 1985 the pace of change was not as fast as had been hoped for. The costs of refurbishment of each house were more than double the original estimate (£6,700 rather than £3,000), the cost of redeveloping the shopping centre had risen from £2 million to £6 million, and only a small number of private houses had been built and sold.

There were two main difficulties. Rents were the main source of income, but these were subject to the fair rents legislation. In addition, incomes in the area were so low that few people could afford to buy houses under the right to buy scheme. The financial position of the trust gave cause for concern, and was made worse by the high interest rates at the time. It was clear that further support from local and national government was necessary or the scheme would collapse.

Progress

The Department for the Environment and Knowsley Council had to provide an extra £5 million between them. By 1987 most of the housing stock had been refurbished. Over 250 new houses had been built for rent, and a private developer had constructed over 125 houses, all of which had been sold. The trust built a new shopping centre, and Knowsley Council started work on a new leisure centre. The demolition of the high-rise flats also began.

During the 1990s the objective of the trust was to continue its housing programme. Housing repairs were to continue and there was a move to broaden the type of tenure. A further 250 new houses were built, some for rent and some for shared ownership. Private developers were also encouraged to add further properties to the area. The management role of the Housing Association is one of ensuring close collaboration between the public and private sectors to provide good-quality housing for all.

build new houses or to buy existing housing stock (e.g. former council housing, NCB housing or even ex-military housing), and seek to make returns on their investments for further reinvestment. As they also receive government subsidy, they are able to provide housing for many people at lower rents.

They are also part of the strategy to encourage greater home ownership — people are initially offered housing for rent, but in the longer term may opt to buy the property. Some Housing Associations in inner cities use this system to initiate the process of home ownership in areas where this is not the norm. In some cases, Housing Associations may offer rental packages on furniture and other household items.

Retail change in the UK

The traditional pattern of retailing is based on two key factors:

- easy, local access to goods such as bread, milk and newspapers which are purchased on a regular basis, often daily and particularly so if perishable
- willingness to travel to a shopping centre for goods with a higher value which are purchased less often, such as household and electrical goods, clothes and shoes

For many years, these factors led to a two-tier structure of retailing. Local needs were met by corner shops in areas of terraced housing, and by suburban shopping parades. Higher-value goods were purchased in the town centre (the central business district or CBD) and required a trip by bus or car. In the last 30 years technology (in the form of the motor vehicle) has had a major influence on the patterns of retailing.

Photograph 7.3
Bluewater shopping
centre, Dartford

Bluewater

**Table 7.4
Out-of-town
shopping centres**

Shopping centre	Location	Number of shops
Braehead	Near Glasgow	100
MetroCentre	Near Newcastle-upon-Tyne	320
Trafford Centre	Near Manchester	320
Meadowhall	Near Sheffield	270
Merry Hill	Near Birmingham	230
Cribbs Causeway	Near Bristol	150
Bluewater	Dartford (south of Thames)	330
Lakeside	Thurrock (north of Thames)	350

In the 1970s **supermarkets** and **superstores** began to be built in residential areas and town centres. These stores sold a full range of food and non-food items, including brand names and shop brands, at the same check-out. This idea expanded into larger **hypermarkets** that also sold electrical goods and clothing and often had smaller specialist retail outlets under the same roof. An important factor in the development of these establishments was the use of the private car to load up once or twice a week with the 'family shop'.

In the 1980s **non-food retail parks** expanded. These housed DIY, carpet and furniture stores such as Do-it-All, B&Q, MFI and Carpetland. Many such parks were constructed on the outskirts of towns or cities, with easy access to main roads, again to attract the car user. The buildings were of a warehouse type construction, often uniform in design, each distinguished by the display on the outside and by the internal design.

In the 1990s huge **out-of-town shopping centres** were built on the periphery of large urban areas and close to major motorways. They often had their own motorway junctions. Some of the best-known shopping areas in the country come into this category: the MetroCentre (near Newcastle), Meadowhall (near Sheffield), the Trafford Centre (near Manchester), and Bluewater (Photograph 7.3) and Lakeside on either side of the Thames east of London (Table 7.4 and Figure 7.8).

**Figure 7.8
Location of out-of-
town shopping
centres**

0   km   200

Braehead
Renfrew,
nr Glasgow

MetroCentre
Edge of Gateshead,
Tyneside

Trafford Centre
Outskirts of
Manchester

Meadowhall
Outskirts of
Sheffield

Merry Hill
Dudley,
West Midlands

Lakeside
Thurrock,
Essex

Cribbs Causeway
Near Bristol

Bluewater
Near Dartford, Kent

Effects on the CBD

During each of these phases of retail change, the traditional town centre has continued to exist. At several times this area of retailing has been said to be dying, and in some small towns there has certainly been a general decline. Evidence that was often quoted in the 1990s was the closure of branches of Marks and Spencer in a number of small towns. Shop closures in some cities did take place, with consequent job losses. An increase in the numbers of charity and low-price shops and a greater emphasis on office developments in town centres are also given as evidence of decline.

The decline of CBDs is a cause for concern. Run-down city centres can become dangerous at night. Dereliction, increased numbers of low-grade shops and lack of investment all encourage a cycle of decay. Planners see the CBD as an important social and cultural meeting point for a city. A declining CBD accelerates the success of the out-of-town shopping centres.

A number of strategies are being devised to help city centres fight back, including:

- the establishment of business and marketing management teams to coordinate activities
- the provision of more attractive shopping environments with pedestrianisation, new street furniture, floral displays, paving and landscaping
- the construction of all-weather shopping malls that are air-conditioned in the summer and heated in the winter, often with integral low-cost parking
- the encouragement of specialist areas, such as open street markets, cultural quarters and arcades
- the improvement of public transport links to the heart of CBDs, with rapid transit systems, park-and-ride schemes and shopper buses
- the extensive use of CCTV and emergency alarm systems to reduce crime and calm the fears of the public, particularly women
- the organisation of special shopping events such as Christmas fairs, late-night shopping and Sunday shopping
- conservation schemes that enhance the ambience of heritage cities such as Chester, York, Bath and Cambridge

Many cities are also encouraging functions other than retailing to increase the attractions of their CBDs, including:

- developing a wider range of leisure facilities that people visit in the evening, such as café bars, restaurants, music venues (e.g. the 'Arenas' in many city centres), cinemas and theatres
- promoting street activity (e.g. Covent Garden in London)
- developing a wide range of nightlife, such as 'clubbing' in Manchester and Leeds (but there are negatives associated with this as a high level of policing is necessary)
- establishing theme areas, such as the gay area in Manchester, or the cultural quarters of Sheffield and Stoke
- developing flagship attractions (e.g. the National Museum of Photography, Film & Television in Bradford)
- encouraging residential use to return to city centres, in the form of either gentrification or new up-market apartments

Effects on the rural–urban fringe

Large areas have been devoted to major retail parks and this has involved the following:

- redevelopment and/or clearance of a large area of cheap farmland or a brownfield site
- the creation of extensive areas of car parking (the Trafford Centre has 10,000 spaces)
- the construction of a link to a motorway interchange or outer ring road
- the development of other transport interchange facilities — bus station, supertram, railway station
- the construction of linked entertainment facilities, e.g. Warner Village cinemas, fast-food outlets

Effects on retailing in general

In the twenty-first century we are seeing the beginning of e-commerce and e-tailers — electronic home shopping using digital and cable television systems. The impact of this form of retailing on other types of shops has yet to be seen. However, it does seem that this type of shopping is unlikely to affect existing shopping locations seriously. People still want to examine items before purchase, and e-tailers depend upon mail delivery services, both road and rail based, none of which can guarantee next-day delivery.

Another recent and more localised change in shopping is the rise in the local petrol-station shop. This is clearly linked to the increase in private car ownership. Local petrol stations are no longer just places people go for fuel. They can also buy newspapers, bread, milk, vegetables, fast food, lottery tickets and often obtain money from a cash machine. The local petrol station has become the corner shop of the twenty-first century.

Factors affecting retail change

A number of factors have combined to produce the changes in retailing described above.

Increased mobility

Nearly all the changes described — from the growing use of the local petrol-station shop to large out-of-town shopping centres — arise from increased ownership and use of the private car. Car parking in city centres is expensive and relatively restricted. Access to such areas is by means of congested roads. Out-of-town retail areas have large, free areas of car parking. Locations next to motorway junctions offer speedy access, which makes reaching them less stressful. Even at a local level, it is often easier to pull into a petrol-station forecourt to make a low-level purchase than to find a parking space outside a suburban shopping parade.

The changing nature of shopping habits

People now purchase many items as part of a weekly, fortnightly, or even monthly shop. The use of freezers in most homes means items that once had to be purchased regularly for freshness can now be bought in bulk and stored. This technology has dovetailed with the changing nature of employment. In many cases, either both income earners or the sole income earner in a family does not have time to shop daily. Retailers have responded to this by developing more 'ready-made meal' products that can be stored in a domestic freezer.

Changing expectations of shopping habits

An increasing number of people use shopping as a family social activity, involving more than just the act of shopping. Consequently, many of the larger shopping areas combine retailers with cinemas, restaurants, fast-food outlets, and crèches and entertainment areas. For example, the White Rose Centre near Leeds has an area set aside for men who accompany their partners but do not wish to 'shop until they drop'! It is claimed that at the Bluewater shopping centre near Dartford you are never more than 100 m away from a coffee bar. Such marketing ploys are used to make the customer feel much more at home.

The changing nature of retailing

There are only a few supermarket/hypermarket companies, each of which strives to be more competitive than the others. They seek to build on cheaper out-of-town locations and to increase their economies of scale. In this way they can afford to reduce prices and provide large car-parking areas.

Case study *An out-of-town retail area: the Trafford Centre, Manchester*

When the Trafford Centre was opened in 1998 many people were concerned about the effect it would have on Manchester's CBD. Nearly 5.4 million people (almost 10% of the UK population) live within 45 minutes' drive of the centre. It was designed to be more than just a shopping centre, with a 1,600-seat food court, an 18-lane ten-pin bowling alley and a 20-screen cinema.

Advantages

The Trafford Centre offers the following:

- 10,000 free car parking spaces
- facilities for the disabled which are regularly spaced within the complex. These include a Shop Mobility Unit offering scooters and wheelchairs
- a weatherproof, air-conditioned and safe environment
- its own security system, with a tannoy and a meeting point for lost children
- a full range of peripheral services, such as a post office, banks and travel agents

Disadvantages

It has the following disadvantages, typical of out-of-town retail areas:

- heavy build-ups of traffic on the access road network at certain times of the year, such as Christmas and Bank Holidays
- the atmosphere within the complex is artificial, although there are themed sections which reflect the styles of Italy, Chinatown and New Orleans
- all the outlets are those of national chain stores — it is too expensive for local or independent businesses to become established in the centre
- public transport services to the centre are restricted, which makes access difficult for elderly shoppers in particular. In 1999, 85% of visitors came by car. However, there are plans to provide a Metrolink connection and a rail link to the centre
- it is difficult for poorer people to gain access to the centre — for example, the homeless are usually kept out by security staff

Changing attitudes of politicians and planners

Politicians believe that retail development in out-of-town locations has gone far enough. New life has to be injected into existing CBDs to avoid the problems of economic decline in these areas. Similarly, traffic congestion in out-of-town locations has to be reduced. A number of the major shopping areas identified above have sought permission to expand and, in most cases, this has been refused. Some supermarket chains are turning their attention back to their existing and new CBD outlets. For example, Sainsbury's has developed new Sainsbury's Local stores which do not sell the full range of goods found in its larger outlets but stock items targeted at local needs.

Attitudes to the development of out-of-town retailing

Those in favour of out-of-town retailing believe it provides greater opportunities to shop without the need to travel into city centres. It also provides greater employment opportunities for local people, especially for students at weekends or young mothers who want to work part time.

Those opposed to out-of-town retailing believe it causes an increase in traffic in the area which creates problems of safety, pollution, noise and parking in local residential streets. All-night shopping means continual movement of both cars and delivery lorries which may cause unacceptable noise levels at night.

Chapter 8

Option T
Historical landscapes of England and Wales

This option examines how the rural and urban landscapes of England and Wales have evolved over time. The key thing to remember here is that a great deal of what we see today in both countryside and towns is the product of what our predecessors did in the past. We tend to take for granted, for example, the layout of fields and lanes in rural areas, and the types of buildings and layout of streets in towns.

Most of the changes that took place in the past were due to technological, social and economic innovations. The products of these changes are now often believed to be worthy of protection and have given rise to a growing tourist industry. A key element of this industry is the conflict between protection and exploitation of historic landscapes. This option provides an opportunity to examine a few of the major changes that have taken place in the rural and urban landscapes of England and Wales.

The evolution of rural landscapes

The first major change to consider is that from open fields to an enclosed landscape. Remnants of previous rural landscapes can still be found across England and Wales. Briefly, these include:

■ Stone Age settlements represented by stone circles, burial mounds (often called tumuli on Ordnance Survey maps) and hill forts. These are scattered across England and Wales, but are concentrated in western and southern England and in the Welsh hills, in places such as Salisbury Plain and the South Downs in England, and the Welsh border lands.

■ Roman settlements, including towns and smaller villas. The Romans had a major influence on the landscape of England, establishing many small towns on communication routes between major forts. The evidence of Roman occupation is frequently seen in the names of towns. The suffixes -caster, -chester and -cester derive from the Latin word *castra* meaning camp. During their occupation of the country, the Romans built an extensive road network, still followed by many A-roads. Later they established a number of agricultural villas, particularly in southern England.

The open-field landscape

During the fifth and sixth centuries AD, Germanic peoples known as the Anglo-Saxons invaded England. They cleared extensive areas of woodland, settled alongside rivers and began to farm the land. They brought a period of self-sufficiency and a system of farming which lasted for over 1,000 years, during which time it had a major influence on the landscape of England and Wales.

During the ninth and tenth centuries the Anglo-Saxons were followed by other invaders from the east and north, including the Vikings and the Danes from Scandinavia. The great majority of villages in England were established during this period.

As with the Romans, much evidence of these later invaders can be found in place names:

■ The early phases of Anglo-Saxon colonisation are indicated by village names ending in -ing, -ton, -ham, -ingham, -borough, -burgh. Barn-, burn- and brin- in place names all indicate the use of fire to clear woodland.

■ Later dispersal of Anglo-Saxon settlements (often called daughter settlements) is indicated by the suffixes -cote, -ley, -field and -stead.

■ Subsequent Scandinavian settlements are indicated by the suffixes -toft, -by, -thorpe and -thwaite.

Photograph 8.1 In Laxton, Nottinghamshire, open-field strip farming is still practised

As stated earlier, the greatest impact was the system of farming that was introduced. It is known as the open-field system and it created a distinctive landscape (Photograph 8.1, Figure 8.1).

Michael St Maur Sheil/Collections

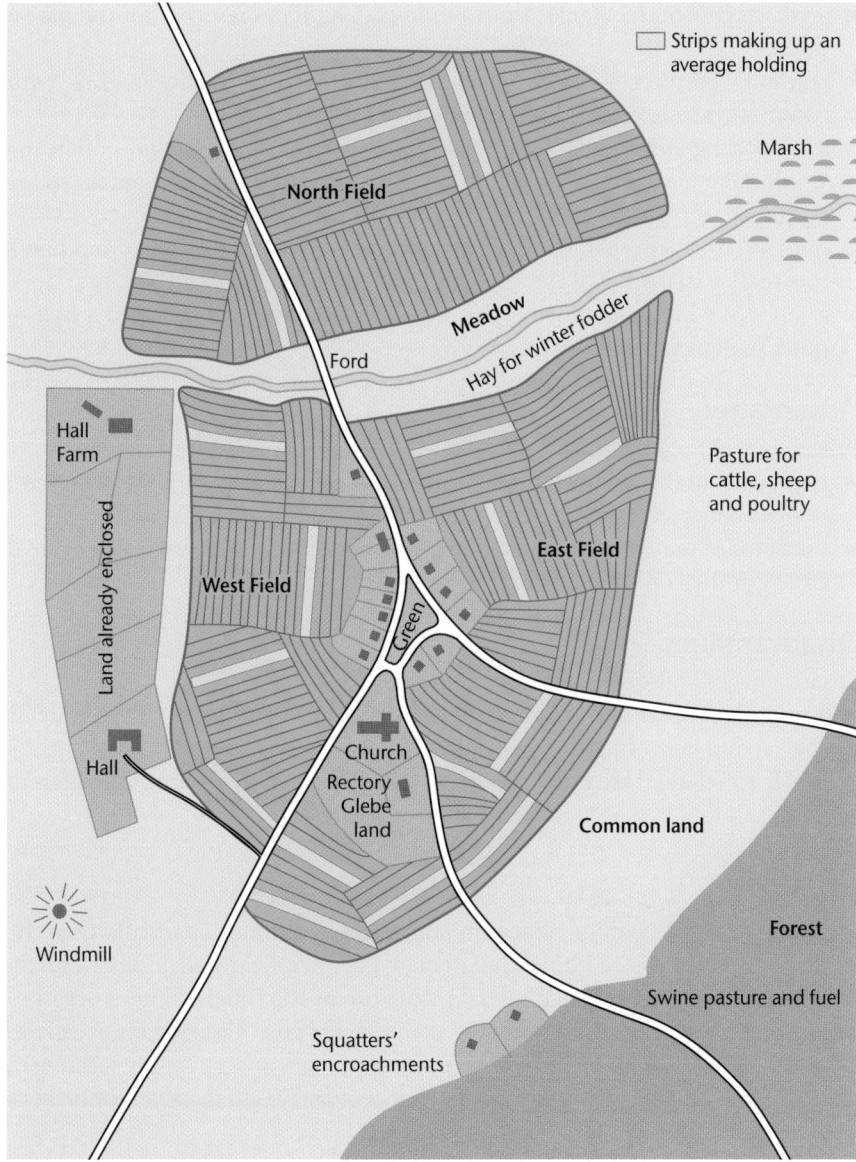

Features of the open-field landscape

The characteristic features of the open-field landscape were:

- two or three very large communal open fields around the village, often called North Field, South Field etc.
- strip cultivation within each field, with farmers having access to a number of strips in each field according to their wealth. Each strip was ideally 22 yards wide and 220 yards long, but this was not always the case. Each field would have a fallow period every 2–3 years
- houses collected together in a central village, with the landowner living in the village or nearby in the manor house

- a village church, rectory and surrounding glebe (land reserved for the parish priest)
- common pastures — often meadows and other low-quality land — with common grazing rights for everyone
- surrounding areas of uncleared woodland where coppicing and pollarding would take place, and pigs and other foraging animals could graze (swine pastures)
- outlying cottages inhabited by squatters and woodmen, each of whom had a role to play in the life of the village
- a series of winding lanes emerging from the village which provided a link to other settlements in the area

Advantages

Advantages of the open-field system were:

- the quality of the land was evenly spread among the farmers of the village. All farmers had a fair share, in theory, of poor- and high-quality land
- pressures of mutual support encouraged communal activities and this made it easier for the Lord of the Manor to control the activities of the local farmers

Disadvantages

Disadvantages of the open-field system were:

- a great deal of land was wasted by walkways and boundaries between fields (balks) and between individual strips
- much time was lost by individuals travelling between strips in different fields

Enclosure

Until the fifteenth century villages in large areas of England, particularly in the Midlands and the east, continued to operate the open-field system of farming. It was closely associated with the social structure of the time. Changes began to take place in the fifteenth century. For example, the great Cistercian monasteries in northern England established large pastoral granges mainly for the farming of sheep. These did away with the open-field system and led to the abandonment of many villages.

Other landowners began to follow this idea, enclosing arable land into smaller fields. Hawthorn hedges and ditches were constructed around these enclosures. Among the first landowners to enclose land were the Spencers of Northamptonshire who later established themselves at Althorp, and the Churchill family at Marlborough. The open-field system had begun to break down. Fields of less than 100 acres were being created. The ideal size was 10–12 acres (4–5 ha).

The concept of enclosure began to spread across the country, and many areas changed from the strip system to enclosed fields by common agreement. Most enclosure did not happen until the eighteenth century, with the parliamentary Enclosure Acts, but counties such as Northamptonshire and Leicestershire were heavily enclosed well before then.

Parliamentary enclosure

The Enclosure Acts dealt with 4.5 million acres (1.8 million ha) of open field, not counting the enclosure of commons and other 'wastes'. This was about half the farmland at the end of the seventeenth century. Parliamentary enclosure began in the 1750s under George II, and continued into the realm of George III. It affected over 3,000 parishes. The complex pattern of strips, with their winding balks, cart roads and grassy footpaths, changed rapidly into a chequer-board pattern of small square or rectangular fields enclosed by hedgerows of hawthorn or by stone walls (Figure 8.2).

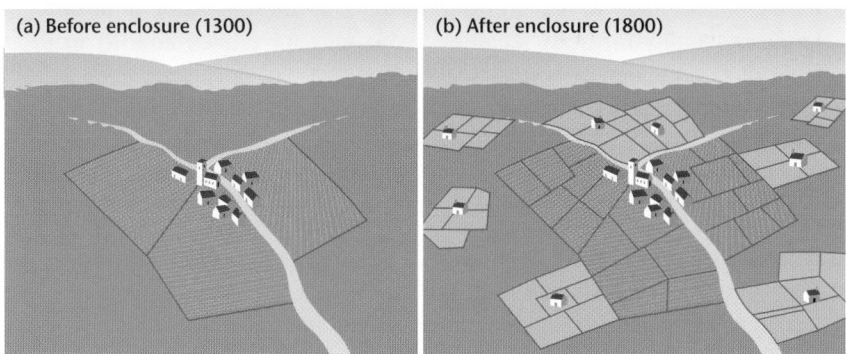

(a) Before enclosure (1300)   (b) After enclosure (1800)

*Figure 8.2
Changes in the
landscape brought
about by enclosure*

People were allocated land according to the amount they had farmed under the old system, but many lost out in the change. All the common land and marginal land in the village also featured in the new field arrangement, and some farmers were allocated unfair proportions of such land.

Landscape changes

Sporadic lines of elm and ash were planted to mark field boundaries. New, straighter roads running across and beyond the parish were created. These frequently had right-angle bends as their route was affected by both the layout of the former open fields and the newly created fields. They were wide, with grass verges available for further widening in the future.

Another key element of the change was the creation of isolated farmsteads in the midst of the new farmlands. It was more economic to live in the middle of the assigned fields than to travel out from the central village.

Residual areas of woodland were preserved to act as areas for coppicing and as fox coverts.

Much of this change took place in only 2–3 years. The change for each village was planned and put into operation by parliamentary commissioners who visited the area over a period of a few days. The rural landscape of today was mostly created at this time.

Commercial farming

Other changes were also taking place in agriculture. New machines were being invented that allowed faster sowing and harvesting, for example Jethro Tull's seed drill. Animal husbandry, for example the work of Robert Bakewell, resulted in new

pedigree breeds of sheep and cattle. Crop rotation advocated by Viscount (Turnip) Townsend also occurred on a large scale. Many of these innovations had royal patronage from George III (Farmer George). In short, farming became much more commercial and the landscape had to reflect this change.

Attitudes to enclosure

Those in favour of the changes were:

- wealthy landowners who were able to purchase the best agricultural land or influence the parliamentary commissioners in the allocation of land. These were the most likely to benefit from the new commercialisation of farming
- landless people, formerly squatters, who may have been able to work as labourers on the newly created large farms
- traders/merchants in towns who welcomed the increased commercialisation of farming

Those against the changes were:

- less wealthy families who were forced to sell or leave land to the wealthier members of the village
- woodsmen and swine herders who had worked the common woodlands which were now being reduced as part of the reallocation
- people (known as burgesses) who had held rights to graze their animals on fallow and common land but now lost their status within the village

The evolution of urban landscapes

All towns and cities in England and Wales, other than New Towns, show considerable evidence of the past in their buildings and street layouts. The best way to analyse this evidence and the changes that have taken place is by examining historical maps and looking at visual evidence from paintings, etchings, engravings and photographs.

In chronological order, the following may well be found in a typical town or city:

- Roman remains: walls, baths, rectangular street layout, amphitheatre
- medieval remains: castle, city walls, cathedral, street names (for example the use of the suffix -gate)
- Georgian and Regency architecture: townhouses, garden squares, coaching inns
- Victorian architecture: lines of terraced houses, mills/factories and breweries, railway stations, canals

Georgian and Regency architecture in Bath

Many town and cities have a Georgian/Regency area, built during the reigns of George II and George III, from approximately 1750 to 1820. Such areas are dominated by large townhouses, garden squares and features such as crescents and terraces. All of these features are best seen in the city of Bath.

Photograph 8.2
The Circus, Bath

Bath is based around a natural spring of mineral waters that were thought to be a cure for rheumatic and other complaints. In the 1700s Bath became a fashionable resort for the well-off. Using local stone from the quarries at Combe Down, a series of palladian mansions, terraces and town houses was built. Architects such as John Wood (the Elder and the Younger) and Robert Adam designed and built prestigious streets such as Queen Square, The Circus, the Royal Crescent and Great Pulteney Street. The Royal Crescent is perhaps the most striking and well-known street in Bath, and has featured in a number of historical films. It consists of 30 houses in a sweeping curve above a sloping lawn.

Figure 8.3
A typical Georgian townhouse

Figure 8.3 shows a typical Georgian town-house. The main features are:

- a rectangular plan, with a symmetrical facade
- sash windows with thin glazing bars
- a crenellated pattern around the edge of the building, made from stone rather than brick
- an ornate portico above the front door, with columns at either side
- a small ring window at the top of the building
- a stone parapet along the edge of the roof

Other common features of Georgian buildings include smaller windows on upper floors (the servants' quarters), decorative wrought iron at the entrances, and steps behind iron railings leading down to the basement. Carved friezes can be found along the tops of some Georgian buildings.

Portico over doorway   Hipped roof   Chimney on the side

String course   Symmetrical, regularly-spaced windows

The industrial town of the nineteenth century

During the late eighteenth and the nineteenth centuries, great industrial progress was made in England. Most notably, iron and steel were mass-produced at locations such as Coalbrookedale, and textiles were produced in large mills in Nottinghamshire, Derbyshire, Lancashire and Yorkshire. Key elements of this progress were the development of cheap power sources, initially water power, followed by steam power using coal, and cheap forms of transport, initially canals and then railways.

Throughout northern and central areas of England industrial towns began to be established, based around existing market towns or villages. The growth of these towns was very rapid. For example, in 1760 Birmingham had a population of 35,000, but by 1801 it had doubled to 73,000. The town of Ashton-under-Lyne in Lancashire grew rapidly and had over 100 cotton mills within a 16 km radius of the town centre.

Growth of towns mainly involved the building of factories or mills and of areas of cheap housing around them. The factories tended to be located on lower, flatter ground, near canals and railways so raw materials could be brought in and products taken away. Rivers were a major locational factor — as a power source in the early days, but also as a place where waste products could be discharged. Water was a significant raw material too in many industrial processes.

Housing

Extensive areas of densely packed houses were constructed close to the factories so that workers had only short distances to walk to work. As land tended to be sold off piecemeal, builders fitted as many houses as possible on to the available space. Some houses were built back-to-back (Photograph 8.3), others in a courtyard

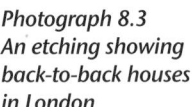

*Photograph 8.3
An etching showing
back-to-back houses
in London*

Topham

design. In both cases, streets with a grid-iron pattern of terraced houses were produced, with the front doors opening directly on to the street. As the land that was being built on was sold field by field (following the pattern of enclosure defined one or two centuries earlier), the streets often represented field boundaries or boundaries between the land of different landowners (Figure 8.4).

Living conditions in these areas were dreadful. In some houses there was no internal staircase — access to upper floors was by ladder. Few houses had an indoor toilet and most people had to use (and even share) an outdoor privy connected to a cesspit. Waste was removed from the pits at night by men with carts and was known as night soil. There was no mains drainage. Initially in back-to-back houses the privy backed on to the wall of the house. The stench was appalling, and it frequently overflowed.

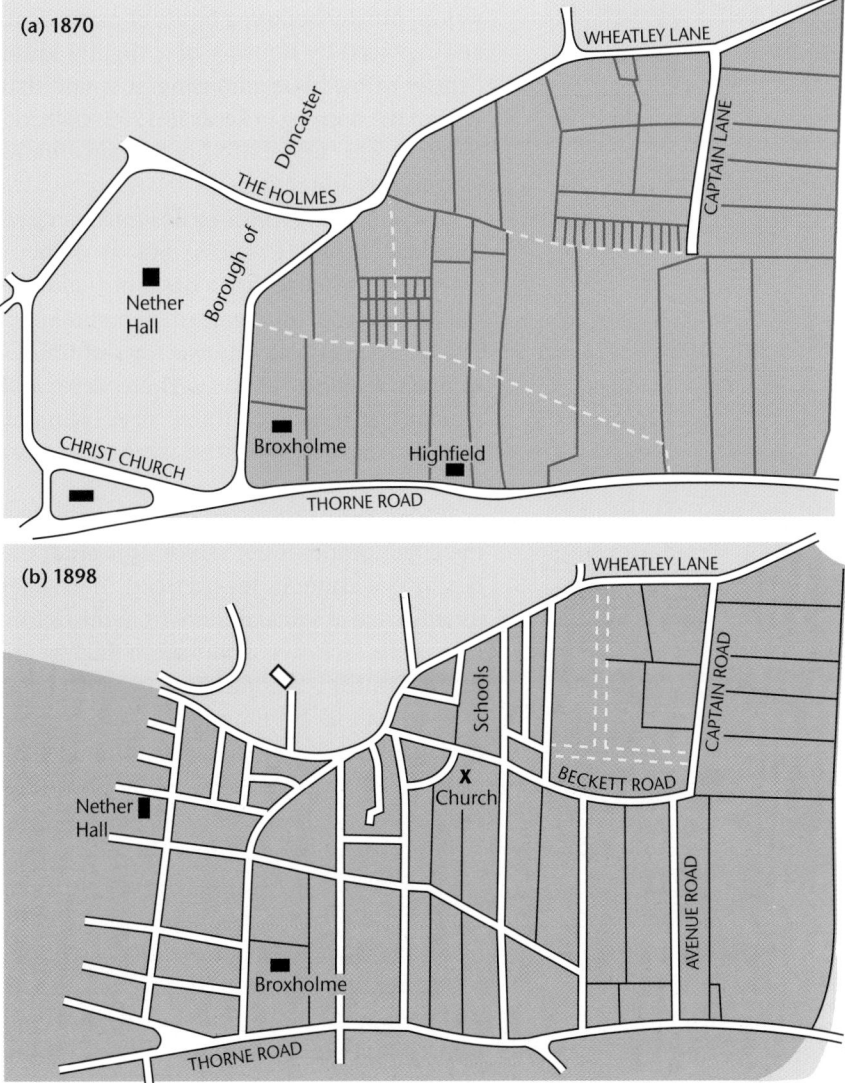

*Figure 8.4
Wheatley in
Doncaster
(a) before and
(b) after housing
development*

Fresh water was taken from nearby wells and pumps, and was contaminated in many cases. Water was also taken from rivers, but as many towns also had stables, manure heaps and pigsties, the water from the river had to be 'stood' in earthenware pots or settlers before it could be drunk. Diarrhoea and cholera were widespread.

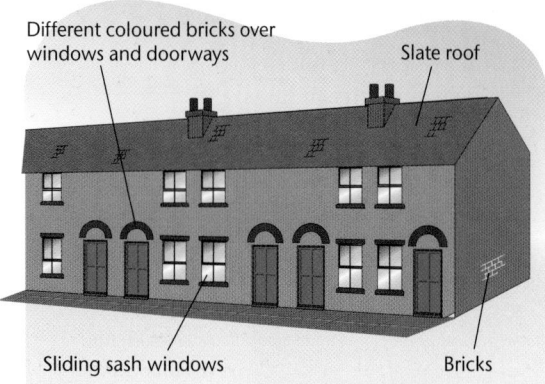

Figure 8.5 Typical Victorian terraced housing

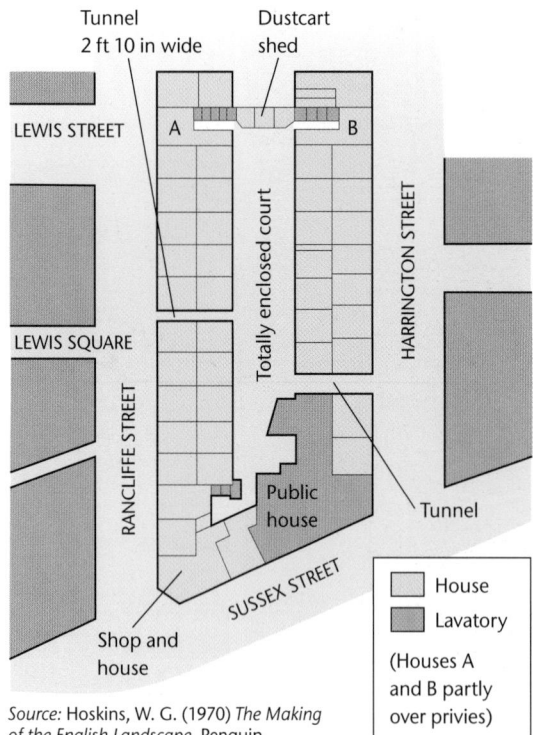

Source: Hoskins, W. G. (1970) *The Making of the English Landscape*, Penguin.

Figure 8.6 Plan of a typical court in Nottingham, 1845

Adequate ventilation was not possible in back-to-back houses, and damp, foul air was common. The quality of the air was made worse by smoke pollution from nearby factories, mills and railways. Bronchitis, pneumonia and other respiratory complaints were common causes of death.

The construction of back-to-back houses was eventually outlawed by act of parliament, and replaced by housing of a slightly lower density called by-law housing. It is said that 70% of the housing in Leeds in 1900 was back-to-back, but none of this remains today (Figures 8.5 and 8.6).

By-law housing consisted of similar terraced housing to back-to-backs, but at a lower density. Most houses had a back yard, with the outdoor toilet in a separate small outbuilding. The back yards between two rows of houses were separated by an alley. There were local regulations regarding room sizes, window sizes and other requirements, such as a pantry and a cast iron grate.

Within the areas of terraced housing were the larger properties of senior officials at the factory (for example, the overseer). The owner usually lived at some distance from the factory in a more exclusive, and usually higher up, part of the town. In some cases areas of land in a town would be untouched, and a large Georgian house would remain surrounded by gardens. Other areas of open space included cemeteries and, later, council parks.

Other buildings featured in the urban landscape — churches, chapels (often of non-conformist origin), washhouses and public baths, public houses and eventually schools. Grand public buildings were also constructed, often financed by wealthy factory and mill owners — libraries, museums and art galleries, town halls and railway stations.

Case study *Saltaire model village*

Saltaire is a model industrial village (Figure 8.7) and former textile mill north of Bradford, built by the industrialist Sir Titus Salt. Having built his fortune on the use of alpaca and mohair, Salt found by the late 1840s that his mill was too small to meet the demands of his new textiles. In addition, in 1849 a major cholera epidemic struck Bradford. Salt was a strict Congregational Christian and believed that 'Cholera was God's voice to the people'. He decided to build a better community for his workers.

Saltaire was built between 1852 and 1872 and was modelled on the buildings of the Italian Renaissance — a period (in Salt's opinion) during which both cultural and social advancement took place. Salt's mill emulated an Italian palace, was larger than St Paul's Cathedral, and was the largest factory in the world when it opened. It was built alongside the Leeds–Liverpool canal, which allowed easy access for raw materials.

It was surrounded by a school, a hospital, a railway station, parks, baths, washhouses, 45 almshouses and 850 houses. The style and size of each house reflected the place of the head of the family in the factory hierarchy. Twenty-two streets were created and all but two (Victoria and Albert Roads) were named after members of Salt's family. The church was the first public building to be completed and there was not a single public house. Workers who lived in Saltaire had to adhere to a strict set of rules set out by Salt, which included personal hygiene and no consumption of alcohol.

Today, the mill has been converted for a variety of uses. Part of the building is used by a microelectronics company, but the mill also contains shops and an art gallery which features work by another famous child of Bradford, David Hockney. The village is well maintained and is a small-scale tourist attraction.

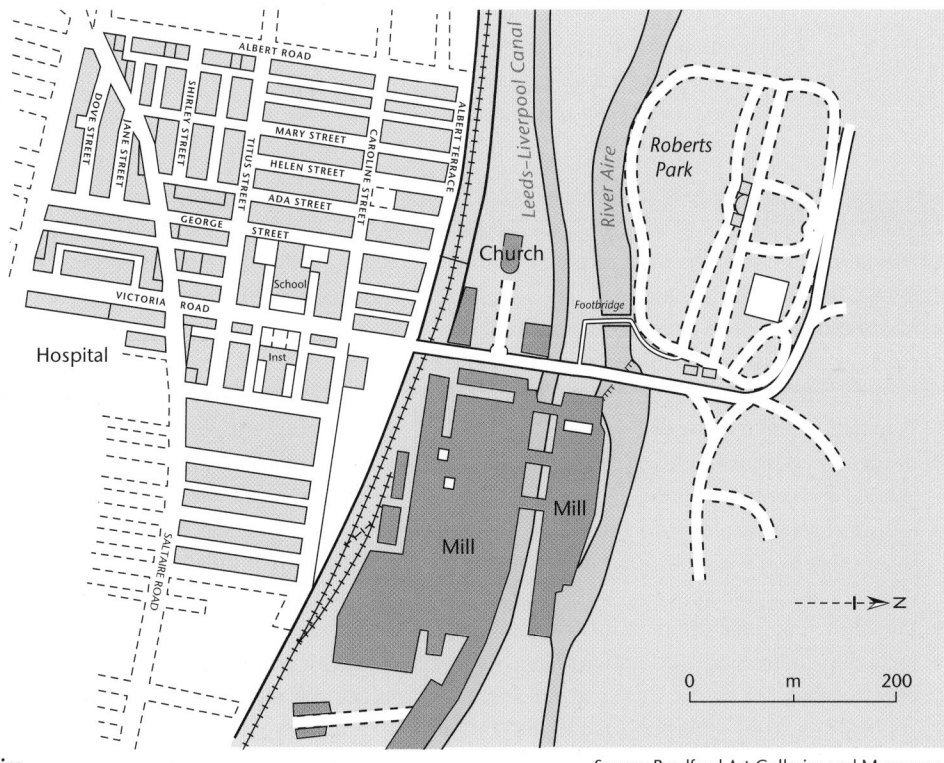

Figure 8.7
Map of Saltaire

Source: Bradford Art Galleries and Museums.

Utopian ideas

Some factory or mill owners were 'enlightened industrialists' determined to provide better-quality housing and other social services for the workforce and their families. Such men were appalled at the poor-quality living conditions faced by ordinary people in industrial towns. They witnessed a number of cholera outbreaks in the mid-nineteenth century, which killed thousands of people.

Some of the enlightened industrialists were motivated by religious principles. The Cadbury brothers and Joseph Rowntree, for example, were Quakers, and abhorred the widespread use of alcohol by workers to make their lives bearable. They were determined to provide better housing and better social values for the workers in their factories. Purpose-built settlements were constructed close to factories and mills, with specially designed housing and services. These are sometimes described as 'utopian' — a term taken from a sixteenth-century book in which Utopia was an imaginary island representing the perfect society.

Examples of enlightened industrialists and the settlements they built are:

- the Cadbury brothers — Bournville, Birmingham
- William Hesketh Lever — Port Sunlight, Wirral
- Sir Titus Salt — Saltaire, near Bradford (see case study on previous page)

New Towns of the twentieth century

New Towns were based on the late nineteenth-century ideas of Ebenezer Howard who was himself influenced by the work of the enlightened industrialists. He envisioned 'garden cities' — completely planned urban settlements with a great deal of open space, designed to combine the best of 'town' and 'country'. He and his associates managed to create two model settlements north of London — Letchworth in 1903 and Welwyn Garden City in 1920.

Case study *Milton Keynes New Town*

Milton Keynes is one of the most recent New Towns in the UK. Its characteristics are:

- housing areas built on a neighbourhood unit principle. Some housing is uniform in design, but there is some individuality of housing types to meet the demands of different families
- modern housing with indoor bathrooms, double glazing and central heating
- industrial areas isolated from the rest of the town but located near major transport routes, including the M1 and the railway
- low-level manufacturing industry buildings, including many single-floored warehouse-type constructions
- service industries (financial, legal) located in purpose-built maisonette-type structures

- planned growth and design of the town, with a block-like structure (Figure 8.8)
- a designed road network connecting housing and industrial areas, with considerable (some would say excessive) use of roundabouts
- transport terminuses (railway station, bus station and major road junctions) all in close proximity
- a variation in residential street style — some curved, others rectilinear, but each confined to a square block within the overall road pattern. Examples of these purpose-designed neighbourhood units include Fishermead and Bradwell
- a central shopping area well served by the transport systems
- extensive planned open spaces with woods, parks and lakeland areas

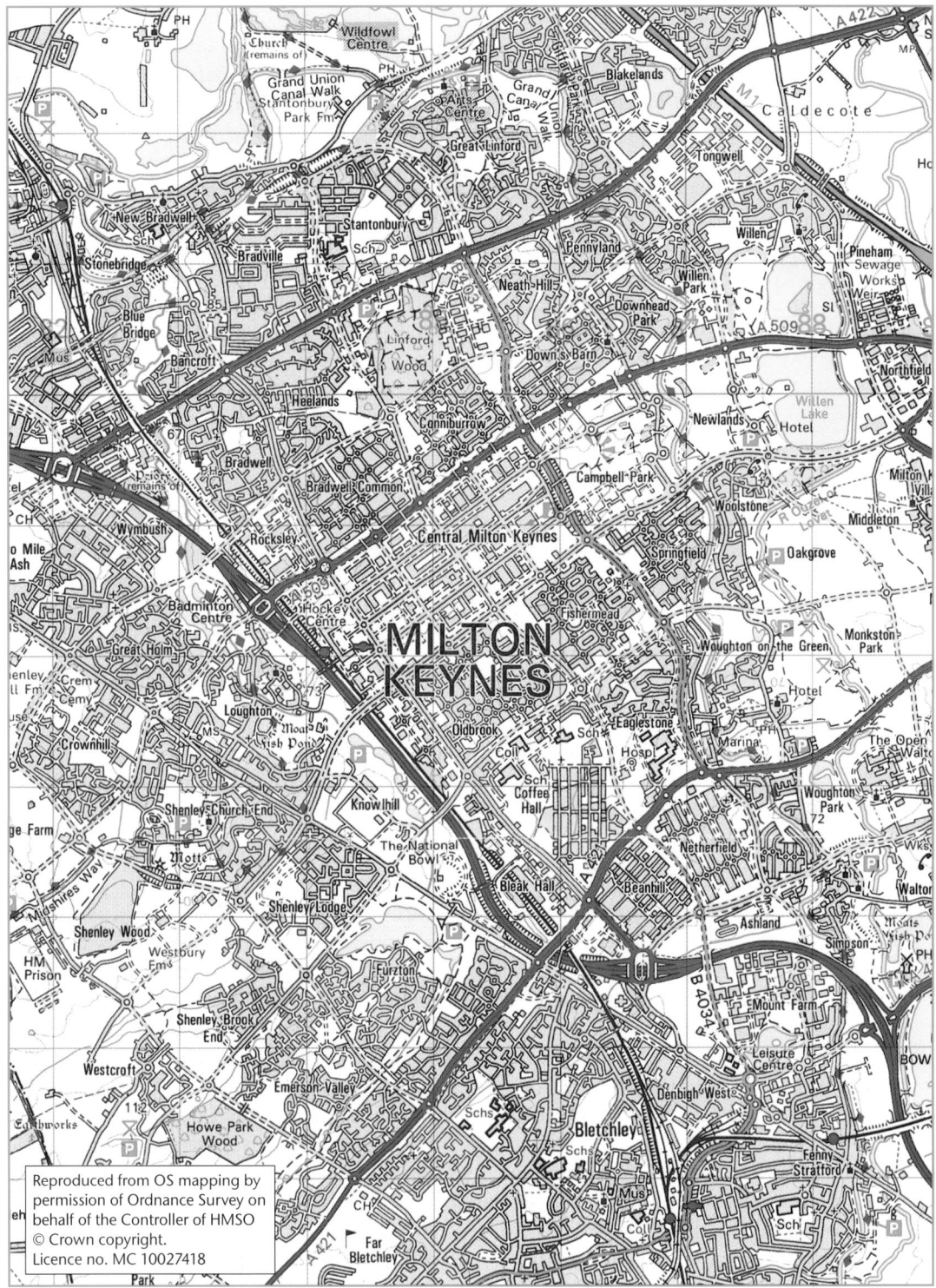

Figure 8.8 Ordnance Survey map showing part of Milton Keynes

New Towns reflected many of the ideas of Howard, but appeared later, following the New Towns Act of 1946. This legislation aimed to relieve over-crowding within major cities such as London, Liverpool and Glasgow, and to regulate the movement of people to such cities. Slum clearance schemes moved people out of inner-city areas, and they were attracted to new houses, with major amenities nearby, in a semi-rural environment. Since 1946 a number of genera-tions of New Towns, designed to be 'self-contained and balanced', have been built, mainly to take overspill populations from expanding cities, but also to act as growth poles in areas of high unemployment. The third generation created Milton Keynes (see case study) and Telford.

New Towns have had varying degrees of success. Several of the earlier settlements were built too near London and were quickly swallowed up by urban growth. Washington, in the northeast, has proved successful and has attracted investment by foreign firms. However, Skelmersdale, near Liverpool, still has high unemployment, few job opportunities and poor local services.

Building materials and technologies since 1900

Traditional building materials for houses, factories and other buildings have been stone, timber, brick and mortar, and metals. During the twentieth century a number of new building materials and technologies became available. These enabled greater flexibility of design, more variation in the use of colour and textures, and increased speed of construction.

Some of the new materials include:
- prestressed concrete
- plate glass and reinforced glass
- felt materials for roofing
- cement wash and cladding for the outside of buildings
- asbestos in wall and ceiling materials
- weather-resistant plastics

Some of the new technologies include:
- the use of steel girders to build high-rise blocks
- the further development of fast-moving lifts (originally designed in the nineteenth century) to facilitate movement within these blocks
- the use of felt materials which have encouraged the development of flat-roofed buildings
- deck access and elevated walkways which have been used in both housing and shopping areas, and the invention of the escalator
- prefabrication stimulated by the use of asbestos
- widespread techniques of infilling and conversion of former industrial premises into housing and service-industry accommodation
- the use of multi-purpose vehicles (e.g. JCBs) for construction

Protection of rural and urban landscapes

It is now commonly believed that both rural and urban landscapes are resources that should be protected and used to generate tourist income. Two aspects of this are the concepts of preservation and conservation.

Preservation is the maintenance of a landscape such that its current state is as close as possible to its original condition. **Conservation** involves the protection and possible enhancement of the landscape for future use. In urban areas buildings may be protected because of their historic interest while in rural areas species, habitats and scenery are conserved for their beauty. There are different attitudes towards preservation and conservation in urban and rural landscapes that can lead to a variety of conflicts.

There are a number of general reasons for the protection of landscapes, both urban and rural:

- to preserve a historical legacy
- to preserve rare species and habitats
- to conserve landscapes that are scenically attractive
- to act as a source of tourist income, and increase employment
- to act as an educational stimulus

The Flamborough case study on the next page illustrates how attitudes vary towards the protection of a landscape.

A historic city: York

The following are issues that arise in a historic city such as York:

- the need for money to prevent the physical deterioration of the buildings
- the importance of safety — another reason to keep buildings in good repair, especially if they are to be opened to the public
- the need for skilled craftspeople and authentic materials which are often more expensive than modern materials
- the question of whether it is possible to recoup all these restoration costs through entry fees, or whether this will make entry charges too high and have a negative effect on tourist numbers. Restricted entry periods can also be considered
- the role of gift and souvenir shops and whether they trivialise the historic aspect of the urban landscapes
- the appropriateness of introducing aspects of modern living, such as satellite dishes and mobile phone masts, in certain parts of the city

Figure B on page 196 shows numbers of visitors to a variety of attractions in York and Figure 8.10 illustrates a variety of ways in which Fairfax House in York can be marketed as a source of tourist income:

Case study *Flamborough, East Yorkshire*

Read the newspaper cutting in Figure 8.9. The following attitudes arose out of this conflict:

- Flamborough Parish Council was very upset at the ruling. It wanted to expand tourist facilities in the village but could not do so because the hedge is protected. It stated that the decision may have detrimental effects on other parts of the area as tourists will park in unsuitable places. In addition, the council will be charged with the future maintenance of the hedge.

- Conservationists were delighted with the ruling, but will clearly need to develop strategies for similar conflicts elsewhere as the judge stated that this ruling should not act as a precedent.
- The local farmer was disappointed with the ruling. He was looking forward to losing the task of keeping the hedge in repair, and would have gained money by selling the land involved.
- Tourists will continue to face parking difficulties and will not have toilet facilities.

Use of 200-year old law helps protect 40,000 miles of greenery

Thousands of miles of hedgerow are in line for stronger protection after a court ruling. A judge has ruled that a 200-year old Enclosure Act still guarantees protection to a wildlife hedge at Flamborough, East Yorkshire which the parish council had hoped to root out in order to extend a tourist car park and toilet area. The Enclosure Act for Flamborough Parish stated in 1765 that all 'ditches and hawthorn hedges shall be maintained for ever'. However, the judge described the56 yards of hawthorn as straggling and unkempt, and said that the ruling should not set a precedent. Each of the 4,000 Enclosure Acts of the eighteeenth and nineteenth centuries would have to be considered on its merits.

Figure 8.9   Adapted from the *Guardian*, January 1997

- there is a 'richly decorated interior' which will entice tourists to come into the building rather than just look at the exterior. Sale of entry tickets will generate an income
- the use of 'liveried footmen' and mention of 'performances' point towards special events and features which will appeal to tourists
- catering facilities are provided, enabling large groups to come for a lengthy visit. This guarantees an economic use of building and facilities
- the provision of guided tours in foreign languages encourages and caters for foreign tourists
- the transport arrangements are clear and precise, facilitating access to the building, and even recognising that access is not straightforward

Figure 8.10

Fairfax House

Location   Fairfax House is in the centre of York

Rail station   10 minutes' walk

Fairfax House was acquired and fully restored by the York Civic Trust in 1983–84. The house, described as a classic architectural masterpiece of its age, and certainly one of the finest townhouses in England, was saved from near collapse after considerable abuse and misuse this century, having been converted into a cinema and dance hall. The richly decorated interior with its plasterwork, wood and wrought iron is now home to a unique collection of furniture, clocks, paintings and porcelain.

Facilities and amenities

Shop. No photography inside the house. Liveried footmen. Musical and dancing performances arranged.

Catering facilities available in the dining room, max. 28 seated. For groups up to 50 a buffet can be arranged.

Disabled visitors may alight at entrance prior to parking. No WCs for the disabled.

A guided tour can be arranged. Evening and daytime guided tours — telephone for details. Available in French and German. Tour time: $1\frac{1}{2}$ hours.

Parking for 300 cars, 50 yards from house. Coach park is $\frac{1}{2}$ mile away — parties are dropped off.

Industrial heritage sites: Ironbridge and Beamish

Deindustrialisation has left many areas with scarred landscapes, derelict factories and workshops, disused canals and railways and unsightly spoil heaps. Such areas are now giving rise to a new enterprise — the heritage industry. Some abandoned coal mines, empty textile mills and deserted iron and steelworks are having a new lease of life. Two examples of this are at Ironbridge in Shropshire and Beamish in County Durham.

Figures 8.11 and 8.12 show the changes in the number and origin of visitors to each of these two heritage sites between 1986 and 1999. There are similarities and differences in the changes. Both show an increase in the late 1980s and gradually fall away in the mid- to late 1990s. The differences are that there was a peak in 1990 for Beamish, with minor peaks in 1993 and 1996. Ironbridge shows minor peaks in 1995 and 1997.

There are also similarities and differences in the patterns of origin of visitors (Figure 8.12). The similarities are that both sites had a minority of visitors from

Figure 8.11
Visitors to
Ironbridge and
Beamish, 1986–99

Figure 8.12
Origin of visitors to
Ironbridge and
Beamish, 1986–99

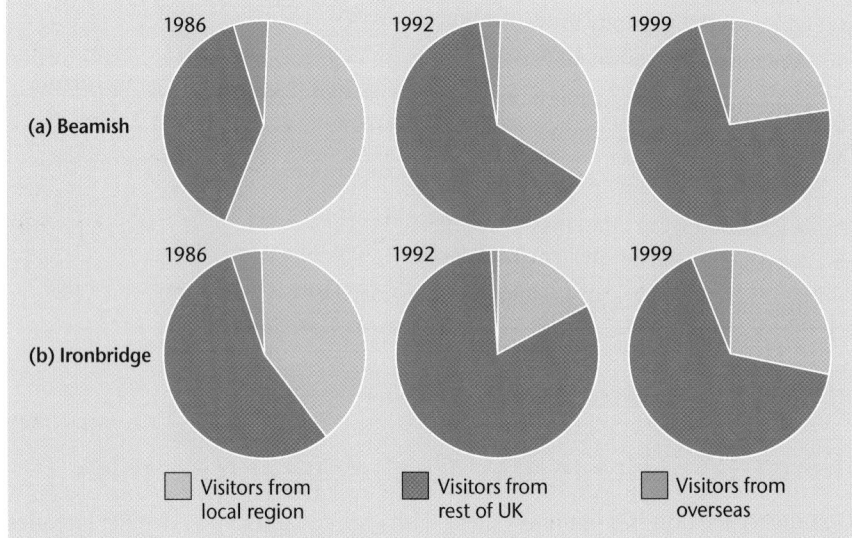

overseas, and both had a marked reduction of these in 1992. Both also demonstrate a declining proportion of visitors from the local region, with a consequent increase in the proportion from the rest of the UK. The differences are that Beamish had a continual fall of local visitors, whereas Ironbridge had an increase in 1999. Beamish also demonstrates less variability in the pattern of overseas visitors.

The Ironbridge industrial heritage site has a range of facilities — educational, historical and entertainment:

- a museum visitor centre in a former warehouse
- the Coalbrookedale Furnace and Museum of Iron which was the site of the original iron-smelting furnace of Abraham Darby
- the Coalbrookedale iron bridge (the first in the world) and its toll house

- the Bedlam furnaces — the first coke blast furnaces in the world on a large scale
- the Blists Hill Open Air Museum featuring the Shropshire Canal, a tile and brick works, a school house, a candlemaker, a doctor's surgery, a pub and a slaughterhouse
- the Coalport China Museum, famous for its historic manufacture of porcelain
- the tar tunnel where natural deposits of bitumen occur

A National Park: the Peak District

National Parks are large areas of beautiful countryside protected so that they can be enjoyed now and in the future. There are currently 12 such parks in England and Wales, the most recently designated being the Norfolk Broads and the New Forest. The main aims of National Parks are:
- to conserve the natural beauty, wildlife and cultural heritage of the area
- to provide a beautiful environment for people to visit in their leisure time
- to look after the interests of local people such as farmers and residents

The Peak District National Park was the first to be created in England, in 1951. Its central location in the country and proximity to large cities such as Manchester and Sheffield mean that it is the most visited of England's National Parks. About 30 million visits are made to the park each year, mainly by car. Most visitors come for the day but there are also over 1 million overnight stays each year.

These visitors cause problems for the park, but they also bring huge amounts of business to local people. In 2002 it was estimated that £152 million was spent in the park by visitors. The Peak Park Planning Board tries to look after the area by providing car parks, information centres and a local ranger service. It has also resurfaced many footpaths damaged by the sheer number of trampling feet.

The Planning Board enforces a variety of protection policies on the landscape, including the following:
- careful control of housing development in the area, although it is important that adequate housing is provided for local people
- restrictions on building materials and other changes to buildings, for example UPVC windows and satellite dishes
- limited access to parts of the countryside popular with walkers and ramblers
- controls on industries such as quarrying — it is recognised that local people need to earn a living but the special nature of the landscape also has to be maintained
- recommended use of designated footpaths, bridleways and cycle paths for individuals and groups who visit the park
- local-scale solutions to traffic congestion; for example, in some areas there is limited road access, one-way systems and designated bus use only
- farm conservation schemes that give grants to farmers to mend and repair walls, preserve meadows, plant deciduous trees and maintain footpaths through their land
- restrictions on the block planting of conifers which took place in the past but can be ugly if badly planned

Assessment exercises

Option S Urban change in the last 30 years

1 What are the economic and social problems caused by the rapid growth of squatter settlements in one major named urban area of a less economically developed country? (7 marks)

2 There have been a number of policies aimed at inner-city improvement in the UK.
  a Name *one* such policy you have studied.
  b Describe the aims of the chosen policy and comment on the effectiveness of that policy in one or more named areas where it was established. (8 marks)

3 The Stockport Road Corridor Initiative is a partnership scheme operating in an area immediately south of Manchester city centre. Figure A and Tables A and B provide information on the area and the partnership scheme.
  a Look at Figure A. Outline the evidence that suggests the Stockport Road Corridor area was undergoing inner-city decline. (5 marks)
  b Look at Tables A and B. Describe the trends in the funding profile of the Stockport Road Corridor Initiative between 1997 and 2004. (4 marks)

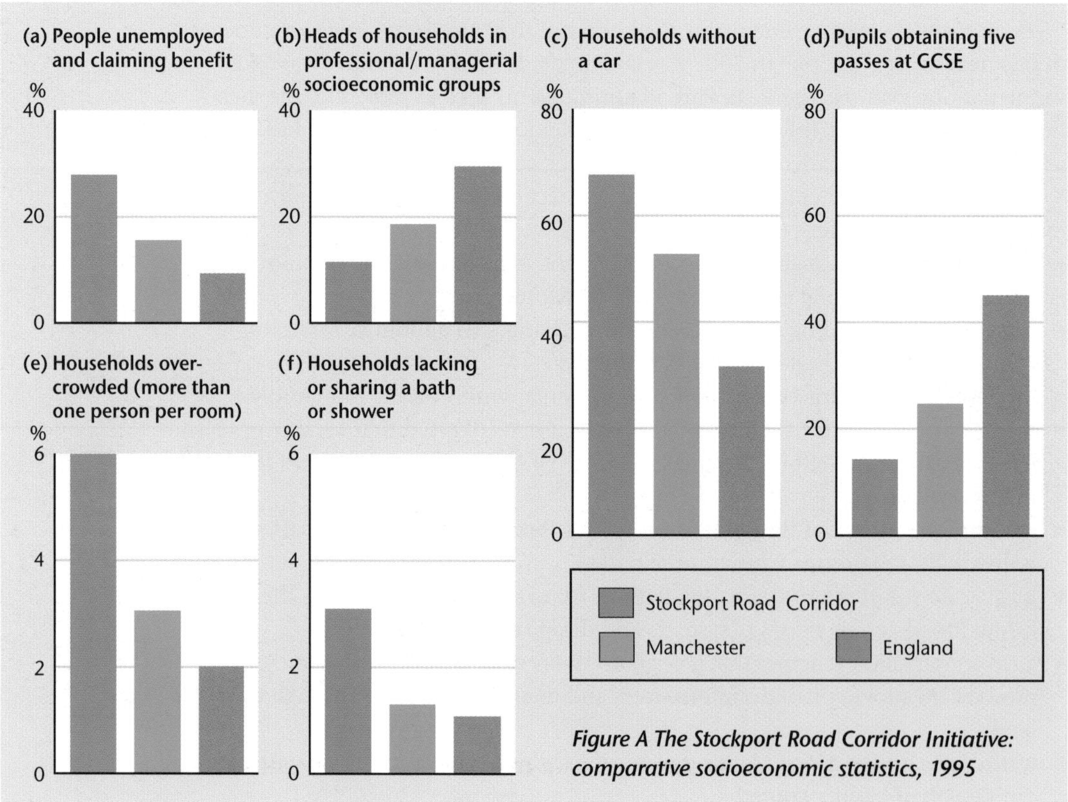

(a) People unemployed and claiming benefit
%
40
20
0

(b) Heads of households in professional/managerial socioeconomic groups
%
40
20
0

(c) Households without a car
%
80
60
40
20
0

(d) Pupils obtaining five passes at GCSE
%
80
60
40
20
0

(e) Households over-crowded (more than one person per room)
%
6
4
2
0

(f) Households lacking or sharing a bath or shower
%
6
4
2
0

- ■ Stockport Road Corridor
- ■ Manchester
- ■ England

Figure A The Stockport Road Corridor Initiative: comparative socioeconomic statistics, 1995

		Table A
Private-sector investment	£51 million	*The Stockport Road*
SRB funding	£12 million	*Corridor Initiative:*
Other public-sector funding (mostly from local organisations)	£21 million	*funding sources,*
		1997–2004

Note: SRB (Single Regeneration Budget) funding derives from central government, administered by the Department for the Environment.

Sources of funding	1997/98	1998/99	1999/00	2000/01	2001/02	2002/03	2003/04
Private-sector investment	0.5	1	2	6.5	10.5	14	16.5
SRB	0.5	1	3	3	3	1	0.5
Other public-sector	2	3	4	3.5	3.5	3	2
Total	3	5	9	13	17	18	19

Table B
The Stockport Road Corridor Initiative: funding profile, 1997–2004 (£millions)

4 Study Photographs A and B which show an area that is undergoing regeneration. Describe the ways in which the area is being redeveloped and regenerated. (8 marks)

Martin Bond/Photofusion

5 Draw an annotated diagram to show the main characteristics of out-of-town retailing developments in the UK. (7 marks)

Option T Historical landscapes of England and Wales

1 Enclosures, technological changes, and social and economic changes took place in the rural landscape between the sixteenth and nineteenth centuries. Describe the effects of these changes. (6 marks)

2 With reference to an example(s), describe the characteristic features of the buildings of the Georgian/Regency period. (8 marks)

3 Study Figure B which shows the numbers of visitors to selected attractions in York. Compare the numbers of visitors to the attractions named, and describe how they changed between 1993 and 1997. (6 marks)

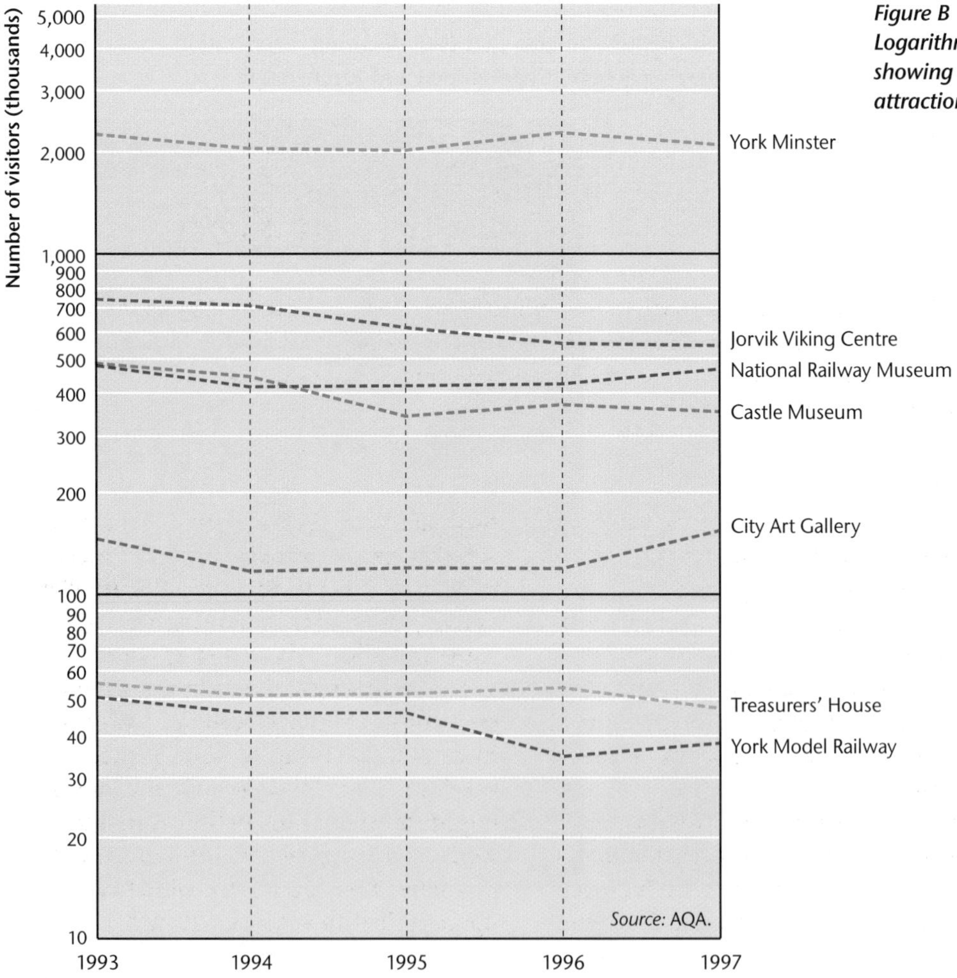

Figure B
Logarithmic graph showing visitors to attractions in York

Number of visitors (thousands)

York Minster

Jorvik Viking Centre
National Railway Museum
Castle Museum

City Art Gallery

Treasurers' House
York Model Railway

Source: AQA.

1993 1994 1995 1996 1997

4 Study Photograph C which shows houses in Saltaire, a town created in the 1870s by the enlightened industrialist Titus Salt. Using only the photograph, describe the features of the housing that make it different from that of a typical industrial town of the nineteenth century.

(6 marks)

Photograph C

5 a Read Figure C. It contains extracts from a Report for the General Board of Health for an area of a town in England in 1850. Using only the information contained in Figure C, describe the consequences of these environmental conditions for the health of the people who lived in the area.

(6 marks)

Figure C

The area shows a population of 3,795 persons resident in 915 tenements, situated chiefly in courts, passages and yards, and not infrequently in close proximity with manure heaps, pigsties, stables, slaughter houses and stagnant pools of foul water. Many of these dwellings have no outlet of any kind at the back, or opening except in the front.

The sleeping accommodation varies from none to four rooms in each house, mostly of small dimensions. In five cases there are no sleeping rooms; in 277 cases each family has but one sleeping room; in 478 they have two; in 51 they have three; in 26 four and upwards. Again the number of living or sitting rooms to each house varies from none to two and upwards.

I know that several members of a family sleep together in a crowded state and never think of such a thing as opening their windows. Sometimes the air in the rooms is quite overpowering. In the courts and yards there is a great deficiency of air, and it is quite impossible for the wind to act upon the bad air and sweep it away.

Source: AQA.

197

b Read Figure D. It contains an extract from a book published in 1897, 12 years after the passing of by-law legislation to improve housing conditions. Using only the information contained in Figure D, outline the ways in which living conditions had been improved in such by-law housing.

(7 marks)

Figure D

The buildings have back doors opening on an enclosed brick-paved yard, 36 feet across. The houses have on the ground floor a living room 13 feet square and a kitchen 12 feet by 9 feet, fitted with an iron sink and a small copper (boiler). There is also a pantry and a coal cupboard.

On the first floor there are two bedrooms and above them a spacious well-lighted attic. Good grates and ovens are provided in every house, and iron is used for mantelpieces and other fittings. Each house has a penny-in-the-slot gas meter and a flushed WC.

Source: AQA.

6 Study the Ordnance Survey map extract in Figure E. It shows part of an industrial town in England in 1891. Using the extract only, describe the characteristic features of the housing areas of this nineteenth-century industrial town.

(6 marks)

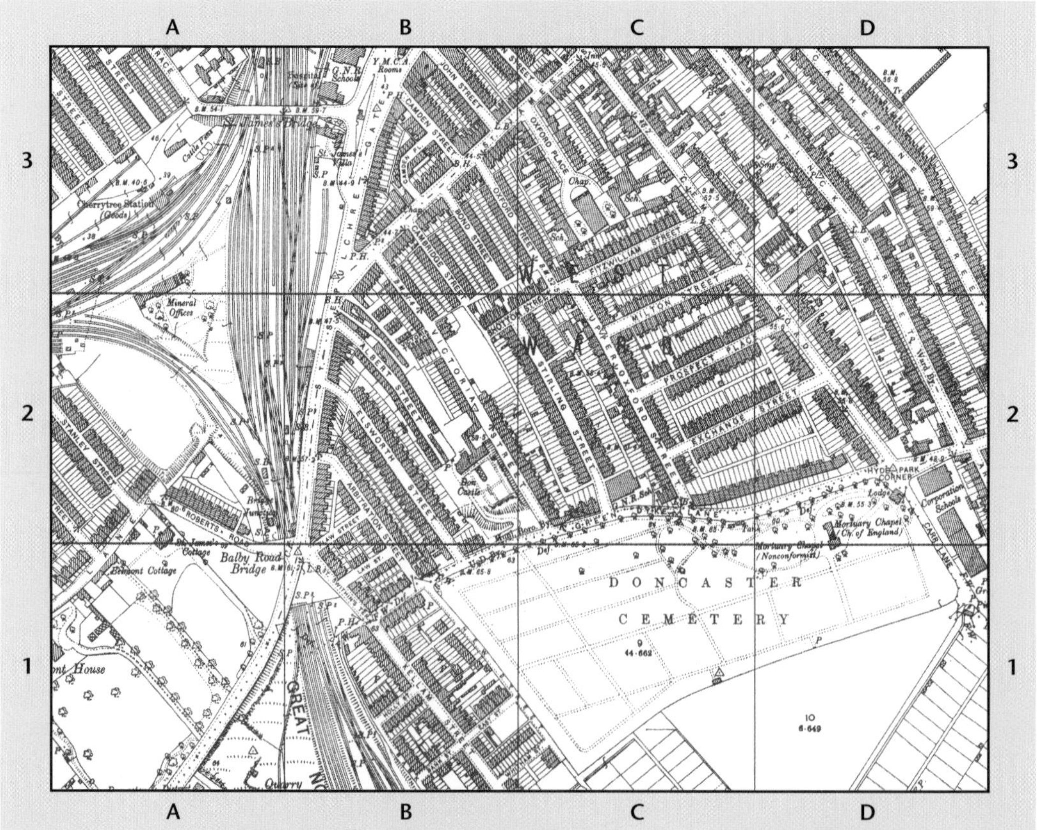

Figure E

Reproduced from OS mapping by permission of Ordnance Survey on behalf of the Controller of HMSO

© Crown copyright. Licence no. MC 10027418

A2

Module 4

Global change

Chapter 9

Physical geography
Seasonal and long-term
change and plate tectonics

The general atmospheric circulation system

The differential heating of the Earth's surface is sufficient to create a pattern of pressure cells. The movement of air within each cell is generally circular and, overall, is responsible for the transfer of surplus energy from equatorial regions to other parts of the Earth. Three cells form in each hemisphere, resulting in major areas of high and low pressure at the surface. High pressure forms where the air is falling and low pressure where it is rising.

The three cells in each hemisphere are known as the Hadley cell, the Ferrel cell and the Polar cell (Figure 9.1).

Figure 9.1
The three-cell model
of atmospheric
circulation

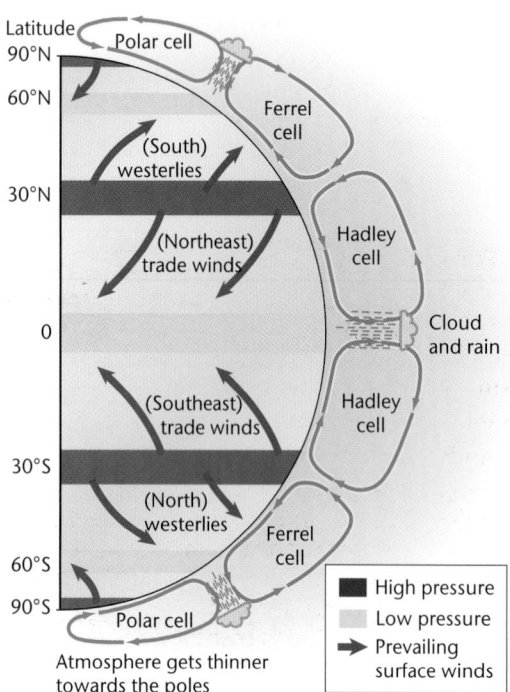

Atmosphere gets thinner
towards the poles

■ High pressure
▨ Low pressure
➡ Prevailing
   surface winds

The Hadley cells

The two Hadley cells, one in each hemisphere, form the basis of tropical air circulation as shown in Figure 9.2, and are responsible for the seasonal changes in climate of those regions that experience a wet and dry climate. Note that the tropopause is the upper limit of the lower part of the atmosphere (known as the troposphere). It is about 16 km above the equator and 8 km above the poles.

Each Hadley cell can be divided into four components:

■ Between the two cells there is an area of low pressure in equatorial latitudes which is known as the **inter-tropical convergence zone** or ITCZ. As the sun is always high in the sky, the ground heats rapidly by day and there is much surface evaporation. As the hot air rises in convection currents, an area of low pressure develops. This rising air cools and the water vapour eventually condenses, giving heavy rainfall.

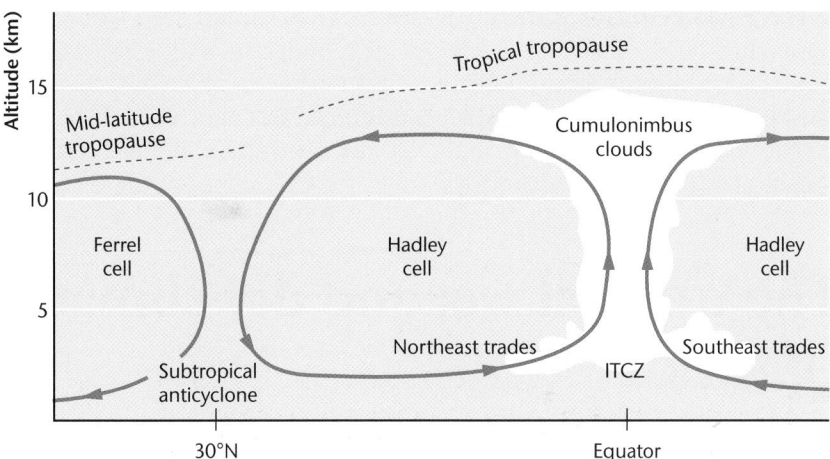

Figure 9.2
The circulation of
the Hadley cells

- At high altitudes the air moves polewards. This air usually circulates as upper westerly winds around the planet due to the deflection effect of the rotation of the Earth, known as the **Coriolis effect**. The net effect, though, is for the air still to move polewards.
- Around 30°N and 30°S the colder air at higher altitudes begins to sink, or subside, back to the Earth's surface. As this air descends, it warms and any residual moisture evaporates. At the surface, high pressure is created, with cloudless skies. These areas are known as the **subtropical anticyclones**.
- On reaching the ground, some of the air returns towards equatorial areas as consistent winds known as the **trade winds**. These air movements are also subject to the Coriolis effect and are deflected to the right in the northern hemisphere and to the left in the south. As a result they blow from a north-easterly direction in the northern hemisphere and from the southeast in the southern hemisphere. The two trade wind systems move air towards the equator where it forms the ITCZ.

Movement of the ITCZ

The position of the Hadley cells does not remain constant. Because the Earth is tilted, the position of the overhead sun moves depending on the season. This movement causes a seasonal shift in the zone of greatest heating, which results in the movement of the ITCZ. On 21 June the sun is overhead at the Tropic of Cancer, and so the ITCZ moves into the northern hemisphere. On 21 December the sun is overhead at the Tropic of Capricorn, and in the intervening time the ITCZ will have moved from the northern into the southern hemisphere. Movements of the ITCZ result in changes in the wind and pressure belt system, particularly the position of the Hadley cells. This movement is responsible for the seasonal changes in the climate of the tropical regions of Africa (see Figure 9.5 on page 203).

Key terms

General atmospheric circulation
The pattern of wind and pressure belts within the atmosphere. The circulation is extremely complex but there are certain movements that occur regularly enough for us to recognise patterns of air pressure distribution and winds.

Tropical wet and dry climate Parts of the tropics experience a distinctly seasonal climate with a hot, wet season followed by one which is cooler and drier. This climate results from changes in the general atmospheric circulation which affect these areas.

There has been some debate in recent years about the system. Some geographers believe that it is the result of the convergence of air within the trade winds and its subsequent rise, rather than the movement of the sun. This is why most texts refer to the equatorial low pressure area as the ITCZ. Whatever the explanation, the system behaves in the way described above, and this creates the climate of certain regions of Africa.

The tropical wet and dry climate of Africa

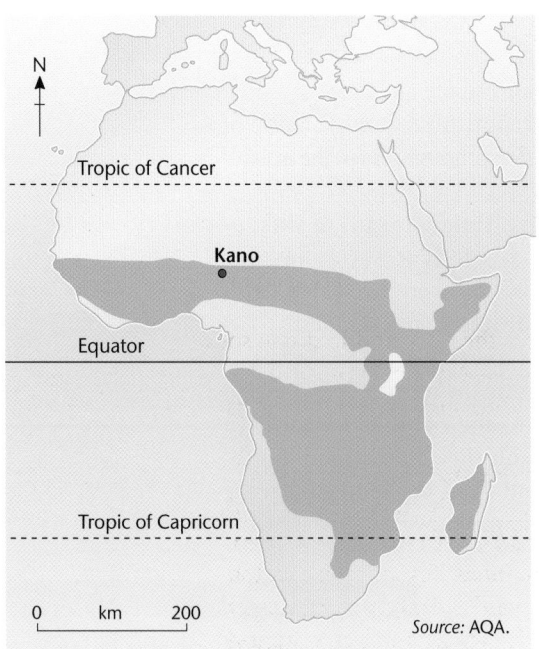

Figure 9.3 The tropical wet and dry regions of Africa

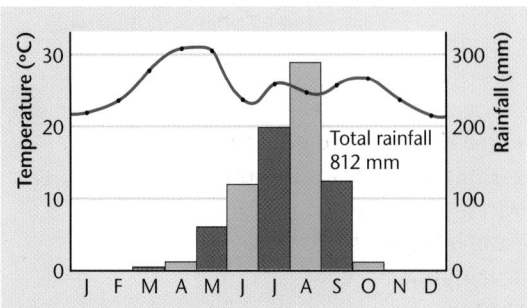

Figure 9.4 Climate statistics for Kano, Nigeria (latitude 12°N, altitude 500 m)

The tropical wet and dry climate of Africa (Figure 9.3) shows seasonal variations in wind direction, precipitation and temperature. It is transitional between the equatorial rainforests, where rain can be expected all year, and the hot deserts, which have minimal precipitation. Variations occur with increasing latitude away from the equator. The climate, however, is generally characterised by having a dry season (in the cooler period) and a wet season during which up to 90% of the annual precipitation falls (in the hotter period) (Figure 9.4).

Precipitation varies as follows:

■ on the equatorial rainforest margins there is over 1,000 mm per year with 10–11 months rainy season

■ on the desert/semi-arid margins there is under 500 mm per year with only 1 or 2 months rainy season away from the equator, so the reliability of the rainfall decreases.

Temperature varies as follows:

■ on the equatorial rainforest margin, temperatures range from 22°C in the wet season to 28°C in the dry season

■ on the desert margins the temperatures range from 18°C in the wet season to 34°C in the dry season

During the dry season, the subtropical anti-cyclone moves over the desert margins of the area. The encroaching high pressure with its subsiding air suppresses convection, giving rise to clear skies and high daytime temperatures.

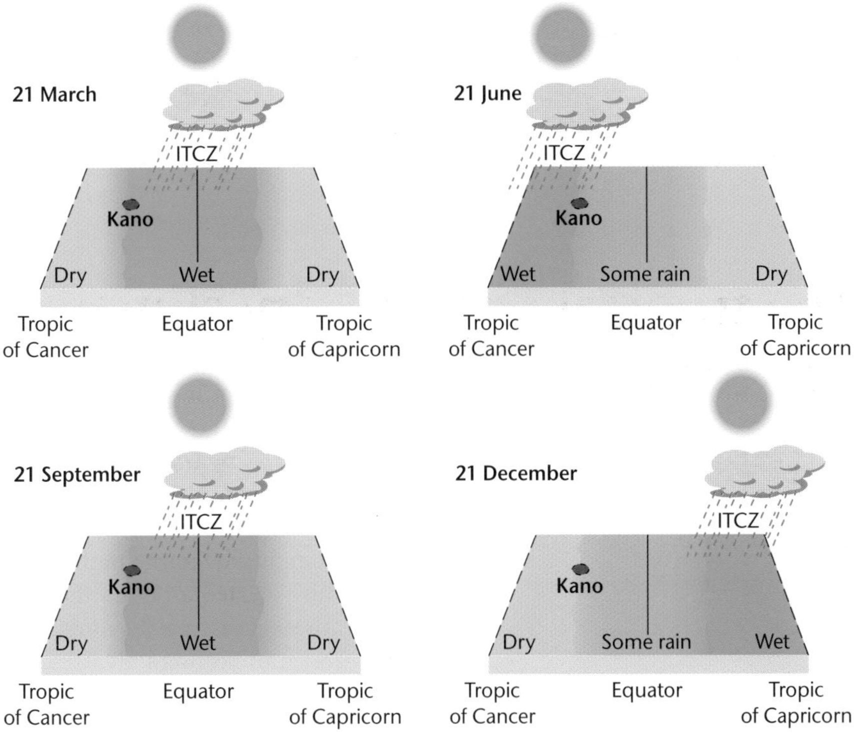

*Figure 9.5
Movements of
the ITCZ*

The trade winds blow from the high pressure towards the ITCZ and in doing so move air from the land towards the coasts. Such air is very dry, and in north Africa often produces an unpleasant wind with a very low moisture content which is known as the harmattan.

In the wet season, the ITCZ migrates polewards (Figure 9.5). As it moves it brings rainfall because uplift and convection are fed by moist, unstable, tropical maritime air. Areas at the poleward limit are only briefly affected, and therefore only have a short wet season with low annual rainfall totals. Towards the equator, the wet season lasts while the ITCZ has moved polewards, giving a much longer rainy season with higher total rainfall amounts. The maximum precipitation period occurs when the ITCZ moves polewards, and when it returns, giving a double maximum of rainfall in some areas.

Responses to seasonal atmospheric changes

Hydrological response

There is a pronounced peak in the discharge of rivers during the late summer through to October (Figure 9.6). This corresponds to the heavy rainfall at this time and to the fact that the soil enters a period of water surplus which results in considerable surface runoff leading to rising river levels (Figure 9.7). During maximum precipitation in June to August there is not an immediate increase in discharge as the soil is experiencing soil moisture recharge.

Once the dry season begins, discharge reduces rapidly and remains low from late October through to May, with some rivers even ceasing to flow for a period.

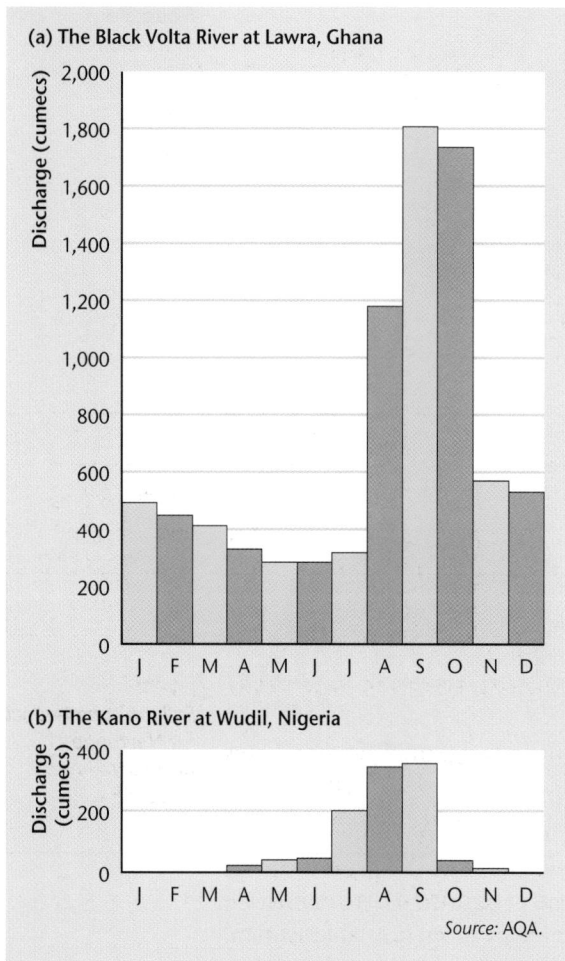

(a) The Black Volta River at Lawra, Ghana

(b) The Kano River at Wudil, Nigeria

Source: AQA.

Figure 9.6
River discharge
in west Africa

The Kano River in northern Nigeria, as shown in Figure 9.6b, has little or no flow from November through to April. For those rivers that flow all year, the minimum discharge often occurs soon after the period of the highest temperatures, which is the peak time for evapotranspiration.

Ecological responses

Soil moisture budgets

Figure 9.7 shows the soil moisture budget for an area in northern Ghana. Precipitation is greater than potential evapotranspiration between July and September, whereas the reverse is true between October and June. There are four distinct periods that can be seen in the soil moisture budget for this area:

- **Soil moisture recharge** occurs through July and early August and is the time when precipitation first becomes greater than evapotranspiration. Rainwater begins to fill the empty pores within the soil. When they are all full the soil is said to have reached its field capacity.

- **Soil moisture surplus** takes place in late August and September. At field capacity, the soil is saturated and rainwater has difficulty infiltrating the ground. This causes surface runoff and explains the high river levels of late summer.

- **Soil moisture utilisation** occurs from October, as evapotranspiration begins to exceed precipitation. There is more water evaporating from the ground surface and being transpired by plants than is falling as rain. Water is also drawn up the soil by capillary action and this leads to further evaporation.

- **Soil moisture deficit** occurs by December when the soil moisture is used up and there is a water deficit. Plants can only survive by being drought resistant or through irrigation. This period lasts until precipitation again becomes greater than evapotranspiration in early July and soil moisture recharge can begin.

This area in northern Ghana therefore shows the following characteristics:

- a lengthy period of moisture deficit
- a short period of moisture surplus
- total annual potential evapotranspiration greater than total annual precipitation

Adaptations by vegetation

The vegetation of the wetter areas consists of tall coarse grasses (elephant grass) with many deciduous trees. It is known as the tree savanna. In drier areas towards the desert margins; shorter tussock grass becomes dominant, with bare soil between the tufts of grass. This is often accompanied by drought-resistant trees such as the acacia and the baobab. These areas are known as the grassland and shrub savannas (Figure 9.8).

Trees in these areas tend to be deciduous, losing their leaves in the dry season, although some evergreens are also present. These have hard leathery leaves to reduce transpiration losses. Other plants are microphyllous (small-leaved), for the same reason.

In the **tree savanna**, a parkland exists. Here isolated trees have low umbrella-shaped crowns which shade root areas and reduce soil moisture evaporation. The trees show xerophytic characteristics, with dense cell fluids, hard waxy leaves, thorns and protected stomata, which all reduce water loss. Two of the main trees of the savanna are:

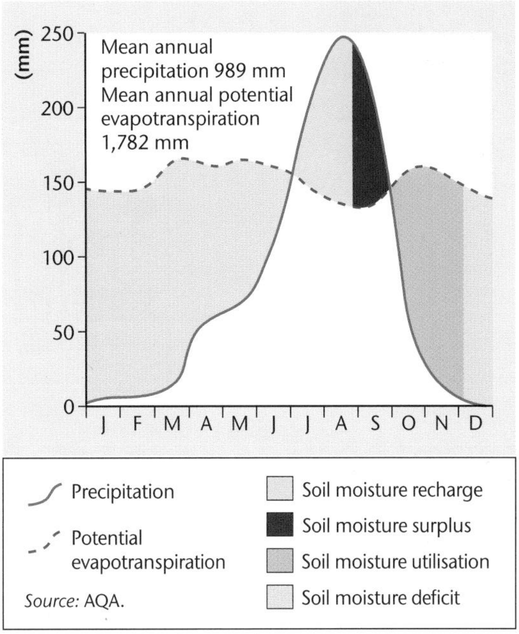

- the acacia, which has a crown structure, often flattened by the trade winds. It loses its leaves in the dry season
- the baobab ('upside-down tree'), which has a thick spongy trunk, long tap roots and bears leaves for only a few weeks. Like the acacia, the baobab is also pyrophytic — it can withstand fire, mainly because of its insulating bark

Figure 9.7
Soil moisture budget for Navrongo, northern Ghana (latitude 11°N)

In the **grassland savanna** (Photograph 9.1) the grasses between the trees become shorter and sparser. They are perennial in that they die back during the dry season and then regrow from root nodules when it rains. The grasses are tussocky,

Photograph 9.1
Savanna, Zimbabwe

Physical geography

Figure 9.8 Section across west Africa showing variations in vegetation

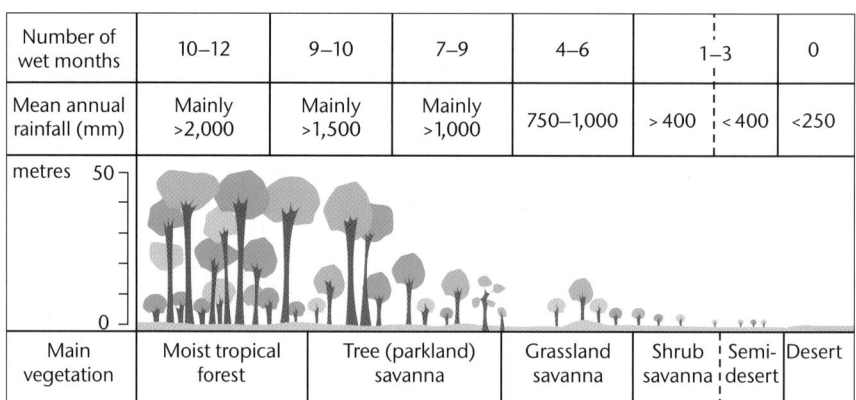

Number of wet months	10–12	9–10	7–9	4–6	1–3	0
Mean annual rainfall (mm)	Mainly >2,000	Mainly >1,500	Mainly >1,000	750–1,000	> 400 ¦ < 400	<250
Main vegetation	Moist tropical forest	Tree (parkland) savanna		Grassland savanna	Shrub ¦ Semi-savanna ¦ desert	Desert

enabling some retention of moisture. The naturally created straw dies down and protects roots.

In the **shrub** or **scrub savanna** there are many acacia trees, thorn bushes and short tufted grasses. Many species generate short stems from a single stock, with deep, branched roots and dormant seeds which compete for water. In some plants, even the stems may be capable of photosynthesis. Some grasses are feathery and wiry, and turn their blades away from the strong sun to reduce water loss.

The effects of human activity

Human activity has two main effects on the vegetation:

- Grass is burnt off to ensure better growth of young grass in subsequent seasons. When fire sweeps through vegetation at frequent intervals, it is very difficult for young trees and bushes to become established. Their place is taken by herbaceous plants and by the few indigenous woody plants whose bark permits them to survive. As the acacia and baobab can survive fires, they are common in the savanna.
- Where numerous cattle graze, woody plants are killed by the cattle eating their foliage. Thorny, animal-repellant trees and shrubs, such as acacia, therefore become numerous.

There is a belief among some biogeographers that humans have had a much greater influence than climate on the development of savanna vegetation. Some have gone as far as to suggest that grassland may therefore not be the climatic climax community.

Long-term climatic change

Climate fluctuations in the British Isles

The climate of the British Isles has changed significantly since the Pleistocene ice age which ended around 10 or 11,000 years ago. At the end of the Pleistocene,

Climatic period	Time before present (years)	Climatic conditions
Sub-Atlantic to present day		Temperatures fluctuate; cooler than the present day
		A cool coastal climate with cooler summers and increased rainfall
		A marked cool period between AD 1300 and AD 1800 — the little ice age
	2,500	A period of warming in the last 200 years
Sub-Boreal		Temperatures falling but rainfall relatively low at the beginning of this period, increasing later
		Period known as the neoglacial in Europe, with evidence of ice advance in alpine areas
	5,000	Warm summers and colder winters
Atlantic		Temperatures reach the optimum for many trees and shrubs — 'the climatic optimum'
	7,500	A warm 'west coast' type of climate, with higher rainfall
Boreal		Climate becoming warmer and drier
	9,000	A continental-type climate
Pre-Boreal		Mainly cold and wet, but becoming warmer and drier
	10,300	Changing from tundra/sub-arctic to more continental

Table 9.1
Climatic periods since the Pleistocene ice age

in the period known as the pre-Boreal, the climate began to warm up and that trend has continued ever since. There have, however, been some fluctuations during which the climate has cooled, the best known being the 'little ice age' from the mid-sixteenth century to around 1800. During this period of global cooling the Thames regularly froze over and fairs were held on the ice. Table 9.1 shows the main climatic periods in the British Isles since the Pleistocene.

Several theories have been put forward to explain climatic change. Suggestions have included the following:

- variations in solar energy (sunspot activity)
- changes in the Earth's orbit and axial tilt (which will affect the amount of solar radiation reaching the surface)
- meteorite impact
- volcanic activity (increasing dust in the atmosphere)
- plate movement (redistribution of land masses)
- changes in oceanic circulation
- changes in atmospheric composition, particularly the build-up of carbon dioxide and other greenhouse gases

Key terms

Climatic change Evidence shows that change has always been a feature of the Earth's climate. Apart from the Pleistocene ice age, recent research has revealed the existence of a whole series of climatic trends on a variety of timescales.

Global warming The recent gradual warming of the Earth's atmosphere due largely to human activity.

Evidence for climatic change

Evidence for climatic change is taken from a variety of sources which can be used to reconstruct past climates. Most of the evidence is indirect — climatic changes are inferred from changes in indicators that reflect climate, such as vegetation.

Pollen analysis

Species have particular climatic requirements which influence their geographical distributions. Each plant species has a distinctively shaped pollen grain and if these fall into oxygen-free environments, such as peat bogs, they resist decay. Changes in the pollen found in different levels of the bog indicate, by implication, changes in climate.

One limitation of this method is the fact that pollen can be transported considerable distances by wind or sometimes wildlife.

Dendrochronology

This is the analysis of tree rings from core samples. Each year, the growth of a tree is shown in its trunk by a single ring made up of two bands: a band reflecting rapid spring growth when the cells are larger; and a narrower band of growth during the cooler autumn. The width of the ring depends on the conditions of that particular year. A wide band indicates a warm and wet year, a narrower one cooler and drier conditions. The change in width from one ring to another is of greater significance than the actual width, as bigger growth rings tend to be produced in the early life of the tree, irrespective of the conditions.

Recent investigations, however, have shown that trees respond more to levels of moisture than to temperature. Dendrochronology has a limitation in that few trees exist that are older than about 4,000 years. It has been possible to extend surveys further back using remains of vegetation preserved in non-oxygen conditions.

Ice-core analysis

Glacial ice can be studied by drilling cores from areas such as Antarctica and Greenland. The carbon dioxide trapped within the ice is a climate indicator — levels tend to be lower during cooler periods and higher when it is warmer. Another method is to look at oxygen isotope levels (see below).

Sea-floor analysis

Core samples from the ocean floor reveal shifts in animal and plant populations which indicate climatic change. The ratio of the isotopes oxygen-18 to oxygen-16 in calcareous ooze can also be measured. This is linked to the ice-core analysis described above. During colder phases, when water evaporated from the oceans and precipitated onto the land eventually forms glacial ice, water containing lighter oxygen-16 is more easily evaporated than that containing heavier oxygen-18. As a result, the oceans have a higher concentration of oxygen-18, while the ice sheets and glaciers contain more of the lighter oxygen-16. During warmer periods, the oxygen-16 held in the ice is released and returns to the oceans, balancing out the ratio. Studies of isotope curves showing the ratio of oxygen-16 to oxygen-18 therefore give a picture of climatic change.

Recent investigations have suggested that isotope variations are an indication of changes in the volume of ice rather than water temperature, but as ice volume itself reflects climatic conditions, such studies have tended to confirm earlier findings.

Radiocarbon dating

Carbon-14 is a radioactive isotope of carbon (normally carbon-12). Carbon is taken in by plants during photosynthesis. As carbon-14 decays at a known rate and carbon-12 does not decay, comparison of the levels of the two isotopes present in plant remains will indicate the age at which a plant died (with an error of up to 5%). The type of vegetation present at any particular time is an indicator of the climate of that period. This method can accurately date organic matter up to 50,000 years old.

Coleoptera

Remains of Coleoptera beetles are common in freshwater and land sediments. Different species of this beetle tend to be found under different climatic conditions. Knowledge of the present climatic range of the different species, and the age of the sediments in which remains are found, allows past climatic conditions to be worked out.

Changing sea levels

The presence of rias and fjords indicates rising (eustatic) sea levels flooding glacial and river valleys. Changes in sea levels are indicators of climate change — the volume of the sea water changes as it warms/cools.

Glacial deposits

These show records of ice advance during colder periods and retreat during warmer times.

Historical records

Historical records include cave paintings, depth of grave digging in Greenland, diaries, documentary evidence of events (such as 'frost fairs' on the Thames) and evidence of areas of vine cultivation. Since 1873 daily weather reports have been documented, and the Royal Society has encouraged the collection of data since the seventeenth century. Parish records are often a good source of climate data.

Recent global warming

Figure 9.9 shows how average world temperature has risen since records were first kept in 1860. Although the overall rise seems small, the top ten hottest years have all occurred since 1980, and the 1990s was the hottest recorded decade. The years 2001 and 2004 continued that trend: 2001 was the second hottest after 1998; and 2004 was the fourth hottest on record, with heatwaves across southern Europe bringing some of the highest temperatures ever recorded in Spain and Portugal. The global mean surface temperature in 2004 was 14°C, which is 0.44°C above the average for the period 1961–90. Global warming has become probably the major environmental issue of our time, and although scientists have often been slow to commit themselves about cause and effect, there is no doubt that the planet is heating up.

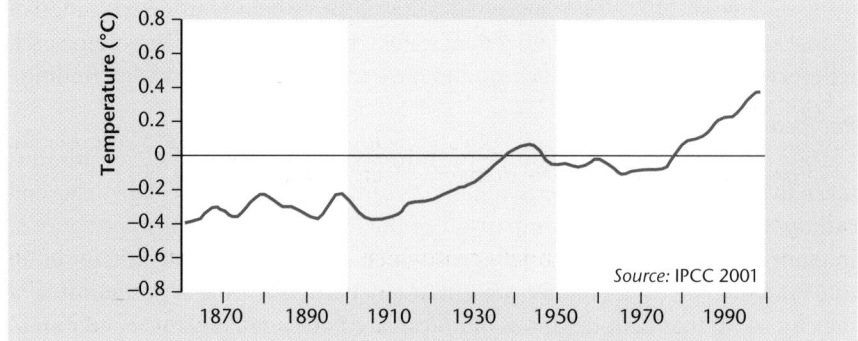

*Figure 9.9
Average variation
in Earth-surface
temperature from
the mean since
1860*

Climatic changes have happened in the past, but present evidence seems to suggest that the recent increase in temperature has been brought about by pollution of the atmosphere, in particular the release of huge amounts of carbon dioxide from fires, power stations, motor vehicles and factories.

Why global warming happens

Carbon dioxide in the troposphere allows incoming short-wave radiation from the sun to pass through and warm the Earth. Some of this radiation is reflected back from the Earth's surface into space at a longer wavelength. Greenhouse gases in the troposphere such as carbon dioxide absorb some of this long-wave radiation and radiate it back again towards the Earth's surface. This trapping of heat is known as the greenhouse effect and is part of the natural process of heat balance in the atmosphere. In fact it is essential for life on Earth — without the greenhouse effect the planet would be about 30°C colder.

Provided the amount of carbon dioxide and water vapour in the atmosphere stay the same and the amount of solar radiation is unchanged, then the temperature of the Earth remains in balance. This natural balance, however, has been influenced by human activity. The atmospheric concentration of carbon dioxide has increased by about 15% in the last 100 years and the current rate of increase is estimated to be 0.4% per year (Figure 9.10). This, together with increases in levels of other greenhouse gases such as methane and nitrous oxide, has upset the natural balance and led to global warming.

It is generally agreed that these continuing atmospheric changes will lead to a further rise in temperature, but it is difficult to predict the extent or speed of

*Figure 9.10
Changes in the
concentration of
carbon dioxide in
the atmosphere,
1720–2000*

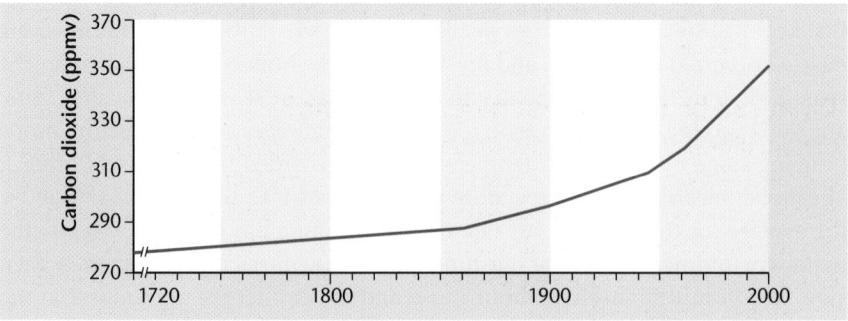

change. If carbon dioxide levels double, then temperatures could rise by a further 2–3°C, with greater rises at higher latitudes, perhaps in the order of 7–8°C.

Effects of global warming

Global warming has serious implications for climate patterns, sea levels and economic activity. Some of the main effects are described below.

Rising sea levels

The sea level rise in the twentieth century has been estimated as 1.5 cm, but in the twenty-first century levels could rise between 5 and 10 cm per decade. This will be sufficient to cause serious flooding in coastal areas and increased erosion in others. For low-lying countries such as the Netherlands and Bangladesh, and for many Pacific and Indian Ocean islands, rising sea levels will have catastrophic consequences. There will also be the huge cost of providing substantial flood defences. Many British estuaries may need defences similar to the Thames Barrier.

Climatic change

Many places will experience warmer summers. One estimate for the UK suggests that, by 2030, average temperatures could rise by over 2°C. Continental areas could have reduced rainfall, producing desert-like conditions in places that were previously good for agriculture.

Some climatologists believe that, if global warming leads to changes in the pattern of ocean currents, the UK could experience a much *colder* climate. It is possible that the Gulf Stream and North Atlantic Drift could be diverted and, without the warming effects of these currents, the UK would have a climate similar to that of Siberia.

Such climatic changes would have widespread effects on vegetation, wildlife and agriculture. The ability of some regions to provide food for the population would diminish, leading to mass migrations as people searched for new areas in which to grow crops. Some areas, though, could benefit. There might be more land suitable for agriculture in countries such as Russia and Canada.

Extreme events

Heatwaves, floods, droughts and storms will all last longer and show an increasing intensity. With higher temperature there will be increased evaporation over the oceans, leading to greater global precipitation.

Figure 9.11 shows a selection of newspaper headlines highlighting some of the possible effects of global warming in the UK.

What can be done about global warming?

Carbon dioxide has an effective lifetime in the atmosphere of about 100 years, so concentrations respond very slowly to changes in emissions. At the 1992 Earth Summit in Rio de Janeiro the developed countries agreed to stabilise carbon dioxide emissions. This will slow down the rate of climate change, but to prevent carbon dioxide concentrations from rising we will need to reduce current global emissions by about 60%.

Case study *Global warming and the British Isles*

As a result of global warming, the UK could experience warmer summers, longer hot spells, droughts and increased storm activity. In contrast, some climatologists believe that changes in the pattern of ocean currents could result in much colder conditions.

Coastal regions

Increases in mean sea levels and in the frequency and magnitude of storms, storm surges and waves would lead to more coastal flooding. Sea levels around Britain are predicted to rise between 12 and 67 cm by 2050, which would make a number of low-lying areas vulnerable, particularly the coasts of East Anglia, Lancashire, the Humber estuary, the Essex mudflats, the Thames estuary, parts of the north Wales coast, the Clyde/Forth estuaries and Belfast Lough. Flooding would lead to disruption in transport, manufacturing and the housing sector. In addition, there would be longer-term damage to agricultural land and coastal power stations, and water supplies could be contaminated by salt infiltration.

Agriculture

Climate changes are likely to have a substantial effect on plant growth and, by extension, plant productivity. Higher temperatures could result in:

- a decrease in yields of cereal crops
- an increase in yields of sugar beet and potatoes
- an increase in the length of the growing season for grasses and trees, bringing a higher productivity
- the introduction of new crops and species — the UK could even become a major wine-producing region
- an increase in some pests, such as the Colorado beetle which causes serious damage to potatoes

Flora, fauna and landscape

A sustained rise in temperatures could have the following effects:

- a significant movement of species northwards and to higher elevations
- the extinction of some native species which are unable to adapt to the increasing temperatures
- the loss of species which occur in isolated damp, cool or coastal habitats
- the invasion and spread of alien weeds, pests, diseases and viruses
- an increased number of foreign species of invertebrates, birds and mammals which may outcompete native species
- the disappearance of snow from the tops of the highest mountains

Soils

Higher temperatures could reduce the water-holding capacity of some soils, increasing the likelihood of soil moisture deficits. The stability of building foundations and other structures, especially in central, eastern and southern England where clay soils with large shrink–swell potential are abundant, would be affected if summers became drier and winters wetter. There could be a loss of organic matter, which would affect the stability of certain soil structures. Soil structure could also be affected if the water table rose with rising sea levels.

Photograph 9.2 Flooding in Sheerness, 1953, following a storm surge which flooded much of the east coast of England. Climate change could lead to an increase in such events

Water resources

Water resources would benefit from wetter winters, but warmer summers with increased evaporation could have the opposite effect.

Energy use

Higher temperatures could decrease the need for heating, but an increase in the demand for air conditioning would increase electricity consumption.

Watch out for the alligators in 'Everglades' Britain

Drought and floods 'may become a way of life'

London will be as hot as Loire valley in 25 years

GLOBAL WARMING IS 'DRIVING FISH NORTH'

'Global warming threat' to 40% of Britain's birds

Figure 9.11

In 1997, at a follow-up meeting in Kyoto, Japan, a Climate Change Protocol was signed by over 100 governments. This set more specific targets for pollution mitigation and proposed schemes to enable governments to reach these targets. Most governments agreed that by 2010 they should have reduced their atmospheric pollution levels to those of 1990. There are three things to note about this:

- Some countries are already polluting at levels significantly above those of 1990. The USA, for example, releases 15% more carbon dioxide at the start of the twenty-first century than it did 10 years ago. Despite this, President Bush refused to ratify the Kyoto proposals, claiming that 'the agreement was fatally flawed' and that the emission targets were unattainable and potentially damaging to the American economy.
- Some countries are disproportionately responsible for releasing greenhouse gases. In 1996, the USA released 21% of global carbon dioxide even though it only had 4% of the world's population.
- Some countries, particularly the least developed countries which have little industry and few vehicles, release very few greenhouse gases into the atmosphere.

Carbon credits

Following the Kyoto meeting, a system of global carbon credits was introduced, under which each country has an annual carbon dioxide pollution limit. Major polluters can buy 'carbon credits' from less polluting countries which are not using up their own quotas. If polluting countries still go over their limit, a number of options might be forced upon them:

- a fine
- investment in ways to reduce domestic carbon dioxide emissions (e.g. wind or solar power)
- paying for improved technologies in other countries or for other countries to

plant trees (in February 2000 the Japanese paid the government of New South Wales in Australia £50 million to plant over 40,000 hectares of trees in the next 20 years)

Critics argue that this system is flawed because it serves the interests of the developed countries which are the major polluters, and enables them to go on polluting. Such people believe the mass industrialisation and consumerism which underpin the economies of the developed world are unsustainable. It is unlikely, though, that the citizens of the developed world will be willing to give up their current lifestyles. The monitoring systems have also been criticised — countries are expected to monitor themselves, leaving much room for cheating and 'massaging' of the figures.

One result of the carbon credit system in the UK has been the introduction of a Climate Change Levy — a tax on energy used by industry, commerce and the public sector. This came into effect in April 2001. It is designed to help meet the target agreed in the Kyoto Protocol which commits the UK to a 12.5% reduction in emissions of six greenhouse gases by 2010. Many people now believe the carbon credit system is the best way to reduce, if not entirely eradicate, atmospheric pollution in the twenty-first century.

Changing vegetation and soils in the UK

Vegetation successions and climax vegetation

The composition of vegetation depends on the interaction between all the components that make up the environment — the plants' habitat. These include natural factors such as climate, relief and soils, and human influence through clearance, fires and livestock grazing. Plants will survive under suitable conditions, depending on environmental factors and competition. Plant populations vary from one area to another and become more complex over time. The change in a plant community through time is called a succession. There are two fundamental types of succession:

- **Primary succession** occurs on ground which has had no previous vegetation, including lava flows, bare rock and sand dunes. There are several types of primary succession. They can be divided into those formed on the land (**xeroseres**) and those formed in water (**hydroseres**), and then into **lithoseres** on bare rock, **psammoseres** on sand dunes, **haloseres** in a salt water environment and **hydroseres** in a freshwater environment.
- **Secondary succession** follows the destruction or modification of existing vegetation, either naturally or by human activity.

Key terms

Climax vegetation The end of a succession, where the characteristics and species of a plant community are in balance with environmental conditions.

Soil The outermost layer of the Earth's crust, providing the foundation for all plant life. Soil is made up of organic, mineral and animal constituents differentiated into horizons (layers) of variable depth. It is the result of the interaction of several physical, chemical and biological processes.

Vegetation succession A series of changes which take place in a plant community through time.

Development of a succession

As a succession develops, it passes through a series of stages, or **seres**, in which the processes of invasion, colonisation, competition, domination and decline operate to influence the composition of the vegetation.

When plants first invade bare ground (through the processes of dispersion or migration), groups of a particular species, or colonies of two or more species, become established. These are known as the **pioneer species** — extremely hardy plants, adapted to survive in the harsh conditions. They compete for available space, light, water and nutrients and, as they die, they help to modify the habitat, affecting the microclimate (wind speed at ground level, shelter, temperature, humidity) and soil conditions (organic content, nutrient recycling, acidity, water retention). The pioneer vegetation also helps to weather the surface and so aid soil formation.

Other plants are then able to colonise the improved ground and change the existing balance of species. Each stage of colonisation produces a better environment for an increasing variety of species. New dominants take over and exert their own influence. The addition of organic matter to the developing soil allows the growth of taller and more aggressive plants that are more demanding of water, nutrients and anchorage. These provide shelter from sun and wind which allows other plants to become established.

Eventually, a period of relative stability is achieved and the **climax vegetation** develops, with dominants excluding rivals that are less suited to the conditions. The number of species continues to increase, but once the major dominants are in place they may cause a decline in the total number of species. Climax is usually dominated by the tallest species that can grow in the given conditions. At this stage the community becomes 'closed', as saturation point is reached with all potential niches occupied. This is known as the **climatic climax community**, the natural vegetation having reached a stable balance with the climate and soils of the area.

Some biogeographers believe that within one climate, local factors such as drainage, geology, relief and even microclimates can create variations in the climatic climax community. This idea is known as **polyclimax theory**.

An example of a primary succession: a lithosere

A lithosere is a succession that begins life on a newly exposed rock surface. This surface may typically have been left behind on the retreat of a glacier, formed on an emerging raised beach (see Figure 9.12) or developed on scree formations. In the UK, such a succession develops as follows:

- The bare rock surface is initially colonised by **bacteria and algae**, which can survive where there are few nutrients. Bare rock tends to be very dry and there is rapid surface runoff.
- The pioneers begin to colonise, starting with **lichens** which can withstand the acute water shortage. They begin to break down the rock and assist water retention.
- As water retention improves, **mosses** begin to grow. These also improve water retention and weathering to produce the beginnings of the soil in which more advanced plants can grow.

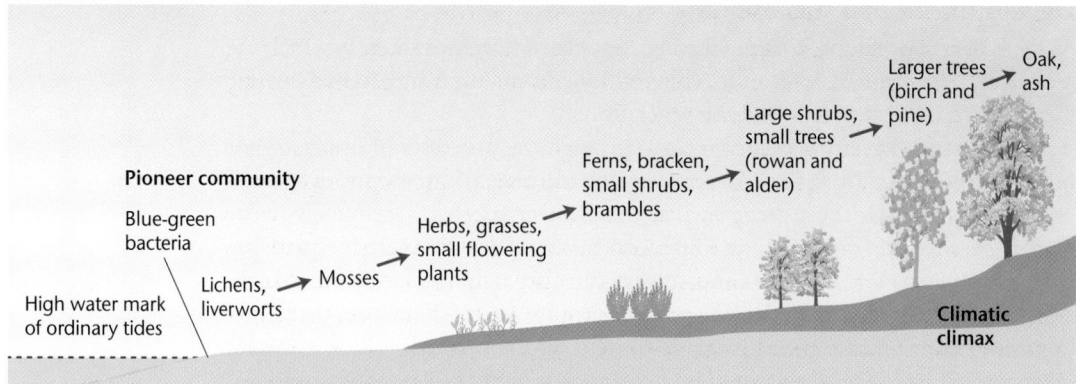

Figure 9.12
The development of a lithosere on a raised beach

- **Grasses, ferns, herbs** and **flowering plants** appear. As these die back, bacteria convert their remains into humus, which helps to recycle nutrients and further improve soil fertility.
- **Shrubs** start to grow, shading out the grasses and herbs.
- Pioneer trees become established. These are mainly fast-growing species such as **willow, birch** and **rowan**.
- Slower-growing tree species begin to develop, such as **ash** and **oak**. Initially they are in the shade of shrubs, so they only appear in the later stages of the succession. They are the dominants of the climatic climax community — temperate deciduous woodland.

Temperate deciduous woodland

Figure 9.13 shows the structure of an oak woodland, one type of temperate deciduous woodland. Such woodland has the following characteristics:

- a net primary production (NPP) of 1,200 grams dry organic matter m^{-2} yr^{-1} (compared with 2,200 for tropical rainforest)
- tall trees are the dominant species. Oak is the tallest (and can reach 30–40 m) followed by elm, beech, sycamore, ash and chestnut
- trees develop large crowns and have broad but thin leaves

Figure 9.13 Structure of a typical English oak woodland

Tree layer	Average height
Oak	20 m
Shrub layer	
Bramble	1.5 m
Rose	1.5 m
Rowan	2.8 m
Herb layer	
Nardus grass	25 cm
Fern bracken	60 cm
Other grass	30 cm
Wood anemones	15 cm
Wild garlic	6 cm
Dog's mercury	20 cm
Bluebells	15 cm
Primroses	10 cm
Ground layer	
Mosses	2 cm

- deciduous trees shed their leaves in winter. This reduces transpiration at a time when water is less available
- relatively few species of dominants. Some woodlands are dominated by only one tree species (such as in Figure 9.13)
- most woodlands show some stratification
- below the canopy is a shrub layer with smaller trees such as holly, hazel, rowan and hawthorn
- just above the forest floor is a herb layer which is dense if the shrub layer allows enough light through. This is made up of grasses, bracken, ferns and some flowering plants such as the bluebell that appear early in the year, before the trees have developed their full canopy
- epiphytes, e.g. lichens and mosses, grow on the trunks and branches of trees
- on the forest floor mosses grow and a thick layer of leaf litter is readily broken down by the the soil microbes and animals

Photograph 9.3 shows beech woodland in the UK.

Photograph 9.3 Deciduous woodland is the UK's climatic climax community

Identifying changes in vegetation over time

Several methods are used to identify past changes in vegetation, including pollen analysis, carbon-14 dating and historical records (see pages 208–209). Pollen grains are frequently used as indicators of previous vegetation distributions and thus of the characteristics of past climates. This is a useful method of investigation because of the following points:

- pollen is produced by a wide range of plants and the pollen grains of each species have a distinctive structure
- pollen grains are resistant to decomposition if they are deposited in an environment with very low oxidation, such as peat bogs, lake floors and sea-floor sediments
- pollen grains are deposited in layers in sediments. If these remain undisturbed, a sequence is retained, with the oldest pollen at the greatest depth

- sampling, by taking a bore through the sediments, will allow identification of a sequence of vegetation through time. Pollen is extracted from the bore sample and counted under a high-power microscope, and the percentage of each type of pollen at each level is recorded

However, wind can move pollen over some distance, so the contents of a peat bog may not entirely reflect the local vegetation at certain times in the past. It is also possible that some of the pollen evidence could have been moved by erosion.

The effects of human activity on successions

Successions can be stopped from reaching the climatic climax, or deflected towards a different climax, by human interference. The resulting vegetation is known as a **plagioclimax**. Examples of human activity that creates plagioclimaxes are:
- deforestation or afforestation
- animal grazing or trampling
- fire clearance

A good example of a plagioclimax in the UK is **heather moorland**. Many of the uplands were once covered by a climax vegetation of deciduous woodland, particularly oak forest. Heather (*Calluna vulgaris*) would have featured, but only in small amounts. Gradually the forests were removed, for a variety of purposes, and as the soils deteriorated without the deciduous vegetation, hardy plants such as heather came to dominate the uplands. Sheep grazing became the major form of agriculture and the sheep prevented the regeneration of climax woodland by destroying young saplings.

Many of these uplands have been controlled by managed burning to encourage new shoots. Burning has eliminated the less fire-resistant species, leading to the dominance of heather. When heather is burnt, one of the aims is to ensure that as much as possible of the available nutrient fund is conserved in the ecosystem. In many areas, heather is burnt on average every 15 years. If a longer time elapses there is too much woody tissue, the fires burn too hot, and nutrients are lost in the smoke.

Figure 9.14 shows the cycle associated with the heather system. If the burning was not continued, the heather moorland would degenerate, eventually allowing

Figure 9.14
The heather system
nutrient cycle

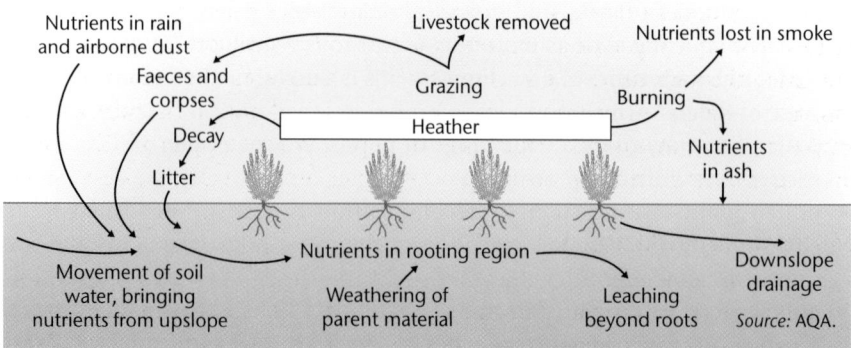

Source: AQA.

Figure 9.15
The heather cycle

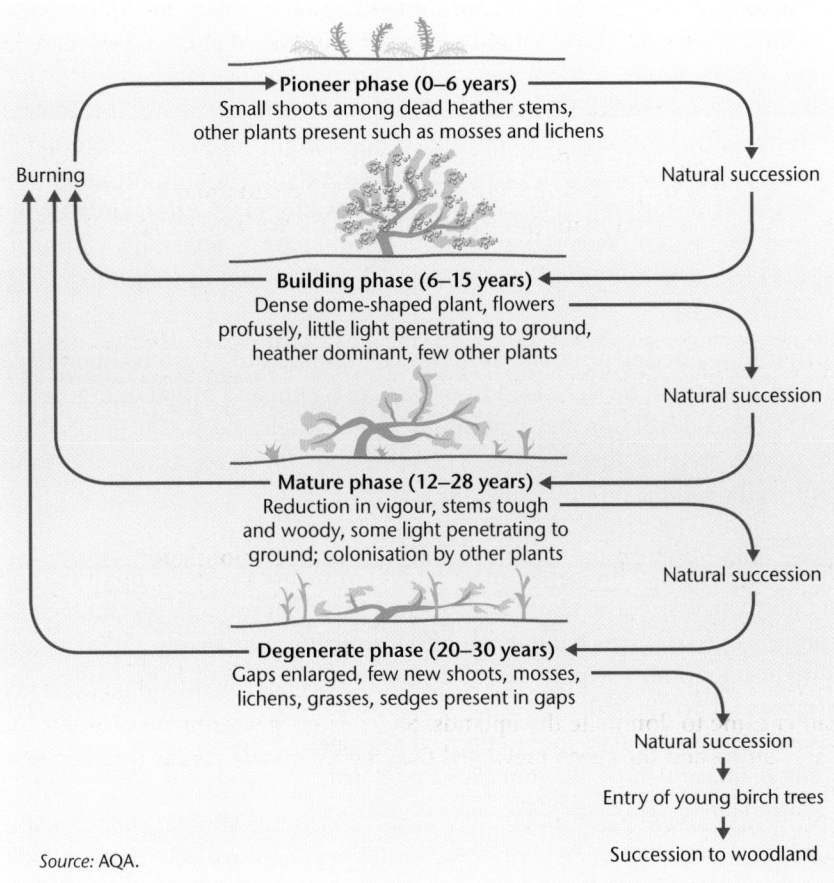

Pioneer phase (0–6 years)
Small shoots among dead heather stems,
other plants present such as mosses and lichens

Burning

Natural succession

Building phase (6–15 years)
Dense dome-shaped plant, flowers
profusely, little light penetrating to ground,
heather dominant, few other plants

Natural succession

Mature phase (12–28 years)
Reduction in vigour, stems tough
and woody, some light penetrating to
ground; colonisation by other plants

Natural succession

Degenerate phase (20–30 years)
Gaps enlarged, few new shoots, mosses,
lichens, grasses, sedges present in gaps

Natural succession

Natural succession
↓
Entry of young birch trees
↓
Succession to woodland

Source: AQA.

the entry of trees and a succession to woodland (Figure 9.15). Much of the present vegetation of the UK is a plagioclimax, largely as a result of clearance from the Roman and Anglo-Saxon periods through to the eleventh century. By this time only about 10% of the original woodland remained in England and Wales.

Soil characteristics

The characteristics of soils are the result of interactions between a number of factors: climate, vegetation (organic matter), relief and parent material, all operating through time. The nature of a soil changes as it develops from an immature skeletal soil to one that is fully developed and in balance with its environment. Human activity can have an important impact on soil. The soil is made up of four main components: mineral matter, organic material, air and water.

Mineral constituents

The mineral constituents of the soil are derived from the weathering of the underlying parent material. They are the products of both physical breakdown (e.g. freeze–thaw) and chemical processes (oxidation, carbonation and hydrolysis).

The weathered products can be divided into two groups:
- primary minerals which remain unaltered from the original parent material and are released by weathering
- secondary minerals which are produced in the soil by chemical reactions. Unlike primary minerals, these are readily soluble and are predominately carbonates, because weak carbonic acid enters soils in rain. Although it is a weak acid, carbonic acid can detach potassium, magnesium and calcium from the parent material to produce carbonates: potassium carbonate (K_2CO_3), magnesium carbonate ($MgCO_3$) and calcium hydrogen carbonate ($Ca(HCO_3)_2$)

Chemical weathering generally produces clay compounds. Even resistant rocks such as granite can be weathered to release potassium and magnesium together with sesquioxides of iron and aluminium. Quartz is released as sand grains when the parent material disintegrates. The sand and clay make up the inorganic fraction; the soluble products enter the soil.

Particle size

The composition of the parent material will influence the make-up of the soil and the particle size. The size of particles in a particular soil is important because it influences texture which ultimately affects structure, pore spaces, and water and nutrient retention. For particles of 2 mm in diameter and below, three size categories are recognised:
- **sand**: diameter 2–0.06 mm (2,000–60 microns)
- **silt**: diameter 0.06–0.002 mm (60–2 microns)
- **clay:** diameter less than 0.002 mm (<2 microns)

Texture

The distribution of grain sizes within a soil is known as its texture. Analysis of texture involves assessing the percentage of sand, silt and clay present in a soil. The usual way of identifying soils according to texture is using a triangular graph (Figure 9.16).
- Soils with a high **sand** content have a large amount of air space which leads to rapid drainage of water and poor nutrient retention. Minerals are easily leached through sandy soils.
- **Clay** soils have a much greater surface area available for nutrient retention but drainage is impeded. Plant root penetration is difficult. Clay soils become easily waterlogged, and shrink on drying.
- **Silty** soils often have a low organic retention and are prone to erosion and waterlogging.

High concentrations of sand, silt or clay in a soil tend to be disadvantageous for agricultural use. The best soils have a mixture of all three, combining the good properties of each particle size. Such soils are known as **loams** (Figure 9.16). The best loam combines 20% clay with 40% of both sand and silt. The clay component retains moisture and nutrients, the sand component aids drainage and reduces the risk of waterlogging, and the silt helps to bind the other particles together.

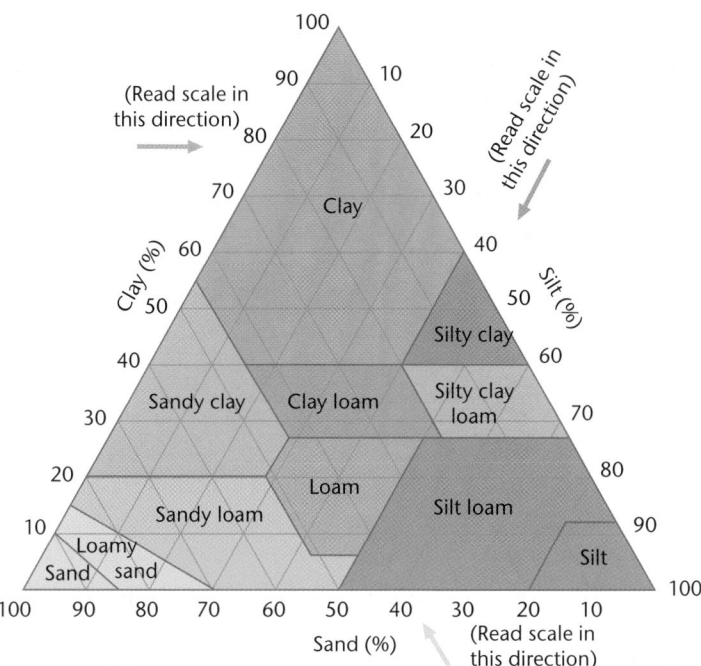

Figure 9.16
Soil textures

Structure

Structure is the way in which mineral particles and humus are aggregated together to form larger units called **peds**. The presence of organic matter and the gums produced during bacterial breakdown help bind the peds together. Spaces between peds allow the passage of water and soil microfauna. Five main types of ped structure are recognised:

- **Crumb (granular)** is soil made up of small breadcrumb-like particles. It contains many air spaces and drains well, which makes it good for agriculture.
- **Platy** soil has particles with horizontal axes longer than the vertical, forming small plates that overlap. It does not provide good drainage, so has poor agricultural potential.
- **Blocky** soil structure consists of larger, irregularly shaped 'blocks'. These are closely packed, but do allow some drainage and a reasonable agricultural productivity.
- **Prismatic (columnar)** structure has particles in a prism-like vertical arrangement with the vertical axis much longer than the horizontal. It is productive because it allows vertical movement of water and roots.
- In **structureless** soil, for example heavy clay soils, it is not possible to recognise any of the above four structures.

Water and air

The texture and structure of a soil influence the amount of space available between the soil particles and peds for the retention of water and for aeration. Clay soils have a large number of small pores (micropores) whereas sandy soils have larger pores (macropores) which allow water to pass through more quickly. Loams generally show a balance between pore sizes.

If all the pores are filled with water there is no space for air in the soil and it is **anaerobic**. As water drains down through the soil under the influence of gravity, it vacates the larger pores but remains within the smaller cavities. When this gravitational water has been lost the soil is said to be at **field capacity**.

Water is held in the finer pores as both hygroscopic and capillary water. Hygroscopic water is tightly bonded to the soil particles and so is not available to plants. Capillary water can move vertically to a height of 1 m above the water table and is available to plants, but is lost to the soil through evaporation and transpiration. When there is no longer sufficient water in the soil to meet the transpiration demands of plants, the soil is said to be at its **wilting point**, and plants will suffer stress. Pores not occupied by water contain air, which is essential for plant growth and organisms within the soil.

Organic constituents

Dead plants and animals, leaf litter and faecal remains are broken down by bacteria and fungi. As this organic matter decays, its previous structure becomes unrecognisable and **humus** is produced. Some of this can be recognised as a distinct layer on top of the soil. As humification continues, organic matter is incorporated into the soil to form an essential part of the clay–humus complex.

Three distinct types of humus can be recognised, each reflecting the environmental and nutrient cycle operating in the area where it is formed. Some plants take up more minerals, such as calcium and potassium, and incorporate them into structures. Such plants are rich in nutrients and, when they decay, the nutrients are returned to the soil and become incorporated into the humus. This humus is neutral or mildly acidic and is known as **mull**. It is soft, crumbly, has a blackish colour and is usually found under lowland areas covered by deciduous woodlands or grasslands. Under more acidic conditions, such as upland heath, bogs or coniferous forests, where breakdown of organic matter is slower, the plants take up, and hence return, fewer nutrients. This produces a raw, fibrous, acidic humus known as **mor**. The third type is a transitional mor-like mull, called **moder**.

Plant nutrients and ionic exchange

Nutrients are elements found in the soil that are essential for plant growth. There are three main sources of nutrients in the soil:

- Nutrients dissolved in the soil water, which are available to plants but easily removed by the loss of gravitational water.
- Humus compounds and clay undergo a linkage within the soil to form the clay–humus complex. Mineral nutrients in exchangeable form (cation exchange) are retained by attachment (adsorbtion) to the clay–humus complex. This is the most important store of soil minerals available to plants.
- Nutrients stored in minerals within the inorganic fraction of the soil are not available to plants until they are released by weathering and become dissolved in soil water.

If the clay–humus complex were not present in the soil, rainwater would easily remove all the soluble salts. Nutrients exist in soil solution as positively charged ions (+) called **cations** and negatively charged ions (–) called **anions**.

The clay–humus complex, which has a negative charge, attracts the positively charged minerals, notably calcium, magnesium, potassium, ammonium and sodium. The cations are said to be adsorbed to the clay–humus complex.

The process of **cation exchange** allows cations to be moved from the clay–humus complex to soil solution or from solution to plant roots. Cation exchange releases hydrogen ions which increase the soil acidity (see section on soil pH below). The measure of the ability of the soil to retain cations for plant use is known as the **cation exchange capacity (CEC)**. Clay–humus soils have a high CEC but sandy soils have a low CEC as they are less able to keep essential soil nutrients and are therefore less fertile.

Soil pH

Soil pH relates to the number of hydrogen cations in the soil. A soil with a low concentration of hydrogen ions has a high pH and is **alkaline**. The higher the hydrogen-ion concentration, the lower is the pH and the more **acidic** the soil. The pH scale runs from pH 1 to pH 14 and is logarithmic. A reading of pH 7 is neutral; most soils fall in the range of pH 4 to 8.

Soil profiles

A vertical section through the soil is known as a profile and shows the different layers, called **horizons**. The profile is the product of the balance between inputs and outputs, and the processes at work in the soil redistributing and chemically altering the various soil constituents. It is common to recognise three main horizons, which are then subdivided as shown in Table 9.2:

- **A horizon**: the upper layer of the soil where biological activity and humus content are at their maximum. This zone may lose soluble salts by drainage and downwash (eluviation)
- **B horizon**: the lower level of the soil where material removed from above is redeposited (illuviation)
- **C horizon**: the weathered parent material (regolith)

O	L	Undecomposed litter; leaf layer
	F	Partially decomposed; fermentation layer
	H	Well decomposed; humus layer
A	Ah	Dark-coloured humic horizon
	Ap	Ploughed layer in cultivated soils
	E	Eluvial horizon from which clay/sesquioxides removed
	Ea	Bleached (albic or ash-like) layer in podzolised soils
	Eb	Brown eluvial layer, depleted of clay
B	Bt	Illuvial clay redeposited (textural B horizon)
	Bh	Illuvial humus layer
	Bf/Bfe	Illuviated iron layer
	Bs	Brightly coloured layer of sesquioxide (iron/aluminium) accumulation
C		Weathered parent material

*Table 9.2
Soil horizons*

Above the A horizon are zones which consist entirely of organic matter (O). Localised conditions may produce a soil that differs from the basic A, B, C pattern. Where soils are waterlogged, for example, a **gley horizon (G)** develops.

Soil processes

There are many processes involved in soil formation and the creation of profiles. **Weathering** breaks down the parent material and releases inorganic matter into the soil. **Humification** occurs when soil organisms break down organic matter to form humus. As organic matter decays, it releases nutrients and organic acids and these can be strong enough to break down clays and other minerals in the A horizon. The resulting chelates (organic–metal compounds) are soluble and easily transported down the profile — a process known as **cheluviation**.

The movement of soil components in any form (solution, suspension, etc.) or direction (up or down) is known as **translocation**. The removal of material from the A horizon and its movement down the soil is called **eluviation**. The deposition of this material in the lower horizon is called **illuviation**. When soluble material is removed in solution from the A horizon, the process is known as **leaching**. It is very common in UK soils. A more extreme form, operating in cool climates, is **podzolisation**, where the oxides of iron and aluminium are removed along with humus. Cheluviation also comes under this heading.

If evapotranspiration begins to exceed precipitation, leaching is limited. Calcium is no longer removed and begins to build up in the process known as **calcification**. Where evapotranspiration becomes much greater than precipitation, salts are drawn upwards in solution by capillary action. These can eventually form a hard salt deposit on the surface. This process of **salinisation** has led to serious problems in irrigated areas around the world.

When soils become waterlogged, reddish-coloured oxidised ferric iron is chemically reduced to grey-blue ferrous iron — a process known as **gleying**.

Examples of soil types

Brown earth

Brown earths (Figure 9.17) form under temperate deciduous forests and are dominated by the process of leaching. Rainwater, a weak carbonic acid, combines with organic acids produced by the breakdown of organic material at the surface to produce a complex solution which causes chemical breakdown in the soil. Soluble bases are dissolved and carried down the soil profile. Clay particles may also be moved downwards in suspension, a process of eluviation known as **lessivage**.

Podzol

Podzols (Figure 9.18) develop where precipitation exceeds evapotranspiration, under coniferous forest and heathland, and where parent material produces coarse-textured soils. They are dominated by the processes of podzolisation and cheluviation which are basically intense forms of leaching operating under these more acidic conditions. Stronger humic acids, or chelating agents, are released by the slower breakdown of the more acidic raw humus (mor) produced under

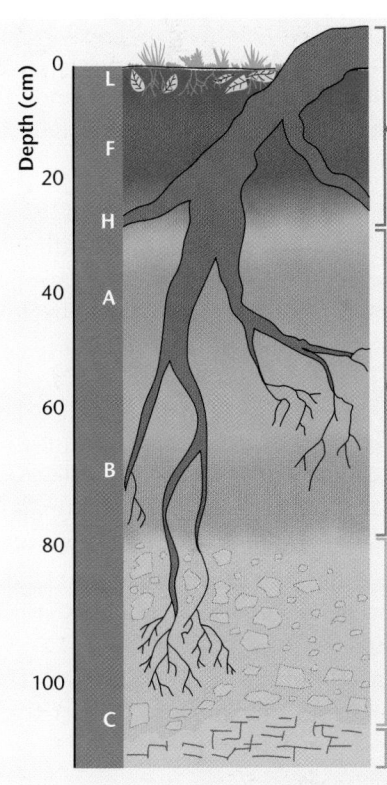

- Brown colour
- Boundaries between horizons blurred
- Roots abstract bases which are recycled through decomposition of leaf litter
- Usually loamy texture throughout profile — crumb structure in A horizon; more blocky lower down

- Litter may be several centimetres thick under deciduous woodland
- Merges with dark brown mild mull humus, pH 5.7; this returns bases to the soil and allows soil fauna to thrive
- Bacterial breakdown is rapid
- Worms incorporate organic material

- Presence of humus which has been moved down; paler at depth
- With precipitation greater than evaporation and mildly acidic conditions, calcium, magnesium, sodium and potassium are leached downwards, along with clay particles
- May be an Eb layer
- Lighter brown; pH 6.75–6.5
- Although leaching is slight, the products from the eluvial A horizon may be moved down and the sesquioxides of aluminium and iron redeposited in the B horizon to give a darker layer than in A
- Lessivage, the movement of clay particles in suspension, may produce 'clay skins' around roots
- May produce a Bt layer

- Weathered parent material

- Parent rock

Figure 9.17
Brown earth profile

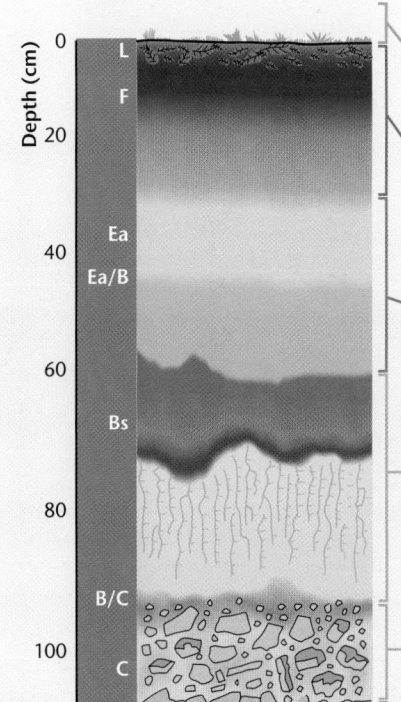

- Clear horizon boundaries
- Acidity restricts fauna
- Less mixing

- Surface vegetation: conifers, heathland, moorland grasses

- Litter of leaves, pine needles; slowly decomposing under colder upland climate
- Dark, acidic mor humus

- Light grey ash colour, clean sand grains, gritty texture; pH 4.5–5.5
- Bleached layer
- Eluviated zone, leached of nutrients; Fe/Al compounds washed down

- Deposition of humus to form Bh layer
- Darker red-brown colour
- Redeposition of sesquioxides of Fe and Al, pH 4.5–6.0
- Clay content increases
- Possible iron pan Bfe
- Transitional B zone

- Weathered parent material

- Parent rock

Figure 9.18
Podzol profile

225

conifers and heathland vegetation. Aluminium and iron sesquioxides are unstable under such conditions and are moved down the profile, leaving a bleached, silica-rich, sandy layer in the A horizon. The aluminium and iron compounds are redeposited in the B horizon and the iron may accumulate to form a pan.

Gley

Gleying (Figure 9.19) occurs when water movement through the soil is restricted by a high water table (groundwater gley) or by an impermeable layer within the soil which impedes drainage (surface-water gley). The formation of an iron pan in a podzol (see above) can restrict water movement and form a gleyed horizon above the pan. Groundwater gleys are usually associated with low-lying areas such as valley floors.

Under waterlogged conditions, all the pore spaces are filled with water, producing anaerobic (oxygen-deficient) conditions. The decomposition of plant debris on the surface is therefore slow and the presence of organic matter and specialised bacteria cause reducing conditions under which the solubility of many constituents is changed. Iron, which is relatively insoluble in its ferric (oxidised) form, becomes much more soluble when reduced to the ferrous form. Under waterlogged conditions, therefore, iron becomes more mobile and is slowly removed from the system.

Waterlogging is often seasonal, as the height of the water table varies. This causes mottling in the soil, with orange-brown coloration produced when oxygen enters the soil and oxidation occurs.

Figure 9.19
Gley profile

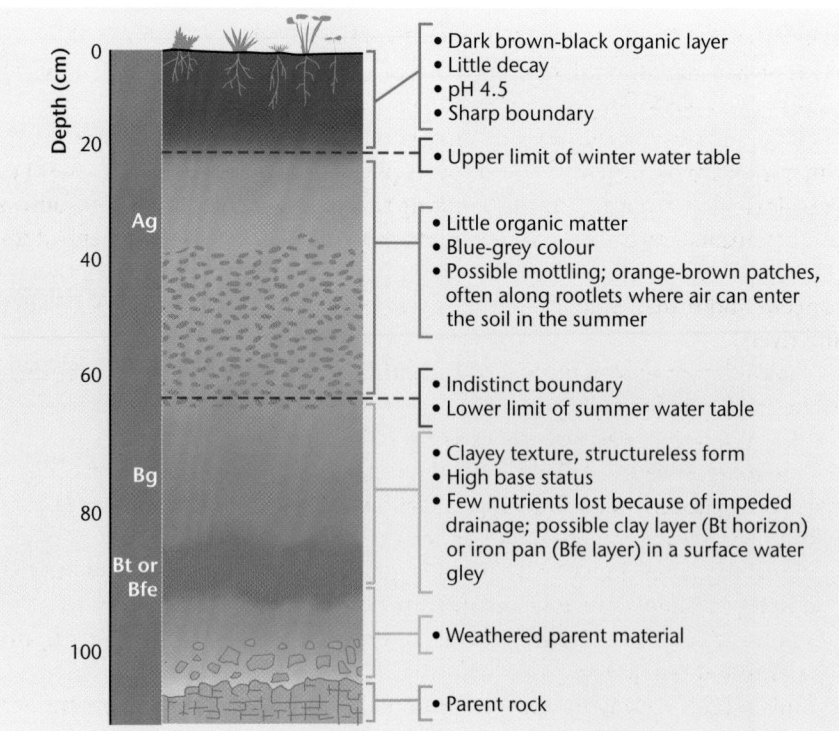

* Dark brown-black organic layer
* Little decay
* pH 4.5
* Sharp boundary

* Upper limit of winter water table

* Little organic matter
* Blue-grey colour
* Possible mottling; orange-brown patches, often along rootlets where air can enter the soil in the summer

* Indistinct boundary
* Lower limit of summer water table

* Clayey texture, structureless form
* High base status
* Few nutrients lost because of impeded drainage; possible clay layer (Bt horizon) or iron pan (Bfe layer) in a surface water gley

* Weathered parent material

* Parent rock

The effects of human activity on soil

Human activity can have a massive impact upon soils, either damaging them (degradation) or improving them (upgrading).

Soil degradation

Degradation of soils is the result of human failure to understand and manage them as a resource.

- Harvesting of crops removes the natural supply of recycled nutrients and organic material.
- Removal of vegetation by either deforestation or overgrazing increases the amount of precipitation impacting on the ground or running over it. This water removes nutrients and organic matter faster than they can be replaced by the weathering of bedrock and the breakdown of vegetation.
- Ploughing under wet conditions can produce a plough plan which restricts drainage. Deep ploughing breaks up stable peds and loosens soil, making it vulnerable to wind erosion. Ploughing up and down the gradient creates channels for water runoff and allows gullying to begin.
- Monoculture (growing the same crop year after year) depletes the soil of certain mineral nutrients.
- Heavy machinery or overstocking with animals causes compaction. The soil develops a platy structure which impedes drainage and can lead to water-logging and gleying.

Soil improvement (upgrading/conservation)

Many areas of the world suffer from soil erosion but there are ways in which it can be prevented. Badly affected areas can be improved by conservation.

Adding fertiliser improves the nutrient content of the soil. Fertilisers can be either inorganic (compounds of nitrogen, phosphorous and potassium, NPK) or organic (mainly farmyard manure and crop residues). Organic fertilisers encourage soil organisms and improve nutrient retention through the development of the clay–humus complex. Artificial inorganic fertilisers do not do this and there is concern about their impact on the environment (e.g. eutrophication of lakes and rivers).

Planting crops or trees helps to stabilise the soil, and organic matter is returned to the soil in the form of leaf litter.

Various farming practices can also help to improve the soil:

- crop rotation with fallow periods allows the soil to replenish nutrients
- replacing hedgerows or building shelter belts reduces wind erosion
- improving field drainage increases aeration
- ploughing *across* slopes helps prevent gullying. Ploughing also aerates the soil, develops a crumb structure and increases the number of pore spaces
- liming raises the pH and provides more nutrients for plant growth and organism development
- mulching, by ploughing in the stubble, increases the organic content and improves nutrient retention

Plate tectonics

Plate tectonic theory revolutionised the study of earth science. As soon as maps of the Atlantic Ocean were produced, people noticed that the continents either side seemed to fit together remarkably well — the bulge of South America fitting into the indent below west Africa. Francis Bacon had noted this fit as early as the seventeenth century but it did not attract any serious attention as no one thought the continents could move about.

The theory of plate tectonics

In 1912 a German, Alfred Wegener, published his theory that a single continent existed about 300 million years ago. He named this super-continent Pangaea, and maintained that it had later split into the two continents of Laurasia in the north and Gondwanaland in the south. Today's continents were formed from further splitting of these two masses. Wegener published this **theory of continental drift** and claimed that it was supported by several pieces of evidence that these areas were once joined.

Geological evidence for the theory included:
- the above-mentioned fit of South America and Africa
- evidence of the glaciation of the late Carboniferous period (290 million years ago), deposits from which are found in South America, Antarctica and India. The formation of these deposits cannot be explained by their present position; they must have been formed together and then moved. There are also striations on rocks in Brazil and west Africa which point to a similar situation
- rock sequences in northern Scotland closely agree with those found in eastern Canada, indicating that they were laid down under the same conditions in one location

Biological evidence for the theory included the following:
- fossil brachiopods found in Indian limestones are comparable with similar fossils in Australia
- fossil remains of the reptile Mesosaurus are found in both South America and southern Africa. It is unlikely that the same reptile could have developed in both areas or that it could have migrated across the Atlantic
- the fossilised remains of a plant which existed when coal was being formed have only been located in India and Antarctica

Development of the theory

Wegener's theories were unable to explain how continental movement could have taken place

---

Key terms

Global structures The major surface relief features of the planet that cover both oceanic and continental areas. They include fold mountains (ancient and young), rift valleys, oceanic ridges, oceanic trenches, island arcs, the abyssal plain (ocean basins) and the continental slope and shelf.

Plates The lithosphere (the crust of the Earth and the upper part of the mantle) is divided into a number of sections called plates. These rigid slabs float on the underlying semi-molten mantle (asthenosphere) and are moved by convection currents within it.

Plate tectonics A theory that explains the formation and distribution of the Earth's major structural features in terms of a series of plates that make up its surface.

and his ideas gained little ground. From the 1940s onwards, however, evidence began to accumulate to show that Wegener could have been correct.

The mid-Atlantic ridge was discovered and studied. A similar feature was later discovered in the Pacific Ocean.

Examination of the ocean crust either side of the mid-Atlantic ridge suggested that sea-floor spreading was occurring. The evidence for this is the alternating polarity of the rocks that form the ocean crust. Iron particles in lava erupted on the ocean floor are aligned with the Earth's magnetic field. As the lavas solidify, these particles provide a permanent record of the Earth's polarity at the time of eruption (palaeomagnetism). However, the Earth's polarity reverses at regular intervals (approximately every 400,000 years). The result is a series of magnetic 'stripes' with rocks aligned alternately towards the north and south poles (Figure 9.20). The striped pattern, which is mirrored exactly on either side of a mid-oceanic ridge, suggests that the ocean crust is slowly spreading away from this boundary. Moreover, the oceanic crust gets older with distance from the mid-oceanic ridge.

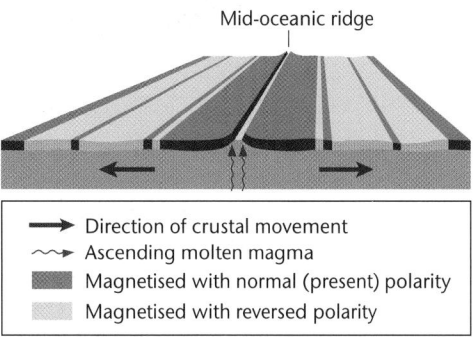

Figure 9.20
Magnetic 'stripes' on the Atlantic Ocean floor

Sea-floor spreading implies that the Earth must be getting bigger. As this is not the case, then plates must be being destroyed somewhere to accommodate the increase in their size at mid-oceanic ridges. Evidence of this was found with the discovery of huge oceanic trenches where large areas of ocean floor were being pulled downwards.

The Earth's layers

Before the development of plate tectonic theory, earth scientists divided the interior of the Earth into three layers: the crust, the mantle and the core. The **core** is made up of dense rocks containing iron and nickel alloys and is divided into a solid inner core and a molten outer one, with a temperature of over 5,000 °C. The **mantle** is made up of molten and semi-molten rocks containing lighter elements, such as silicon and oxygen. The **crust** is even lighter because of the elements that are present, the most abundant being silicon, oxygen, aluminium, potassium and sodium. The crust varies in thickness — beneath the oceans it is only 6–10 km thick but below continents this rises to 30–40 km. Under the highest mountain ranges the crust can be up to 70 km thick.

The theory of plate tectonics has retained this simple threefold division, but new research has suggested that the crust and the upper mantle should be

	Continental crust	Oceanic crust
Thickness	30–70 km	6–10 km
Age	Over 1,500 million years	Less than 200 million years
Density	2.6 (lighter)	3.0 (heavier)
Composition	Mainly granite; silicon, aluminium and oxygen (SIAL)	Mainly basalt; silicon, magnesium, oxygen (SIMA)

Table 9.3
Differences between continental and oceanic crust

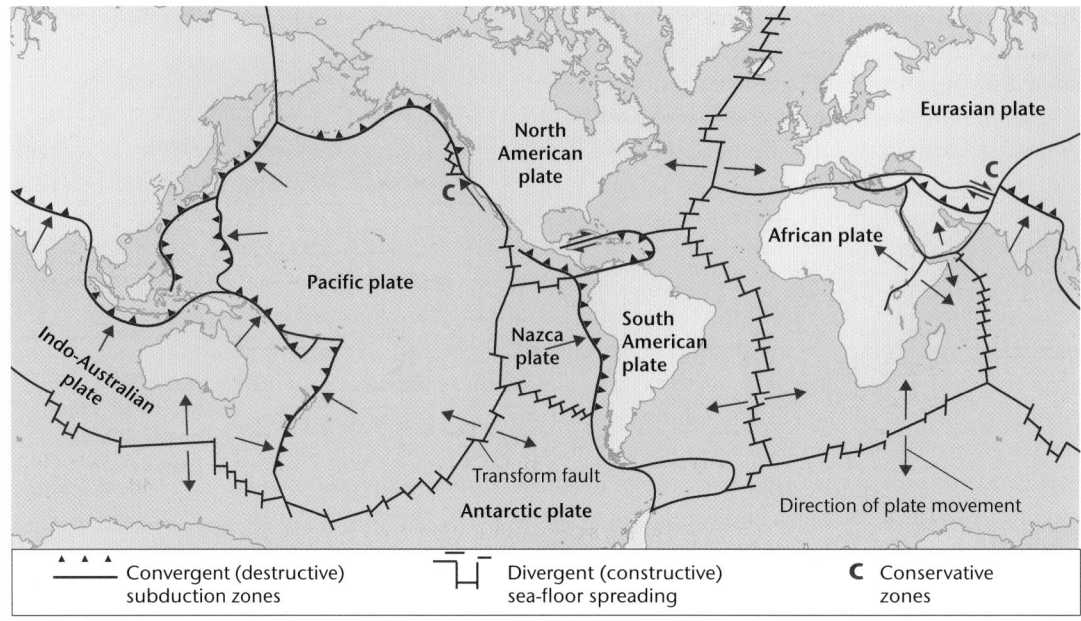

| ▲▲▲ Convergent (destructive) subduction zones | ⊤⌐ Divergent (constructive) sea-floor spreading | **C** Conservative zones |

Figure 9.21
Tectonic plates and their margins

Figure 9.22
Convection currents and plate movement

divided into the lithosphere and the asthenosphere. The **lithosphere** consists of the crust and the rigid upper section of the mantle and is approximately 80–90 km thick. It is divided into seven very large plates and a number of smaller ones (Figure 9.21). Plates are divided into two categories, oceanic and continental, depending on the type of material from which they are made (see Table 9.3). Below the lithosphere is the semi-molten **asthenosphere**, on which the plates float and move.

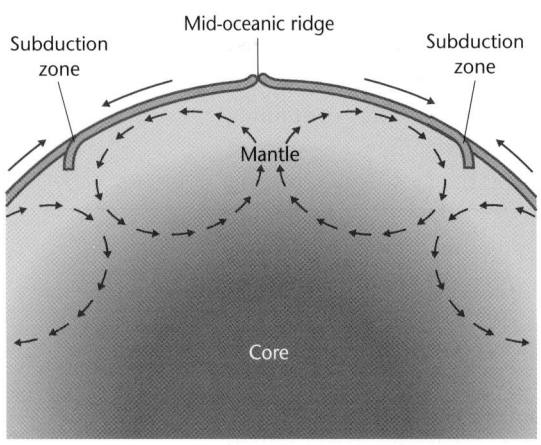

Hot spots around the core of the Earth generate thermal convection currents within the asthenosphere, which cause magma to rise towards the crust and then spread before cooling and sinking (Figure 9.22). This circulation of magma is the vehicle upon which the crustal plates move. The crust can be thought of as 'floating' on the more dense material of the asthenosphere. This is a continuous process, with new crust being formed along the line of constructive boundaries between plates (where plates move away from each other) and older crust being destroyed at destructive boundaries (where plates are moving towards each other).

Features of plate margins

Constructive (divergent) margins

Where plates move apart in oceanic areas they produce mid-oceanic ridges. Where they move apart in continental crust they produce rift valleys. The space between the diverging plates is filled with basaltic lava upwelling from below. Constructive margins are therefore some of the youngest parts of the Earth's surface, where new crust is being continuously created.

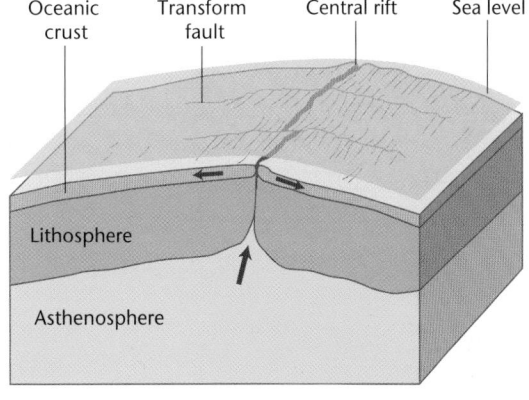

Figure 9.23
Cross section of the mid-Atlantic ridge

Oceanic ridges

Oceanic ridges (Figure 9.23) are the longest continuous uplifted features on the surface of the planet, and have a total length of 60,000 km. In some parts they rise 3,000 m above the ocean floor. Their precise form appears to be influenced by the rate at which the plates separate:

- a slow rate (10–15 mm per year), as seen in parts of the mid-Atlantic ridge, produces a wide ridge axis (30–50 km) and a deep (3,000 m) central rift valley with inward-facing fault scarps
- an intermediate rate (50–90 mm per year), such as that on the Galapagos ridge (Pacific), produces a less well-marked rift (50–200 m deep) with a smoother outline
- a rapid rate (>90 mm per year), such as on the east Pacific rise, produces a smooth crest and no rift

Volcanic activity also occurs along the ridge, forming submarine volcanoes, which sometimes rise above sea level, e.g. Surtsey, to the south of Iceland (Iceland itself was formed in this way and is the largest feature produced above sea level on a divergent margin). These are volcanoes with fairly gentle sides due to the low viscosity of basaltic lava. Eruptions are frequent but relatively gentle (effusive).

As new crust forms and spreads, transform faults occur at right angles to the plate boundary. The parts of the spreading plates on either side of these faults may move at differing rates, leading to friction and ultimately to earthquakes. These tend to be shallow-focus earthquakes, originating near the surface.

Rift valleys

At constructive margins in continental areas, such as Africa, the brittle crust fractures as sections of it move apart. Areas of crust drop down between parallel faults to form **rift valleys** (Figure 9.24). The largest of these features is the African

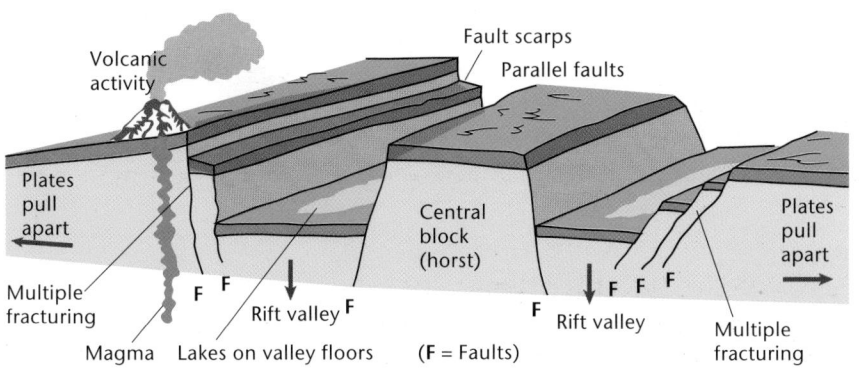

Figure 9.24
Cross section of a rift valley

231

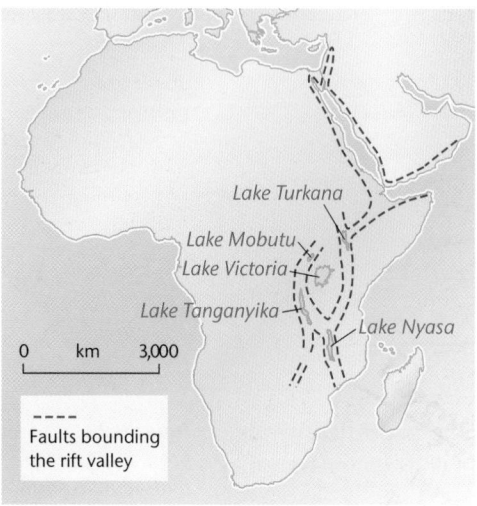

Lake Turkana

Lake Mobutu
Lake Victoria
Lake Tanganyika
Lake Nyasa

0 km 3,000

- - - -
Faults bounding
the rift valley

Figure 9.25
The African rift
valley

rift valley which extends 4,000 km from Mozambique to the Red Sea. From the Red Sea it extends north into Jordan, a total distance of 5,500 km (Figure 9.25). In some areas, the inward-facing scarps are 600 m above the valley floor and they are often marked by a series of parallel step faults.

The area is also associated with volcanic activity (for example the highest mountain in Africa, Kilimanjaro, Photograph 9.4). The crust here is much thinner than in neighbouring areas, suggesting that tension in the lithosphere is thinning the plate as it starts to split. The line of the African rift is thought to be an emergent plate boundary, the beginning of the formation of a new ocean as eastern Africa splits away from the rest of the continent.

Destructive (convergent) margins

There are two types of plates, so there are three different convergent situations:

- oceanic plate moves towards continental plate
- oceanic plate moves towards oceanic plate
- continental plate moves towards continental plate

Oceanic/continental convergence

Where oceanic and continental plates meet, the denser oceanic plate (see Table 9.3 on page 229) is forced under the lighter continental one. This process is known as **subduction**. The downwarping of the oceanic plate forms a very deep part of the ocean known as a **trench** (Figure 9.26). A good example of an ocean trench is off the western coast of South America where the Nazca plate is subducting under the South American plate, forming the Peru–Chile trench.

Photograph 9.4
Mt Kilimanjaro,
Tanzania, in the
African rift valley
area

Corel

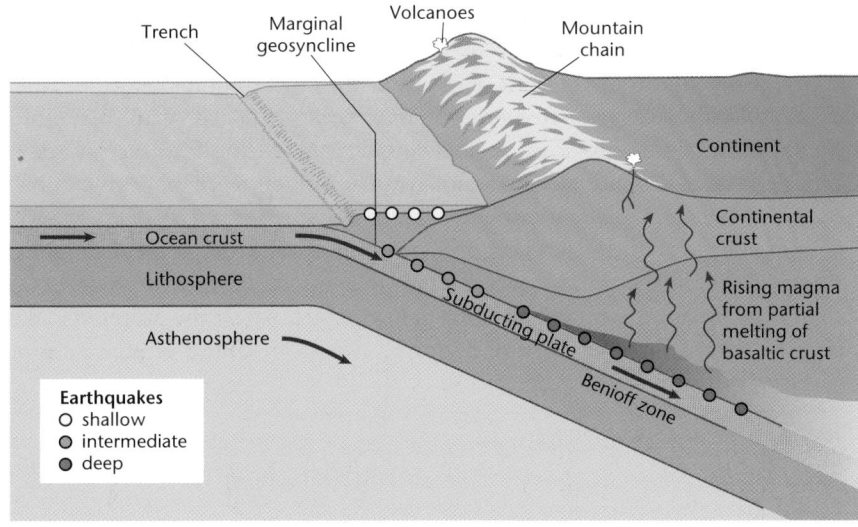

Figure 9.26
Cross section of oceanic/continental plate convergence at a destructive plate margin

Sediments that have accumulated on the continental shelf on the margin of the land mass are deformed by folding and faulting. Along with the edge of the continental plate, these are uplifted to form **fold mountains** (Figure 9.26), such as the Andes along the Pacific side of South America. As the oceanic plate

Table 9.4
Relationship of tectonic activity to plate margins

Plate margin	Movement of plates	Tectonic features	Examples
Constructive	Divergent: two plates moving away from each other	New crust is formed from upwelling magma: mid-oceanic ridges, effusive ridge (shield) volcanoes, shallow focus earthquakes, median rift valleys	Mid-Atlantic ridge
		Continental rift valleys	East African rift valley
Destructive (1) Subduction	Convergent: two plates moving towards each other	(1a) Oceanic to oceanic: trenches, island arcs, explosive volcanoes, earthquakes (shallow, intermediate and deep)	On the margins of Pacific plate, with subduction under other, separate sections of the plate — Tonga trench
		(1b) Oceanic to continental: trenches, fold mountains, explosive volcanoes, earthquakes (shallow, intermediate and deep)	Andean type: Nazca plate subducting under South American plate
(2) Collision		(2) Continental to continental: fold mountains, shallow-focus earthquakes	Himalayan type: Indian plate colliding with Eurasian plate
Conservative	Two plates shearing past each other	Shallow-focus earthquakes	San Andreas fault: Pacific plate and North American plate
Not at plate boundaries	Hot spots: may be near the centre of a plate	Plume volcanoes	Hawaiian islands: Emperor seamount chain

descends, the increase in pressure can trigger major earthquakes along the line of the subducting plate; these may be shallow-, intermediate- or deep-focus.

The further the rock descends, the hotter the surroundings become. This, together with the heat generated from friction, begins to melt oceanic plate into magma in a part of the subduction zone known as the Benioff zone. As it is less dense than the surrounding asthenosphere, this molten material begins to rise as plutons of magma. Eventually, these reach the surface and form volcanoes. The andesitic lava, which has a viscous nature (flows less easily), creates complex, composite, explosive volcanoes (contrast this to the basaltic emissions on constructive margins which tend to be gentle eruptions). If the eruptions take place offshore, a line of volcanic islands known as an **island arc** can appear, e.g. the West Indies.

Oceanic/oceanic convergence

Where oceanic plates meet, one is forced under the other and the processes involved with subduction begin. Ocean trenches and island arcs are the features associated with this interaction, as it takes place well offshore. A good example is on the western side of the Pacific Ocean where the Pacific plate is being subducted beneath the smaller Philippine plate. Here the ocean floor has been pulled down to form the very deep Marianas trench. A line of volcanic islands, including Guam and the Marianas, has been formed by upwelling magma from the Benioff zone (Figure 9.27).

Figure 9.27
Cross section of oceanic/oceanic plate convergence at a destructive plate margin

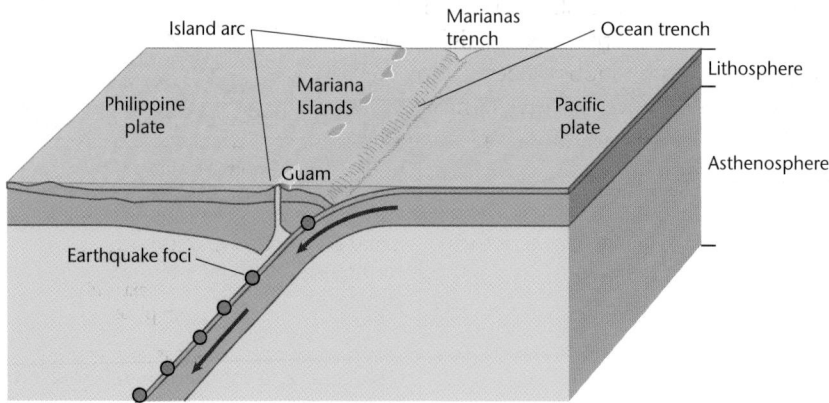

Continental/continental convergence

The plates forming continental crust have a much lower density than the underlying layers, so there is not much subduction where they meet. Instead, as the plates move towards each other, their edges and the sediments between them are forced up into fold mountains. As there is little subduction, there is no volcanic activity, but the movement of the plates can trigger shallow-focus earthquakes. Material is also forced downwards to form deep mountain roots (Figure 9.28).

The best example of such a margin is where the Indo-Australian plate is being forced northwards into the Eurasian plate. The previous intervening ocean, known as the Sea of Tethys, has had its sediments forced upwards in large

*Photograph 9.5
The Himalayas have
been formed at a
convergent margin*

*Figure 9.28
Cross section of
continental/
continental plate
convergence
(collision boundary)*

overfolds to form the Himalayas, an uplift that is
continuing today. The Himalayan range of fold
mountains, containing the highest mountain on
the planet (Everest 8,848 m, Photograph 9.5), is
up to 350 km wide and extends for 3,000 km.

Conservative margins

Where two crustal plates slide past each other and
the movement of the plates is parallel to the plate
margin, there is no creation or destruction of crust.
At these conservative margins (sometimes called
passive) there is no subduction and therefore no
volcanic activity.

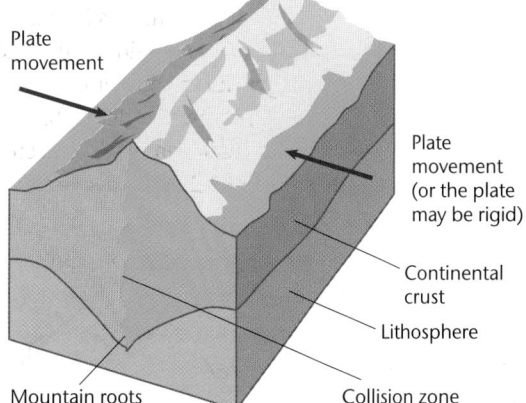

Plate
movement

Plate
movement
(or the plate
may be rigid)

Continental
crust

Lithosphere

Mountain roots

Collision zone

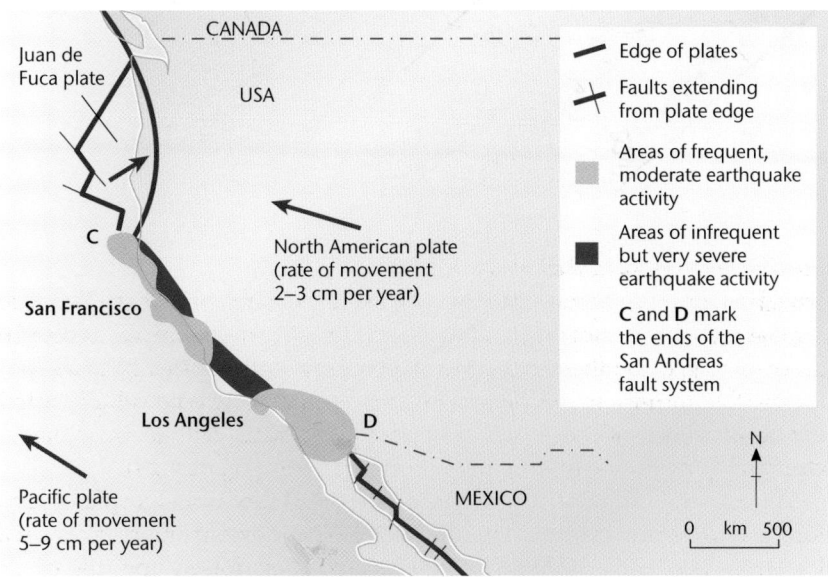

Juan de
Fuca plate

CANADA

USA

North American plate
(rate of movement
2–3 cm per year)

C

San Francisco

Los Angeles

D

Pacific plate
(rate of movement
5–9 cm per year)

MEXICO

Edge of plates

Faults extending
from plate edge

Areas of frequent,
moderate earthquake
activity

Areas of infrequent
but very severe
earthquake activity

C and **D** mark
the ends of the
San Andreas
fault system

N

0   km   500

*Figure 9.29
The San Andreas
fault system: a
conservative plate
margin*

The movement of the plates, however, creates stresses between the plate edges and, as sections of the plates rub past each other, the release of friction triggers shallow-focus earthquakes (e.g. San Francisco 1906 and 1989, Los Angeles 1994). These earthquakes occurred at the best-known example of a conservative margin — the San Andreas fault in California, where the Pacific and North American plates move parallel to each other (Figure 9.29). Both plates are moving in the same direction but not at the same speed. Stresses set up by this movement cause transform faults to develop, running at right angles to the main San Andreas fault.

Hot spots

Vulcanicity is normally associated with plate margins but, in the centre of the Pacific Ocean, we find the volcanic Hawaiian islands which are not connected with any plate boundary. It is believed that this volcanic area is caused by a localised **hot spot** within the Pacific plate. A concentration of radioactive elements inside the mantle may cause such a hot spot to develop. From this, a plume of magma rises to eat into the plate above. Where lava breaks through to the surface, active volcanoes occur above the hot spot.

The hot spot is stationary, so as the Pacific plate moves over it, a line of volcanoes is created. The one above the hot spot is active and the rest form a chain of islands of extinct volcanoes. The oldest volcanoes have put so much pressure on the crust that subsidence has occurred. This, together with marine erosion, has reduced these old volcanoes to seamounts below the level of the ocean. Figure 9.30 shows the line of the Hawaiian islands and their ages. From this evidence it is clear that the Pacific plate is moving northwest. This is further proof that the Earth's crust is moving, as originally suggested by Alfred Wegener.

Figure 9.30
The Hawaiian hot spot

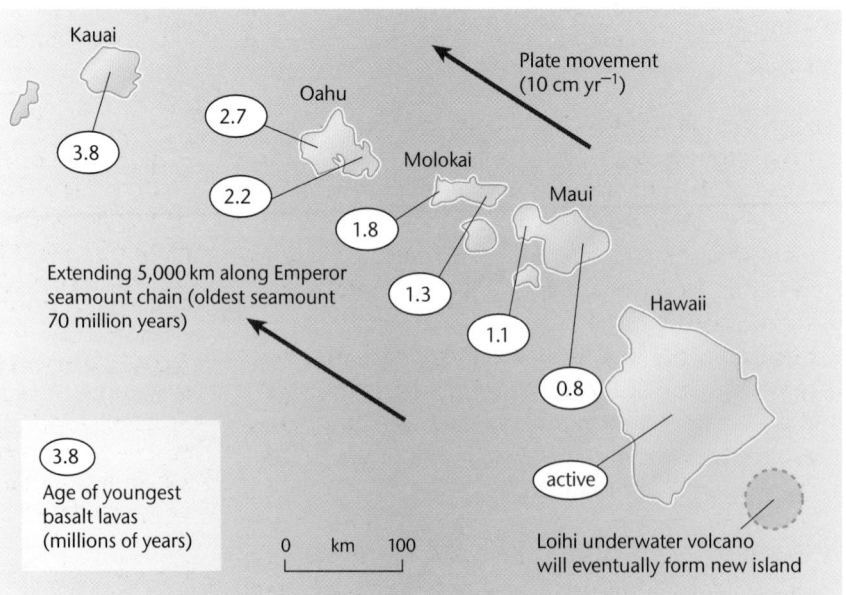

Chapter 10

People and the environment Hazards and resource conflicts

Responses to hazards

Response to a hazard can be made at both a collective and an individual level. It very much depends upon how people or organisations perceive the hazard and this depends on factors such as past experience, values, personality and expectations. The response is influenced not only by perception but also by the economic ability to take various courses of action and the technological ability to carry them out.

The way in which hazards are perceived can be classified as follows:

- **Fatalism**: the acceptance that hazards are natural events which are part of living. Some communities go as far as saying that they are 'God's will'. Losses are accepted as inevitable and people remain in the area.
- **Adaptation**: people see that they can prepare for, and therefore survive, a hazard event by prediction, prevention and/or protection, depending upon the type of hazard and the economic and technological circumstances of the area.
- **Fear**: people feel so vulnerable to the event that they are no longer able to face living in the area and move away to regions perceived to be unaffected by the hazard.

Management of hazards takes a number of forms:

- **Prediction**: it is possible to give warnings of some hazards. Action can then be taken to reduce their impact. Improved monitoring, information and communications technology have made the prediction of hazards and the issuing of warnings more useful in recent years.
- **Prevention**: the ideal form of management is to prevent the occurrence of the hazard event. For most hazards this is entirely unrealistic, and the best that can be achieved is some form of control, often through modification of the environment.
- **Protection**: the aim is to protect people and property from the impact of the hazard, but protection can also take the form of insuring against losses (particularly in MEDCs) and the supply of aid (usually in LEDCs).

> ### Key term
>
> **Hazard** A hazard is a natural event that threatens life and property, but it can also result from human action, for example burglary in urban areas. Remote volcanic or earthquake events that pose no threat to life or property are not hazard events — without people they are simply natural processes. John Whittow, in his book *Disasters: The Anatomy of Environmental Hazards*, suggested that 'A hazard is a perceived natural event which threatens both life and property — a disaster is the realisation of this hazard.' It is the interaction of people and the environment that defines a hazard.

In MEDCs, areas affected by hazard events are able to draw on central government funds for protection and relief.

Attempts to manage hazards must be evaluated in terms of their success in prediction, protection and prevention. Successful schemes, for example, have included the use of dynamite to divert lava flows on the slopes of Mt Etna in Italy, and pouring sea water on lava fronts in Iceland to solidify the flow. On the other hand, the Japanese felt they were well-prepared for earthquakes and yet the Kobe earthquake in 1995 left more than 100,000 buildings damaged and over 6,000 people dead (with 35,000 injured).

Volcanic activity

Causes and distribution

Most volcanic activity is associated with plate tectonic processes and is mainly located along plate margins (Figure 10.1). Such activity is therefore found:

- along oceanic ridges where plates are moving apart. The best example is the mid-Atlantic ridge — Iceland represents a large area formed from volcanic activity
- associated with rift valleys. The African rift valley has a number of volcanoes along it including Mt Kenya and Mt Kilimanjaro

Figure 10.1
Global distribution of active volcanoes

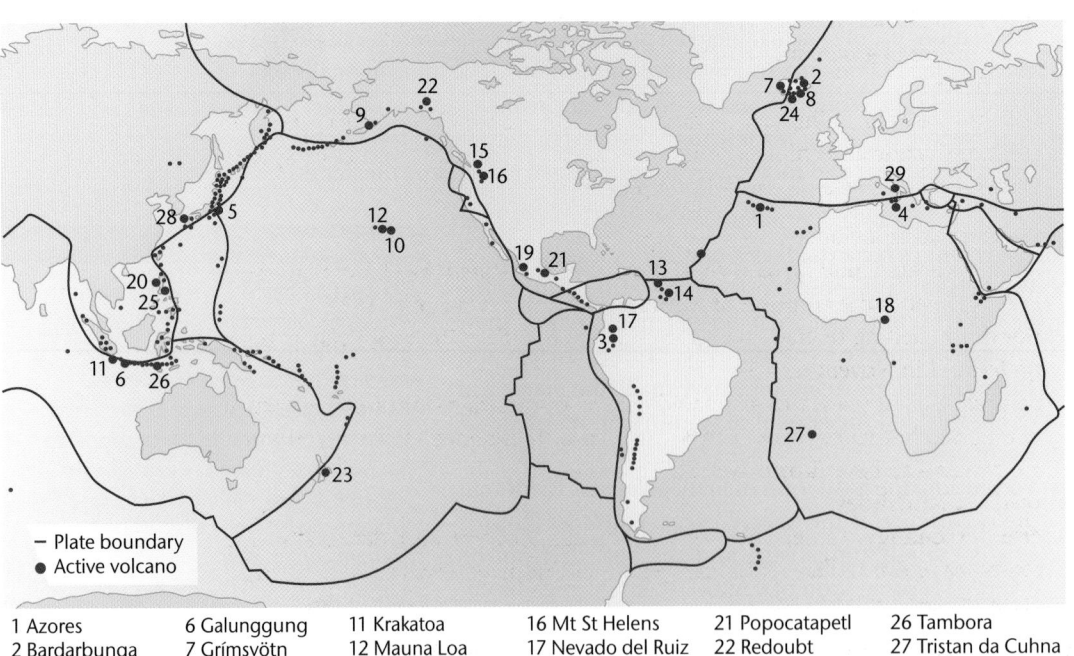

1 Azores	6 Galunggung	11 Krakatoa	16 Mt St Helens	21 Popocatapetl	26 Tambora
2 Bardarbunga	7 Grímsvötn	12 Mauna Loa	17 Nevado del Ruiz	22 Redoubt	27 Tristan da Cunha
3 Cotopaxi	8 Haeimaey	13 Soufrière Hills	18 Nyos	23 Ruapehu	28 Unzen
4 Etna	9 Katmai	14 Mt Pelée	19 Parícutin	24 Surtsey	29 Vesuvius
5 Fujiyama	10 Kilauea	15 Mt Rainier	20 Pinatubo	25 Taal	

- on or near subduction zones. The line of volcanoes, or 'ring of fire', that surrounds the Pacific Ocean is associated with plate subduction. This tends to be the most violent of all activity
- over hot spots such as the one in the middle of the Pacific Ocean which has given rise to the Hawaiian islands

Magnitude and frequency of the events

Attempts have been made to classify volcanic activity by noting the physical differences in eruptions and by using a table which runs from 0 (non-explosive) to 8 (very large). The impact of the hazard event is usually measured in terms of loss of life or the cost of damage to the built environment. The frequency of eruption of a volcano can be determined by vulcanologists and geologists examining the volcanic deposits in the surrounding area and working out an eruption history.

The effects of volcanic activity

A volcanic event can have a range of impacts, affecting the area immediately around the volcano or the entire planet. Effects can be categorised into primary and secondary.

Primary effects consist of:
- **tephra:** solid material of varying grain size, from volcanic bombs to ash particles, ejected into the atmosphere
- **pyroclastic flows:** very hot (800°C), gas-charged, high-velocity flows made up of a mixture of gases and tephra
- **lava flows**
- **volcanic gases** including carbon dioxide, carbon monoxide, hydrogen sulphide, sulphur dioxide and chlorine. Emissions of carbon dioxide from Lake Nyos in Cameroon in 1986 suffocated 1,700 people

Secondary effects include:
- **lahars:** volcanic mud flows such as those that devastated the Colombian town of Armero after the eruption of Nevado del Ruiz in November 1985
- **flooding:** melting of glaciers and ice caps such as the Grímsvötn glacial burst on Iceland in November 1996
- **tsunamis:** giant sea waves generated after violent caldera-forming events such as that which occurred on Krakatoa in 1883 — the tsunamis from this eruption are believed to have drowned 36,000 people
- **volcanic landslides**
- **climatic change:** the ejection of vast amounts of volcanic debris into the atmosphere can reduce global temperatures and is believed to have been an agent in past climatic change

Volcanic effects become a hazard when they impact upon the human and built environments, killing and injuring people, burying and collapsing buildings, destroying the infrastructure and bringing agricultural activities to a halt.

Case study *Soufrière Hills, Montserrat, 1995–97*

During and after the eruption of the Soufrière Hills volcano (Figure 10.2), the British government had to provide, or assist in, the following, as Montserrat was still a dependent territory:

- evacuation of 7,000 of the island's 11,000 population to neighbouring islands such as Antigua, or resettlement in the UK, and financial help with all resettlement
- resettlement of some of the population from the volcanic south to the 'safer' north of the island
- setting up of temporary shelters in the north
- re-establishment of air and sea links with the island
- building of new permanent housing
- moving the capital from Plymouth (now destroyed) to Salem
- providing farming areas for those resettled in the north

Hazard zones (April 1997)

- ■ No access
- ■ Limited access for essential visitors
- □ Prepare for possible evacuation
- □ Full occupation (possible evacuation)
- □ Full occupation

0 km 2

N

St John's

Salem
Airport ✈
Spanish
⊗ Montserrat Volcano Observatory
Point

Soufrière Hills

Plymouth ●
New Dome

*Figure 10.2
Montserrat:
danger zones*

South Soufrière Hills

St Patrick's

The British government spent over £100 million in total on mitigating the effects of the eruption and on a 3-year development plan for the island.

Popperfoto/Reuter

*Photograph 10.1
The former capital of
Montserrat, Plymouth,
covered by dust and
ashes from the
Soufrière Hills volcano,
August 1997*

Management

Prediction

It is easy to locate volcanoes, but it is very difficult to predict exactly when activity will take place, particularly a major eruption. The Colombian volcano, Nevado del Ruiz, came to life in late 1984 with small-scale activity. Vulcanologists knew the danger a major eruption could pose to the surrounding area, but were unable to predict when the major event would take place. Small-scale volcanic activity continued for several months and people were not prepared to evacuate their homes on the basis of this threat. When the violent eruption came on 13 November 1985, almost all the population had remained in the area. Devastating lahars, resulting from melting snow and ice, swept down the valleys, killing over 20,000 people.

A study of the previous eruption history of a volcano is important in prediction, along with an understanding of the type of activity produced. At present, research is being conducted to see if it is possible to predict the time of an eruption accurately using the shock waves that are produced as magma approaches the surface, expanding cracks and breaking through other areas of rock. There was some success in predicting the recent eruption (2000) of Popacatapetl in Mexico, but it remains to be seen if such techniques can be applied to all volcanoes.

Protection

With volcanic activity, protection means preparing for the event. Monitoring of the volcano may suggest a time when the area under threat should be evacuated. Such monitoring includes observations of land swelling, earthquake activity, changes in groundwater level and chemical composition, emission of gases, magnetic field studies and the shock wave analysis mentioned above. Several governments of countries in volcanic areas have made risk assessments and from them produced a series of alert levels to warn the public. In New Zealand the government has produced a five-stage table that includes the following:

1 Signs of volcanic activity. No significant volcanic threat.
2 Indications of intrusive processes. Local eruption threat.
3 Increasing intrusive trends indicate real possibility of hazardous eruption.
4 Large-scale eruption now appears imminent.
5 Destruction within the permanent danger zone (as identified). Significant risk over a wider area.

Geological studies of the nature and extent of deposits from former eruptions and associated ashfalls, lahars and floods may also provide evidence for hazard assessment. Figure 10.3 shows the hazards posed by Mt Rainier (Cascade Range, USA), one of the most studied volcanoes in North America.

Following assessments, it is possible to identify the areas at greatest risk, and land use planning can be applied to avoid building in such places.

Figure 10.3
Risk assessment of the Mt Rainier area

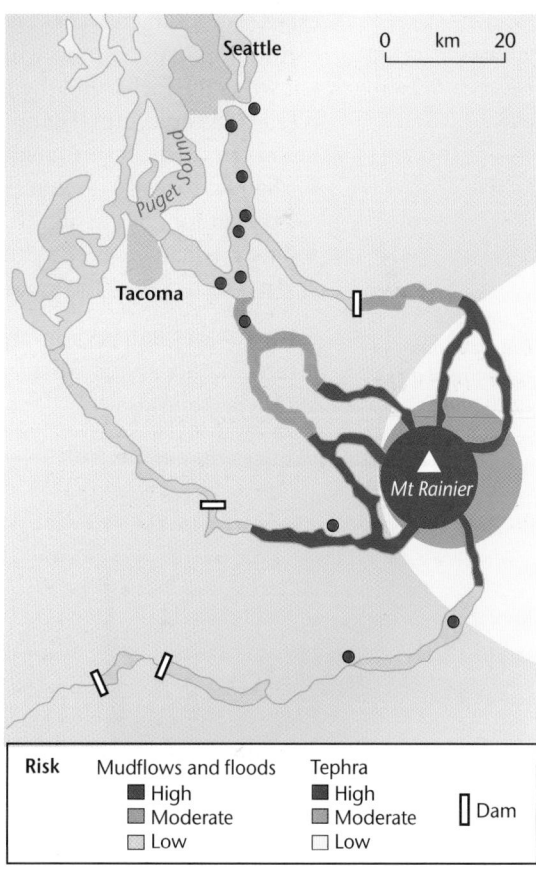

Risk	Mudflows and floods	Tephra	
	■ High	■ High	
	▨ Moderate	▨ Moderate	▯ Dam
	☐ Low	☐ Low	

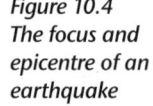

Once the lava has started to flow, it is possible, in certain circumstances, to divert it from the built environment by:

- digging trenches (Mt Etna, Sicily)
- explosive activity (Mt Etna, 1983)
- artificial barriers, which also protect against lahars (Hawaiian islands)
- pouring water on the lava front (Haeimaey, Iceland, 1973)

Foreign aid to LEDCs suffering volcanic eruptions may be required for considerable periods of time as volcanic events can be prolonged and devastating to the local economy. Such aid is needed for monitoring, evacuation, emergency shelters and food, long-term resettlement of the population and restoration of the economic base and the area's infrastructure (see case study, page 240).

Earthquakes

Causes

As the crust of the Earth is mobile, there tends to be a slow build up of stress within the rocks. When this pressure is suddenly released, parts of the surface experience an intense shaking motion that lasts for just a few seconds. This is an earthquake. The point at which this pressure release occurs within the crust is known as the **focus**, and the point immediately above that on the Earth's surface is called the **epicentre** (Figure 10.4). The depth of the focus is significant and three broad categories of earthquake are recognised:

- shallow-focus (0–70 km deep): these tend to cause the greatest damage and account for 75% of all the earthquake energy released
- intermediate-focus (70–300 km deep)
- deep-focus (300–700 km deep)

*Figure 10.4
The focus and
epicentre of an
earthquake*

Seismic waves radiate from the focus rather like the ripples in water when a rock is thrown into a pond. There are three main types of seismic wave, each travelling at different speeds:

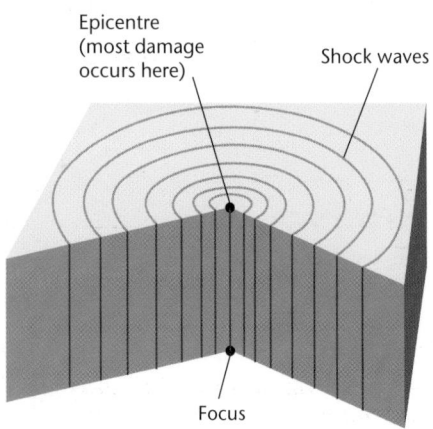

Epicentre
(most damage
occurs here)

Shock waves

Focus

- primary (P) waves travel fastest and are compressional, vibrating in the direction that they are travelling
- secondary (S) waves travel at half the speed of P waves and shear rock by vibrating at right angles to the direction of travel
- surface waves travel slowest and near to the ground surface. Some surface waves shake the ground at right angles to the direction of wave movement and some have a rolling motion that produces vertical ground movement

P and S waves travel through the interior of the Earth and are recorded on a seismograph. Studying

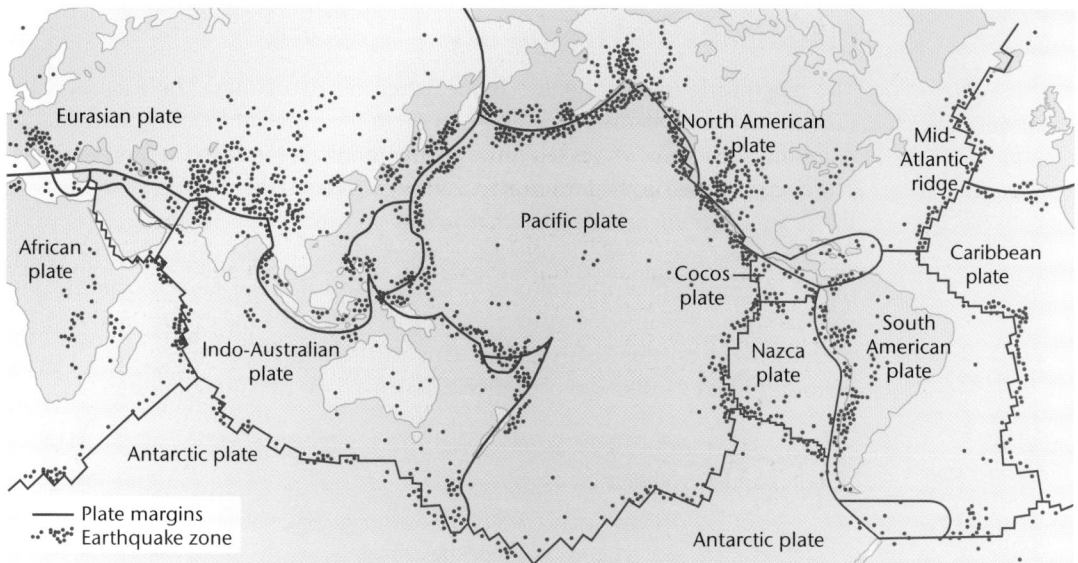

earthquakes and the seismic waves they generate has made it possible to build up a picture of the interior of the Earth.

Figure 10.5
Global distribution
of earthquakes

Distribution

The vast majority of earthquakes occur along plate boundaries (Figure 10.5), the most powerful being associated with destructive margins. At conservative margins, the boundary is marked by a fault, movement along which produces the earthquake. Perhaps the most famous of these is the San Andreas fault in California which represents the boundary between the North American and Pacific plates. In reality, the San Andreas system consists of a broad complex zone in which there are a number of fractures of the crust (Figure 10.6).

Some earthquakes occur away from plate boundaries and are associated with the reactivation of old fault lines. An example is the event that occurred on 23 September 2002 in the UK Midlands. This earthquake measured 4.8 on the Richter scale, and the epicentre was located in Dudley, west of Birmingham. It is believed that the cause was movement along an old fault line known as the Malvern lineament.

It has been suggested that human activity could be the cause of some minor earthquakes. Examples are the building of large reservoirs in which the water puts pressure on the surface rocks, or subsidence of deep mine workings.

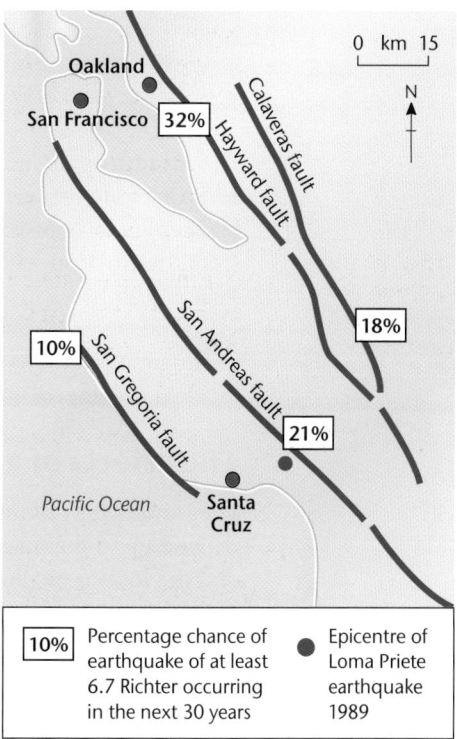

Figure 10.6 Earthquake probability in the San Francisco area

243

Magnitude and frequency

The magnitude of earthquakes is measured on two scales. The **Richter** scale (Table 10.1) is a logarithmic scale — an event measured at 7 on the scale has an amplitude of seismic waves ten times greater than one measured at 6 on the scale. The energy release is proportional to the magnitude, so that for each unit increase in the scale, the energy released increases by approximately 30 times.

The largest event ever recorded was measured at 8.9 on the scale. The earthquake in Dudley described above, at 4.8 on the scale, was large for the UK but small compared to major earthquakes such as the 1999 Turkish earthquake which measured 7.4 on the Richter scale. This earthquake killed more than 14,000 people, injured 25,000 and completely destroyed over 20,000 buildings.

Table 10.1
The Richter scale

Number (logarithmic)	Effects
1–3	Normally only detected by seismographs, not felt
4	Faint tremor causing little damage
5	Widely felt, some structural damage near epicentre
6	Distinct shaking, less well-constructed buildings collapse
7	Major earthquake causing serious damage (e.g. Kobe 1995, Turkey 1999)
8	Great earthquake causing massive destruction and loss of life (e.g. Mexico City 1985, San Francisco 1906)
9–10	Very rare great earthquake causing major damage over a large region. Ground seen to shake

The **Mercalli** scale measures the intensity of the event and its impact. It is a 12-point scale that runs from Level I (detected by seismometers but felt by very few people — approximately equivalent to 2 on the Richter scale) to Level XII (total destruction with the ground seen to shake — approximately 8.5 on the Richter scale).

Seismic records enable earthquake frequency to be observed, but these records only date back to 1848 when an instrument capable of recording seismic waves was first developed.

The effects of an earthquake

The initial effect of an earthquake is **ground shaking**. The severity of this will depend upon the magnitude of the earthquake, the distance from the epicentre and the local geological conditions. In the Mexico City earthquake of 1985, for example, the seismic waves that devastated the city were amplified several times by the ancient lake sediments upon which the city is built.

Secondary effects are as follows:

- **soil liquefaction**: soils with a high water content lose their mechanical strength when violently shaken, and start to behave like a fluid
- **landslides/avalanches**: slope failure as a result of ground shaking

Case study **Northridge, Los Angeles**

There have been five earthquakes in the Los Angeles area since 1933 that measured at least 5.8 on the Richter scale (Figure 10.7). They were:

- 1933 Long Beach, 6.4 Richter, 120 people died
- 1987 Whittier Narrows, 5.9 Richter, 8 people died
- 1971 San Fernando, 6.6 Richter, 65 people died
- 1991 Sierra Madre, 5.8 Richter, 2 people died
- 1994 Northridge, 6.7 Richter, 57 people died

The Northridge earthquake occurred at 4.30 a.m. on Monday 17 January 1994 and was the result of movement along a thrust fault, whose presence was not known to geologists. The focus of the earthquake was at a depth of 18.4 km. The low death toll has been attributed to the fact that the earthquake occurred in the early morning. If it had happened several hours later, far more people would have been away from their homes, many of them on the roads of the area.

The main effects of the earthquake were:

- 57 people killed, over 1,500 seriously injured
- 12,500 structures suffered moderate to serious damage
- 11 major roads were seriously damaged (Photograph 10.2) and had to close; roads were damaged up to 32 km from the epicentre
- over 11,000 landslides were triggered

Photograph 10.2 Damage caused by the Northridge earthquake, Los Angeles

- 20,000 people were immediately made homeless
- several days after the event 9,000 premises had no electricity, 20,000 had no gas, 48,500 had little or no water
- nearly 6,000 aftershocks were felt in the days following the event, causing damage to already weakened buildings
- the cost of the damage exceeded $30 billion
- around 700,000 applications were made to federal and state assistance programmes for financial help

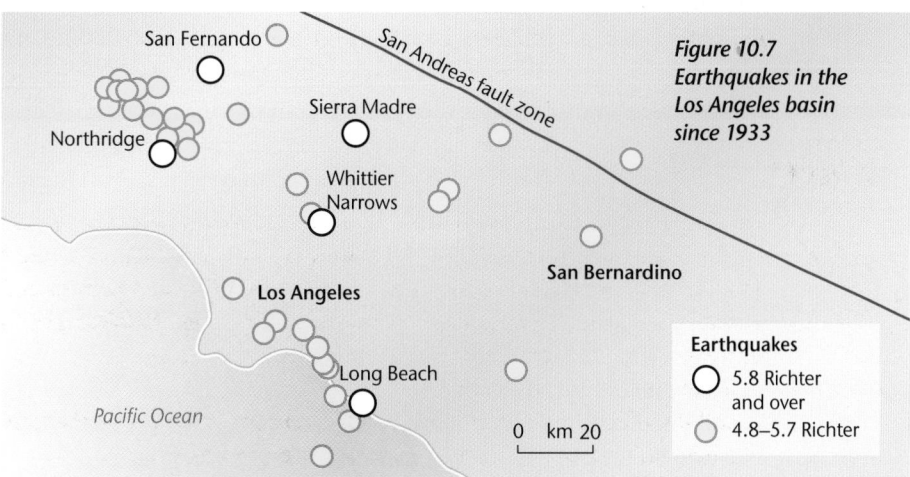

Figure 10.7 Earthquakes in the Los Angeles basin since 1933

Earthquakes
- ◯ 5.8 Richter and over
- ◦ 4.8–5.7 Richter

0 km 20

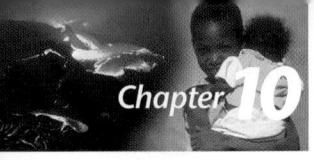

Case study *The Indian Ocean tsunami, 26 December 2004*

Pressure had been building up for some time where the Indo-Australian plate subducts beneath the Eurasian plate south of Myanmar, and on Boxing Day 2004 there was a slippage along the plate edge some 25 km beneath the Indian Ocean. A section of seabed on the Eurasian side of the fault rose several metres, generating a powerful earthquake which measured about 9.0 on the modified Richter scale. This makes it one of the biggest earthquakes ever recorded.

The epicentre of this earthquake was just off the northwestern tip of the island of Sumatra (see Figure 10.8). The earthquake triggered a tsunami that raced across the Indian Ocean, devastating islands (the Maldives, and the Andaman and Nicobar Islands) and the coastlines of the countries bordering the ocean, particularly Indonesia (Sumatra), Malaysia, Thailand, Myanmar, India and Sri Lanka. In some places the wall of water that came ashore was over 25 m in height. In the Pacific basin tsunami warning systems are in place but in the Indian Ocean no such system had been set up. The populations of these countries had no idea of what was about to arrive.

The main effects of the tsunami were:

- an estimated 300,000 people were killed by the waves
- tens of thousands of people were injured by the force of the waves and the debris that they carried
- many of these coastal areas, particularly in Thailand, Sri Lanka and the Maldives, are popular tourist destinations, so many hundreds of the dead and missing were from Europe — tourist figures were high as these areas are popular winter destinations, especially over the Christmas holiday
- whole towns and villages were swept away, particularly in northern Sumatra, the nearest land to the epicentre — it has been estimated that over 1,500 villages were destroyed in this area alone
- destruction of property resulted in millions of people being made homeless
- there was massive damage to the tourist infrastructure, particularly hotels, bars, restaurants and shops

Photograph 10.3 Devastation caused by the tsunami: a lone mosque left standing in a flattened village in Aceh, Sumatra

Topham

- there was widespread damage to coastal communications, particularly bridges and railway lines — in one instance in Sri Lanka, a train was swept off the tracks resulting in over 1,000 deaths

- damage to the economies of these coastal areas, particularly agriculture and fishing, left hundreds of thousands of people unable to feed themselves — the damage was so severe in places that coastal economies will be seriously affected for many years

- many hospitals and clinics were washed away or damaged, so a great deal of medical aid had to be brought in from outside the affected areas

- despite the enormous human cost, the insurance industry estimated that the disaster could cost it less than $5 billion

On the western side of the Indian Ocean, countries did receive a warning of what was to come and were able to take action. Kenya, for example, reacted quickly, moving thousands of tourists off beaches to safety.

One positive result of this tsunami is that a warning system is to be set up among the countries that border the Indian Ocean. This would have been of little use in northern Sumatra as the area was so close to the epicentre of the earthquake, but other countries would have benefited from some warning.

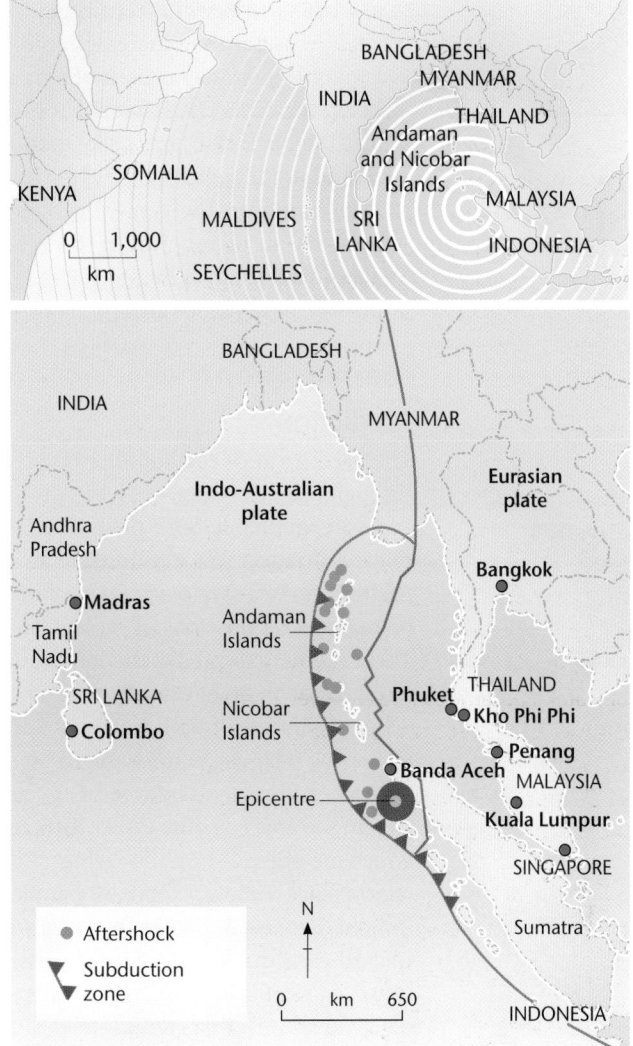

Figure 10.8 Maps showing the area affected by the tsunami

- **effects on people and the built environment**: collapsing buildings; destruction of road systems and other forms of communications; destruction of service provision such as gas, water and electricity; fires resulting from ruptured gas pipes and collapsed electricity transmission systems; flooding; disease; food shortages; disruption to the local economy. Some of the effects on the human environment are short term; others occur over a longer period and will depend to a large extent on the ability of the area to recover

- **tsunamis**: giant sea waves (tsunami means 'harbour wave' in Japanese) generated mainly by shallow-focus underwater earthquakes; they can also be caused by volcanic eruptions, underwater debris slides and large landslides into the sea. They travel quickly over the ocean (possibly in excess of 700 km h^{-1})

as shallow waves with very long wavelengths (about 100 km). When they enter shallow water they increase rapidly in height, reaching in excess of 25 m. A tsunami event may consist of a number of waves — the largest not necessarily being the first. The wave washes boats and wooden coastal structures inland and the backwash may carry them out to sea. People are drowned or injured by the event as both the water itself and the debris it contains are hazards. Depending on the coastal geography, tsunamis can have an effect at least 500–600 m inland and in some circumstances even further. Around 90% of all tsunamis are generated within the Pacific basin as result of the tectonic activity taking place around its edges

Management

Prediction

The prediction of earthquakes is very difficult. Regions at risk can be identified through plate tectonics, but attempts to predict earthquakes a few hours before the event are unreliable. Such prediction is based upon monitoring groundwater levels, release of radon gas and unusual animal behaviour. Fault lines such as the San Andreas can be monitored and local magnetic fields can be measured. Areas can also be mapped on the basis of geological information and studies of ground stability. These can help to predict the impact of earthquakes and can be used to produce a hazard zone map that can be acted upon by local and even national planners.

Close studies of fault lines can sometimes indicate the point along the fault where the next earthquake might be due. A study of the pattern of events along the San Andreas fault between 1969 and 1989 revealed the existence of a 'seismic gap' in the area of Loma Prieta. This area suffered an earthquake in October 1989 which measured 7.1 on the Richter scale and was the worst to hit the San Francisco region since 1906. In total, 63 people died and more than 3,700 were seriously injured. Because of the seismic survey, this event was not entirely unexpected, but, like all earthquakes, it was not possible to predict it precisely. Such a system, however, would not work for events such as the one at Northridge, which took place on an unknown fault line.

Prevention

Trying to prevent an earthquake is thought by most people to be impossible. This, however, has not stopped studies into the feasibility of schemes to keep the plates sliding past each other, rather than 'sticking' and then releasing, which is the main cause of earthquakes. Suggestions so far for lubricating this movement have focused on water and oil. Some people have even gone as far as to suggest nuclear explosions at depth!

Protection

Since earthquakes strike suddenly, violently and without warning, preparation cannot be put off until the event. Being prepared for an earthquake involves everyone from civil authorities to individuals. In the USA, the Federal Emergency Management Agency's earthquake program has the following objectives:

- to promote understanding of earthquakes and their effects
- to work better to identify earthquake risk
- to improve earthquake-resistant design and construction techniques
- to encourage the use of earthquake-safe policies and planning practices

Protection therefore means preparing for the event by modifying the human and built environments to decrease vulnerability. It also includes attempts to modify the loss by insurance and aid. Some of the means of protection are described below.

Hazard-resistant structures

Buildings can be designed to be aseismic or earthquake-resistant. There are three main ways of doing this:

- putting a large concrete weight on the top of a building which will move, with the aid of a computer program, in the opposite direction to the force of the earthquake to counteract stress
- building large rubber shock absorbers into the foundations to allow some movement in the building
- adding cross-bracing to the structure to hold it together when it shakes

Older buildings and structures such as elevated motorways can be **retro-fitted** with such devices to make them more earthquake-proof. A comparison between the 1989 Loma Prieta earthquake in California (7.1 Richter) and the 1988 event in Armenia (6.9 Richter) shows the effects of different types of building structures. In California, with its earthquake-proof buildings, there were only 63 deaths, whereas in Armenia more than 25,000 people died, many inside buildings that collapsed as a result of soft foundations and no earthquake-proofing features. In the town of Leninakan, for example, over 90% of the modern 9–12 storey buildings with pre-cast concrete frames were destroyed.

Education

Education is a major way of minimising loss of life in the event of an earthquake. Instructions issued by the authorities explain how to prepare for an earthquake by securing homes, appliances and heavy furniture, and getting together earthquake kits. Schools, offices and factories may have earthquake 'drills'. Government offices and many companies in Japan observe Disaster Prevention Day (1 September) which marks the anniversary of the Tokyo earthquake.

Following the Loma Prieta earthquake (1989), the American Red Cross issued a list of supplies that people should keep at hand in case of an earthquake. These included:

- water: at least 3 days' supply for all persons and pets in the house
- a whole range of foodstuffs, particularly canned and high-energy foods
- clothing and bedding
- first-aid kit
- tools and supplies, to include radio, torch, batteries, can opener, matches, toilet paper, small fire extinguisher, pliers, aluminium foil

Figure 10.9 shows the instructions issued by the metropolitan government of Tokyo advising people what to do if an earthquake occurs in the city.

Figure 10.9

Tokyo Metropolitan Government
What to do if a big earthquake hits

The worst shake is over in about a minute, so keep calm and do the following:

1 Quickly turn off all stoves and heaters. Put out fires that may break out. Do not become flustered by the sight of flames, and act quickly to put out the fire.

2 Get under a table or desk to protect yourself.

3 Do not run outdoors where you are liable to be hit by falling objects.

4 Open the door for an emergency exit. Door frames are liable to spring in a big quake and hold the door so tight it cannot be opened.

5 If you are outdoors keep away from narrow alleys, concrete block walls and embankments, and take temporary refuge in an open area.

6 During evacuation from department stores or theatres do not panic. Do as directed by the attendant in charge.

7 If driving in the street, move the car to the left and stop. Driving will be banned in restricted areas.

8 Evacuate to a designated safety evacuation area if a big fire or other danger approaches.

9 Walk to emergency evacuation areas. Take the minimum of personal belongings.

10 Do not be moved by rumours. Listen to local news over the radio.

Fire prevention

'Smart meters' have been developed which can cut off the gas if an earthquake of sufficient magnitude occurs. In Tokyo, the gas company has a network transmitting seismic information to a computer which informs employees where to switch off major pipelines, reducing the number of fires.

Emergency services

Use of the emergency services in the event of an earthquake needs careful organisation and planning. Heavy lifting gear needs to be available. Civilians must be given first-aid training as trained medical personnel can take some time to arrive. Much of the preparation in California involves the establishment of computer programs that will identify which areas the emergency services should be sent to first.

Land-use planning

The most hazardous areas in the event of an earthquake can be identified and then regulated. Certain types of buildings such as schools and hospitals should be built in areas of low risk. It is also important to have sufficient open space, as this forms a safe area away from fires and aftershock damage to buildings.

Insurance and aid

In MEDCs, people are urged to take out insurance to cover their losses. This can be very expensive for individuals. Only 7% of the people affected by the Kobe earthquake in Japan (1995) were covered by earthquake insurance.

Most aid to LEDCs has been emergency aid in the few days after the event — providing medical services, tents, water purification equipment, and search and rescue equipment. Aid over the longer term, to reconstruct the built environment and redevelop the economy, is much less readily available.

Tropical cyclones (hurricanes)

Origins and formation

Tropical cyclones are violent storms between 200 and 700 km in diameter. They represent the end product of a range of weather systems that can develop in the tropics. They begin with an area of low pressure into which warm air is drawn in a spiralling manner. Such small-scale disturbances enlarge into tropical depressions with rotating wind systems and these may grow into a much more intense and rapidly rotating system — the cyclone. It is not entirely clear why tropical storms are triggered into becoming cyclones, but there are several conditions that need to be present:

- an oceanic location with sea temperatures over 26°C — this provides a continuous source of heat to maintain rising air currents
- an ocean depth of at least 70 m — this moisture provides the latent heat, released by condensation, which drives the system
- a location at least 5° north or south of the equator in order that the Coriolis force can bring about the maximum rotation of air (the Coriolis force is weak at the equator and will stop a circular air flow from developing)

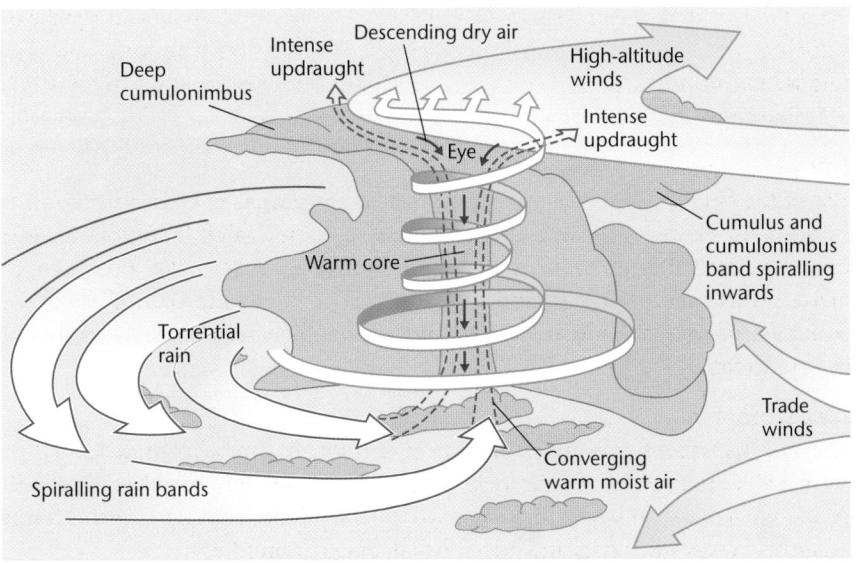

Figure 10.10
The structure of a tropical cyclone

- low-level convergence of air in the lower atmospheric circulation system
- rapid outflow of air in the upper atmospheric circulation

The tropical cyclone exists while there is a supply of latent heat and moisture to provide energy and low frictional drag on the ocean surface. Once the system reaches maturity, a central eye develops. This is an area 10–50 km in diameter in which there are calm conditions, clear skies, higher temperatures and descending air. Wind speeds of over 300 km h^{-1} have been observed around the eye. Figure 10.10 shows the structure of a typical mature tropical cyclone. Once the system reaches land or the colder waters polewards, it will decline as the source of heat and moisture is removed.

Distribution

Tropical cyclones occur between latitudes 5° and 20° north or south of the equator (Figure 10.11). Once generated they tend to move westwards and are at their most destructive:

- in the Caribbean Sea/Gulf of Mexico area where they are known as **hurricanes** (11% of all tropical cyclones)
- on the western side of central America (east Pacific) (17%)
- in the Arabian Sea/Bay of Bengal area where they are known as **cyclones** (8%)
- off southeast Asia where they are known as **typhoons** (main area, with one third of all cyclones)
- off Madagascar (southeast Africa) (11%)
- in northwestern Australia, where they are known as **willy-willies,** and the southwestern Pacific (20%)

Figure 10.11
Global distribution and seasons of tropical cyclones

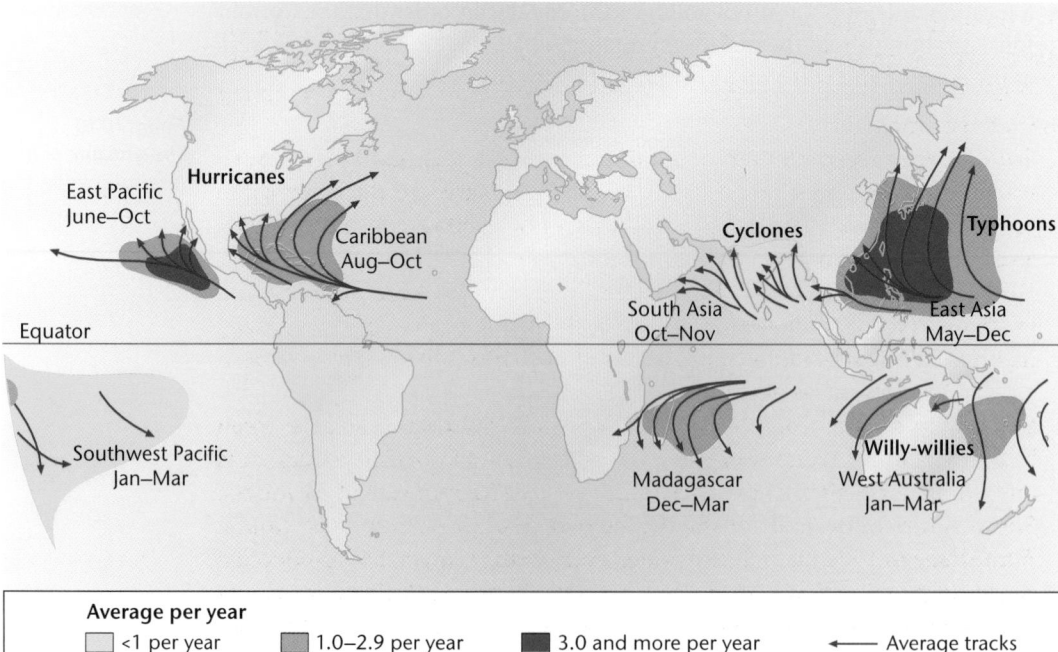

Magnitude and frequency

Tropical cyclones are measured on the Saffir-Simpson scale which has five levels based upon central pressure, wind speed, storm surge and damage potential. Scale 5, for example, has:

- central pressure at 920 mb or below
- wind speed at 69 m s^{-1} (250 km h^{-1}) or greater
- storm surge at 5.5 m or greater
- damage potential that refers to 'complete roof failure of many buildings with major damage to lower floors of all structures lower than 3 m above sea level. Evacuation of all residential buildings on low ground within 16–24 km of coast is likely'

The average lifespan of a tropical cyclone is 7–14 days. Every year about 70–75 tropical storms develop around the world, of which around 50 will intensify to become tropical cyclones.

Effects

People's vulnerability to this hazard depends upon a whole range of factors, both physical and human. The main physical factors that determine the impact of a cyclone include:

- intensity of the cyclone (Saffir-Simpson scale 1–5)
- speed of movement, i.e. length of time over area
- distance from the sea
- physical geography of the coastal area — width of coastal plain/size of delta, location of any mountain ranges relative to the coast

The human factors concern the preparations a community has made to resist the effects of a cyclone, and these are considered below in the section on management.

There are several ways in which tropical cyclones pose a hazard to people and the built environment:

- **Winds** often exceed 150 km h^{-1} and have reached over 300 km h^{-1}. Such winds can bring about the collapse of buildings, cause structural damage to roads and bridges, bring down transmission lines and devastate agricultural areas. They can also hurl large pieces of debris around and this may cause deaths.
- **Heavy rainfall**, often over 100 mm per day, causes severe **flooding** and **landslides** (very common in Hong Kong). High relief can exaggerate already high rainfall figures and totals in excess of 500–700 mm per day have been recorded in some areas of the world.
- **Storm surges** result from the piling up of wind-driven waves and the ocean heaving up under reduced pressure. These can flood low-lying coastal areas and, in flat areas such as the Ganges delta, flooding may extend far inland. Storm surges cause most of the deaths that result from tropical cyclones. A hurricane in 1970 that affected Bangladesh on the Ganges delta produced a storm surge of 6 m that killed more than 300,000 people. Agriculture in areas affected by such flooding often takes a long time to recover as the soil becomes contaminated with salt.

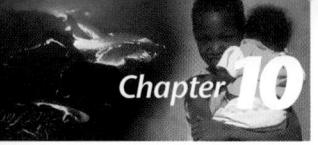
Management

Prediction

The prediction of tropical cyclones depends upon the monitoring and warning systems available. Weather bureaus such as the National Hurricane Center in Florida (USA) are able to access data from geostationary satellites and from both land and sea-based recording centres. The USA also maintains round-the-clock surveillance by weather aircraft of tropical storms that have the potential to become hurricanes and affect the Caribbean/Gulf of Mexico area. Such information is compared with computer models to predict a path for the storm and warn people to evacuate the area.

It is essential that such warnings are accurate, as there is a high economic cost associated with evacuation, and false alarms can cause people to become complacent and refuse future advice. It has been estimated that the cost of evacuating coastal areas in the USA is roughly $1 million per kilometre of coastline due to losses in business and tourism and the provision of protection. As cyclones tend to follow an erratic path, it is not always possible to give more than 12–18 hours warning. LEDCs, where communications are not as developed as those in MEDCs, may not be as well prepared and this can lead to a higher death toll. Some coastal areas of LEDCs, however, including the Bay of Bengal and some Central American coasts, do have adequate warning systems. In 1997 a tropical cyclone warning in the Cox's Bazaar area of Bangladesh allowed the evacuation of 300,000 people, resulting in fewer than 100 deaths. Some progress is therefore being made in warning systems in LEDCs.

Prevention

Like other natural hazards, tropical cyclones cannot really be prevented, but there has been research into the effect of cloud seeding to cause more precipitation before the cyclone hits land. The theory is that if the cyclone can be forced to release more water over the sea, this will result in a weakening of the system as it approaches land. There has been some concern over the effect of this on the global energy system. Because of this, research has not continued.

Protection

As with all natural hazards, protection means being prepared. Predicting the landfall of a storm enables evacuation to take place and the emergency services to be put on full alert. If evacuation does take place, then protection units, such as the National Guard in the USA, have to be called in to prevent homes and commercial properties from being looted. Local authorities in the USA are required to have evacuation and post-disaster redevelopment plans.

People in hurricane areas are given information about strengthening their homes and commercial properties against storms. Cyclone/hurricane drills similar to those carried out in earthquake areas are practised. In Florida, for example, Project Safeside is a hurricane awareness programme of precautionary drills for use in schools and emergency operation centres.

Case study *Hurricane Mitch, Central America*

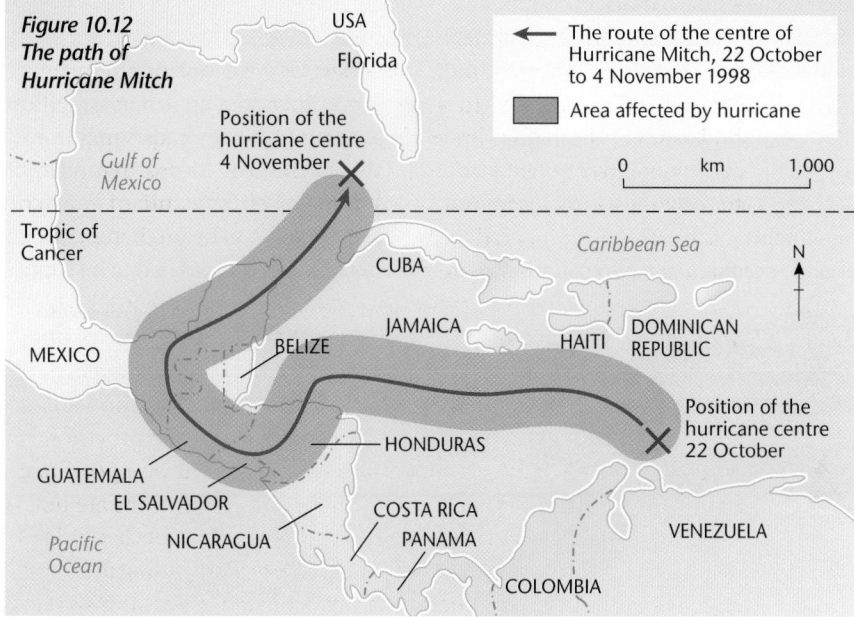

**Figure 10.12
The path of
Hurricane Mitch**

← The route of the centre of Hurricane Mitch, 22 October to 4 November 1998

Area affected by hurricane

Hurricane Mitch began as a tropical storm on around 22 October 1998 in the southern Caribbean Sea. In its early days it moved slowly westwards, on an unpredictable track, typical of many such disturbances in the region. Wind speeds built, however, and as the system deepened through the last days of the month, it became classified as a hurricane, building into one of the most severe to hit this part of the world, a category 5.

After following its predicted westward track, Mitch suddenly turned southward to hit the north Honduras coast, moving through that country and then Nicaragua before departing through El Salvador and Guatemala (Figure 10.12). What made Hurricane Mitch so devastating was not the winds, although they did reach 300 km h^{-1}, but the fact that the system was slow moving and produced torrential rain. It has been estimated that, in some places, the hurricane produced over 1,000 mm of rainfall in 5 days — as much as the region normally receives in a year. This led to severe flooding, as rivers burst their banks. The main effects of the hurricane were:

- more than 19,000 people dead or missing
- many settlements were completely wiped out, particularly on the northern coast of Honduras
- transport links were obliterated — in Nicaragua, 50 main bridges were destroyed, along with countless minor ones, including all those on the roads leading to the capital, Managua
- 2.7 million people were made homeless — in Honduras it is estimated that 20% of the population lost their homes (Photograph 10.4)
- the centre of the capital city of Honduras, Tegucigalpa, was turned into a vast lake as the Rio Choluteca overflowed
- in Nicaragua the crater lake in the dormant volcano Casita burst through the crater walls, sending a 6-m high wall of mud and water onto the villages that surrounded it and killing 1,500 people
- land worked by many farmers was covered by a deep layer of mud which baked hard after the hurricane had passed
- food supplies were destroyed, leading to serious shortages — food prices increased rapidly
- 70% of the Honduran economic output was lost,

255

mainly in agriculture, where the coffee and banana plantations were severely affected. Even where crops could be harvested, the lack of communications meant that produce could not be marketed. Agricultural losses for 1998 and 1999 were put at $1.5 billion

- the limited export base in Nicaragua and Honduras, mainly agricultural, was severely cut, and this had implications for servicing large foreign debt commitments
- Honduras and Nicaragua, the two countries most affected, were already among the poorest in the Americas with GDPs, at the time, of under $700 per head
- crime vastly increased as abandoned homes and businesses were looted
- power transmission systems were destroyed and fuel sources quickly became depleted (immediately after the disaster, the Honduran Air Force had only 5 days' fuel left, important in a country where most of the roads had been left impassable)
- large amounts of aid had to be supplied

Photograph 10.4 Devastation in Honduras following Hurricane Mitch

Nigel Dickinson/Still Pictures

Case study *Hurricanes in Florida, 2004*

In a 6-week period in 2004, Florida was struck by four deadly hurricanes (Figure 10.13) which left trails of devastation across the state.

Hurricane Charley

Hurricane Charley passed through the Caribbean in mid-August causing some damage to Jamaica, Cuba and the Cayman Islands. It intensified as it moved over Florida, when it reached category 4 on the Saffir-Simpson scale with winds in excess of 240 km h^{-1}. The hurricane crossed central Florida, cutting a swathe of destruction across the state, resulting in 27 deaths and estimated damage of $14 billion in Florida and the states to the north. This made Charley the second costliest hurricane in US history behind Hurricane Andrew (1992).

Hurricane Frances

Two weeks later, Hurricane Frances ran in from the Atlantic Ocean as a category 4 hurricane. It inflicted widespread damage on the Turks and Caicos Islands and the Bahamas before making landfall on the Florida coast as a category 2. Crossing the peninsula, with winds exceeding 190 km h^{-1}, it gradually weakened as it moved up through the eastern states of the USA. Although not as powerful as Charley, Frances covered a huge area — it reached a size equivalent to that of the state of Texas at its peak.

It also brought vast amounts of rainfall to many areas resulting in severe flooding — several districts reported that over 300 mm of rain had fallen in a 24-hour period. At least 2.5 million people were ordered to evacuate the east coast in advance of the hurricane, which resulted in blocked roads and petrol shortages. Frances also disrupted the holiday plans of over 7,000 Britons as large numbers of flights to and from the UK were cancelled. Frances was responsible for six deaths and an estimated $8 billion of damage.

Hurricane Ivan

Just over a week after the devastation of Frances, Hurricane Ivan moved through the Caribbean causing widespread devastation and loss of life on Grenada, Jamaica and the Cayman Islands (many islanders referred to it as 'Ivan the Terrible'). At times, it was rated as category 5, the top of the scale, although by the time it reached the USA it had been downgraded to a category 3. The strip of coastline between Mobile (Alabama) and Pensacola (Florida) took the brunt of Ivan's winds (which reached over 240 km h^{-1}) and the storm surges built up by the hurricane. Ivan was believed to be responsible for 26 deaths across the southern USA and caused an estimated $13 billion worth of damage — the third costliest hurricane in US history after Andrew and Charley.

Hurricane Jeanne

A week after Ivan devastated parts of Florida, another tropical storm was spawned in the southern Caribbean. It moved northwards across the Leeward and Virgin Islands, Puerto Rico and Hispaniola (Haiti and the Dominican Republic) where it caused devastating floods — at least 3,000 deaths are believed to have occurred in Haiti from inland flooding. As it moved northwards, it gradually intensified. It was a category 2 hurricane when it crossed the Bahamas, but it made landfall on Florida as a category 3, in the same area affected by hurricane Frances a few weeks earlier. Two people were reported dead as a result of the storm and it is estimated that damage in the USA totalled $6.5 billion.

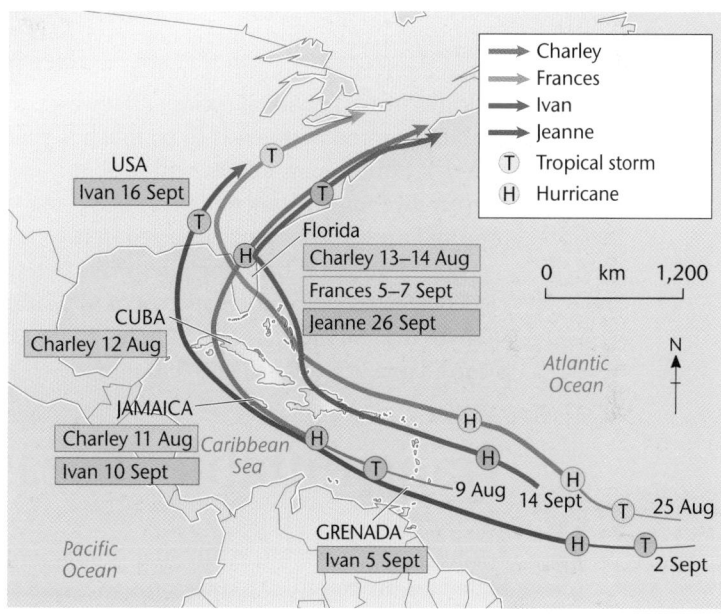

Figure 10.13 Hurricane paths, August–September 2004

Where there is a threat of storm surges, land-use planning can identify the areas at greatest risk and certain kinds of developments can be limited. In the USA, local authorities are required to limit expenditure on developments in high-risk areas and direct population away from them. In LEDCs, however, the need for land usually outweighs such considerations. In addition, sea walls, breakwaters and flood barriers can be erected in high-risk areas and houses can be built on stilts. The sea wall that was built to protect Galveston (Texas) from further flooding after the storm surge of 1900 that killed over 6,000 people was so expensive that it is unlikely to be repeated elsewhere.

In areas of LEDCs prone to tropical cyclones, concrete storm shelters have been built. Such countries are seeking international aid to build more shelters. Bangladesh, for example, has fewer than 500, although it is estimated that more than 10,000 are needed.

As with other natural phenomena, people in MEDCs are urged to take out insurance, whereas in LEDCs it is important that aid is available after a cyclone, in both the short and the long term. Damage to the economic base of the area is likely to last for a number of years.

The economics of disaster management

Human ability to resist natural hazards such as cyclones depends upon a range of political and economic factors. LEDCs suffer more because land-use planning, warning systems, defences, infrastructure and emergency services all cost money and they therefore have fewer of them. The death toll of such events in LEDCs therefore tends to be much higher.

MEDCs have planning systems, sophisticated warning systems, better defences and infrastructure, and emergency services that are much more comprehensive and well prepared. Countries such as the USA thus suffer much smaller numbers of deaths from individual cyclones. However, the financial cost of the damage to the built environment tends to be much higher in MEDCs. Having said this, the loss of a house in an MEDC will probably be covered by insurance, while the simpler dwelling of someone in an LEDC may well be irreplaceable — given the time and uninsured savings invested in it.

The cyclone that hit Bangladesh in 1991 was responsible for an estimated 131,000 deaths and a monetary cost of $1.7 billion whereas a hurricane of similar strength that hit Florida in 1992 (Hurricane Andrew) killed only 60 people but caused damage estimated at $20 billion.

Transmittable diseases

A transmittable disease is one that can be passed from one person to another. Malaria is an example of such a disease; cancer is not. Insects and animals that carry a disease are known as vectors and transmission can occur in a number of ways, including:

- inhalation of infected droplets
- ingestion of contaminated food or water
- skin penetration by insect bite or contaminated needles
- exchange of body fluids

When the number of people with a disease is significantly measurable at a national or regional level it is said to be an **epidemic**. If the disease is a problem at an international level then it is said to be **pandemic**.

Geographers are concerned with transmittable diseases because they do not spread evenly throughout the world and the resources to cope with the management of them vary between areas. Understanding the geography of a disease helps make sense of the social, cultural and economic impact it is likely to have. Some of the major transmittable diseases are:

- **trypanosomiasis (sleeping sickness)**, caused by a single-celled parasite spread by the tsetse fly
- **malaria,** caused by single-celled organisms spread by the anopheles mosquito
- **Ebola,** caused by a virus and caught by eating infected monkey meat or through contact with blood, faeces or body fluids of infected people
- **cholera**, a bacterial water-borne disease which spreads quickly because of its rapid incubation rate
- **tuberculosis (TB),** caused by a bacterium and spread through droplet infection or by drinking untreated milk from infected animals

Acquired immunodeficiency syndrome (AIDS)

HIV is the human immunodeficiency virus which causes the illness AIDS. HIV is a slow retrovirus, which means that it takes years to show symptoms and that it invades the white cells in the blood by literally writing the structure of itself backwards (retro) into them and reproducing itself inside. White cells produce the antibodies that are the body's main defence against disease and without them the body becomes the target of everyday infections and cell changes that cause cancers. This is what happens to someone suffering from AIDS.

There is some controversy over the source of the disease. Some people believe that it is man-made and was produced by a chemical weapons laboratory or by medical research gone wrong. However, the generally accepted view is that it evolved in sub-Saharan Africa, crossing over from the chimpanzee population in contaminated meat or by a bite from a pet, possibly in the 1930s. Some scientists have suggested that the virus was present in the human population of central Africa for a long time, but on a very local scale. Developments in the twentieth century such as international travel, blood transfusions and intravenous drug use brought it onto the world stage and produced a growing pandemic.

The spread of HIV

Today, the virus is spread in the following ways:

- exchange of body fluids during sexual intercourse
- contaminated needles in intravenous drug use (IDU)
- contaminated blood transfusions
- from mother to child during pregnancy

Evidence shows that the disease started in small high-risk groups such as gay men, drug users and prostitutes, and then spread into the population as a whole. This meant that in the early stages of the disease in MEDCs, AIDS was regarded by many as a 'gay plague'. The heterosexual community took little notice of it and education about safe sex was disregarded. In LEDCs, mainly in Africa, the transfer of the virus was more commonly through heterosexual sex, but this information was not known in most of the developed world. Three distinct patterns of distribution therefore developed:

- **Pattern 1** covers countries which began to see a spread of HIV in the late 1970s, first among the homosexual, bisexual and drug-using communities and later in the general population. This includes North America, western Europe, Australia and some parts of Latin America.
- **Pattern 2** includes those countries where the spread has been essentially through heterosexual contact. This covers the bulk of sub-Saharan Africa.
- **Pattern 3** covers those countries where the disease appeared later (in the 1980s) and was brought in by travellers and sometimes by blood imported for transfusions. This includes eastern Europe (including the former USSR), Asia, the Middle East and north Africa.

By 2002, it was estimated that over 40 million people worldwide were living with HIV or full-blown AIDS. Over 70% of these (28.5 million) were in the countries of sub-Saharan Africa. North America had 900,000 people with the disease and western Europe 520,000 (1.5% of the world total). It has been estimated that 8.6% of the adult population of sub-Saharan Africa is HIV positive. Nine countries have 10% or more of their adult population infected: Botswana, Kenya, Malawi, Mozambique, Namibia, Rwanda, South Africa, Zambia and Zimbabwe. The United Nations has estimated that, by 2020, 70 million people will have died from AIDS.

Evidence for frequency and scale

Figures can be obtained from medical records (from doctors and hospitals), national government health department records, the World Health Organization (WHO) and the media. These statistics, however, can never be totally accurate for the following reasons:

- medical records are confidential
- many people with HIV are not aware that they are infected
- the social stigma of AIDS means that many sufferers do not report the illness until it is well into its later stages
- AIDS is not always given as the cause of death because the sufferer usually dies of another disease, such as pneumonia, that they have succumbed to because of the effects of HIV on their immune system
- it has been alleged that the disease has been overestimated in parts of Africa so countries can obtain more overseas aid

Effects

The United Nations has estimated that only one in ten sufferers know they have the virus in the early days of infection. Apart from the physical effects of the disease, sufferers may experience **prejudice** in their employment and social life. This may even extend to the immediate family: in some societies children are ostracised if one of their parents has the disease.

Life expectancy in much of sub-Saharan Africa will soon fall to levels not seen since the nineteenth century (Table 10.2). Some authorities estimate that by 2010

*Table 10.2
Estimated life
expectancy in
selected African
countries (age in
years), 2010*

	Without AIDS	With AIDS
Angola	41.3	35
Botswana	74.4	26.7
Lesotho	67.2	36.5
Malawi	59.4	36.9
Mozambique	42.5	27.1
Namibia	68.8	33.8
Rwanda	54.7	38.7
South Africa	68.5	36.5
Swaziland	74.6	33
Zambia	58.6	34.4
Zimbabwe	71.4	34.6

Source: US Census Bureau.

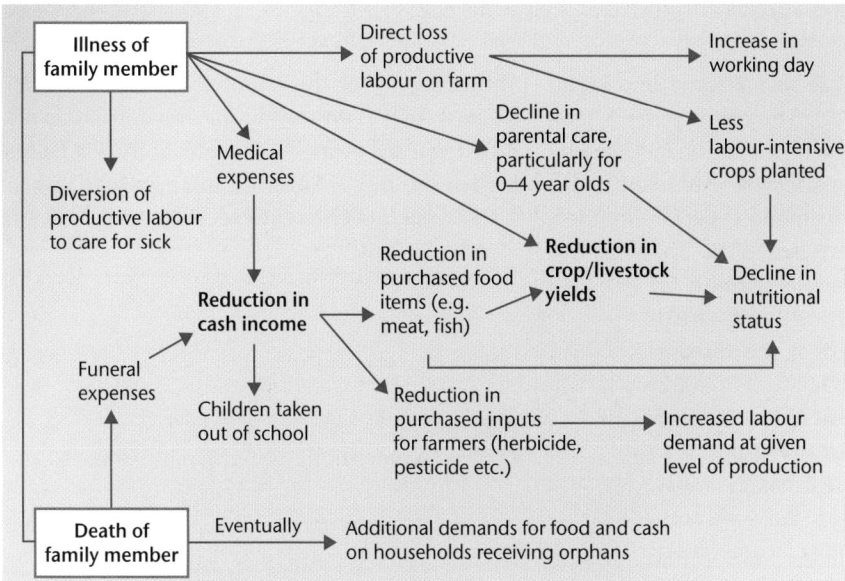

Figure 10.14
Effects of AIDS on families in Africa

people in many southern African countries will not be expected to live beyond their 30s. Populations of some of these countries will have started to shrink. AIDS has also reversed the decline in infant mortality that was seen across southern Africa in the 1980s and early 1990s. Rates in certain countries are now close to double what they would have been without the pandemic.

In Africa, the effects on families have included loss of income-earning opportunities, the diversion of effort and income into care and medicine, and the withdrawal of children from school, either because of lack of money or because they are needed for agricultural work. There has been a huge effect on farming, with AIDS threatening food-growing and income-earning potential in areas already facing food shortages. Some countries have large numbers of orphans as a result of AIDS, and this has put a great strain on local resources (Figure 10.14).

Management

In the case of AIDS, management takes the following forms:

- **Trying to find a vaccine**: the hope of this seems remote, but research is continuing, particularly trying to find groups that might possess some degree of natural immunity through their white cells.
- **Prolonging life through drugs**, particularly in MEDCs. Such drugs are available but expensive — a typical course of AZT costs $10,000 a year per individual. Even Nevirapine, a slightly cheaper retroviral drug, is beyond the reach of the governments of most LEDCs. In 2001, however, a court case against the South African government by a group of multinational pharmaceutical companies was dropped because of massive public pressure on the companies. This means that the South African government can now manufacture and import cheap generic versions of anti-retroviral drugs instead of having to buy the expensive brand-name products. Similar cases have occurred in other LEDCs.

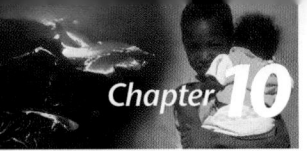

Case study **Botswana**

The AIDS pandemic has had a huge impact on Botswana (Figure 10.15). The country has a total population of 1.6 million and it has been estimated that 39% of adults are infected with HIV. In the north-east of the country and among expectant mothers in urban areas rates are over 50%. The virus has run rampant despite various attempts to control it. The government has tried to manage the spread of the disease by focusing on prevention, but in 2002 Botswana became the first African country to provide free anti-retroviral drugs. It is able to do this because it has the most lucrative diamond mines in the world and, as a result, has a per capita income seven times the average for sub-Saharan Africa.

Life expectancy in the country has dipped to below 40 years of age for the first time since 1950 and in 2002 stood at just under 34 years. It would have been expected to rise to 74 years and 5 months by 2010 if there had been no AIDS pandemic, but the current projected figure for that year is only 26 years and 8 months.

The economy of the country has been affected because AIDS is destroying the workforce. It is predicted that the economy of the country will be one third smaller by 2021 than it would have been without AIDS, while government expenditure will have to increase by 20%.

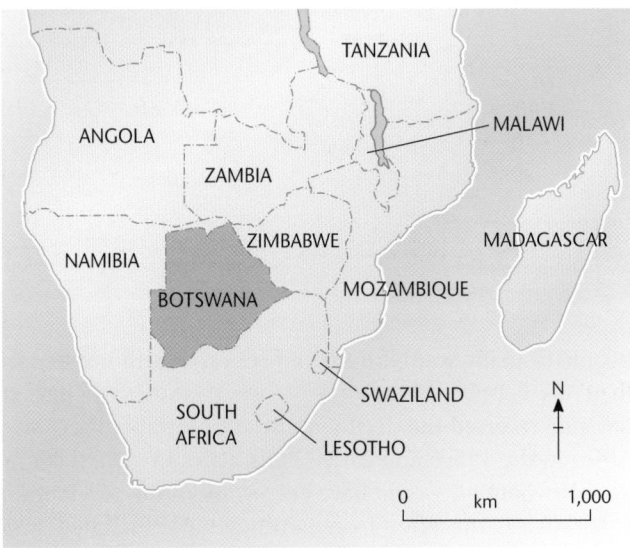

Figure 10.15 The location of Botswana

- **Plotting the course** of an outbreak, making it possible to predict the future spread of the disease and to identify areas where resources should be concentrated.
- **Screening of blood** for HIV antibodies before it is used for transfusions in MEDCs, leading to a negligible risk of infection, although this was not always the case. Blood plasma products, such as factor eight (for haemophiliacs), are also subject to treatment.
- **Education and advertising** aimed at vulnerable groups such as homosexuals and intravenous drug users to try and prevent the spread of the disease. Raising the profile of the disease in schools through sex education has been a major feature in the UK government programme. Other campaigns in the UK have included free needles for drug users, free condoms, and warnings to travellers about their sexual behaviour in foreign countries.
- **Caring for victims and families**, which involves many agencies including charities. In the UK these include the Terrence Higgins Trust and London Lighthouse.

Burglary in urban areas

Burglary is a crime involving the removal of goods from property, usually dwellings. The person carrying out the crime commits trespass in order to carry it out.

Distribution and frequency

There is no doubt that the incidence of burglary tends to be higher in inner-city areas than in the outer suburbs. Insurance company premiums reflect this and are highest in urban areas with low-numbered postcodes — the most central areas. In the West Midlands, the 1999 figures for reported burglary offences were:

- 12.4 per 1,000 households for the Halesowen/Stourbridge/Lye/Cradley/ Kingswinford division (outer suburbs)
- 36.9 per 1,000 households for the Birmingham Central division

On the urban fringe, large, detached houses that are often screened from view offer opportunities for burglars, who can benefit from the less well developed social network in such areas. Younger, less experienced criminals are more likely to offend in the inner city because of familiarity and opportunism near their home territory. Information about potential targets is readily available in the immediate vicinity of their homes but decreases rapidly with distance — there is a strong distance-decay effect.

Information about the distribution and frequency of burglary is often incomplete. Several sources can be used:

- **Police records**, but not all crimes are reported, for various reasons: fear of retaliation, loss too small to bother about, goods uninsured, unwillingness to lose no-claims bonus or receive higher premium after claim, desire to sort it out without recourse to authority (burglar known to victim), lack of confidence in police to solve the crime.
- **Insurance company records**, but not all people are insured. Premiums are often highest and therefore unaffordable in inner-city areas, and some addresses may find it difficult to obtain insurance.
- **British Crime Survey**, conducted by the Home Office, which measures crime levels in England and Wales by asking members of the public about the crime they have experienced. Critics maintain it underestimates crime levels, but it does allow those at greatest risk to be identified. It is used in planning crime-prevention programmes.
- **The media,** but only certain crimes are reported, distorting the overall picture.

In terms of frequency, several studies have suggested that the incidence of burglary tends to rise at holiday times (particularly summer) and Christmas.

The nature of the hazard (effects)

The hazard of burglary can take several forms:

- removal of goods from property

- damage to property caused when trying to gain entry
- invasion of privacy
- psychological damage (fear of crime)

This last effect may become so severe that people barricade themselves in their property or move away from the area. People have died in house fires because their homes were so tightly secured that the fire services could not reach them and they could not escape. In Dover (Kent) in May 2000, a case was reported in which two elderly people were overcome by fumes when trapped by a fire in their bungalow. The police reported that they had three locks on each door and another three locks on each internal door.

Figure 10.16
Map of Newcastle-upon-Tyne showing reported fear of crime in 26 wards

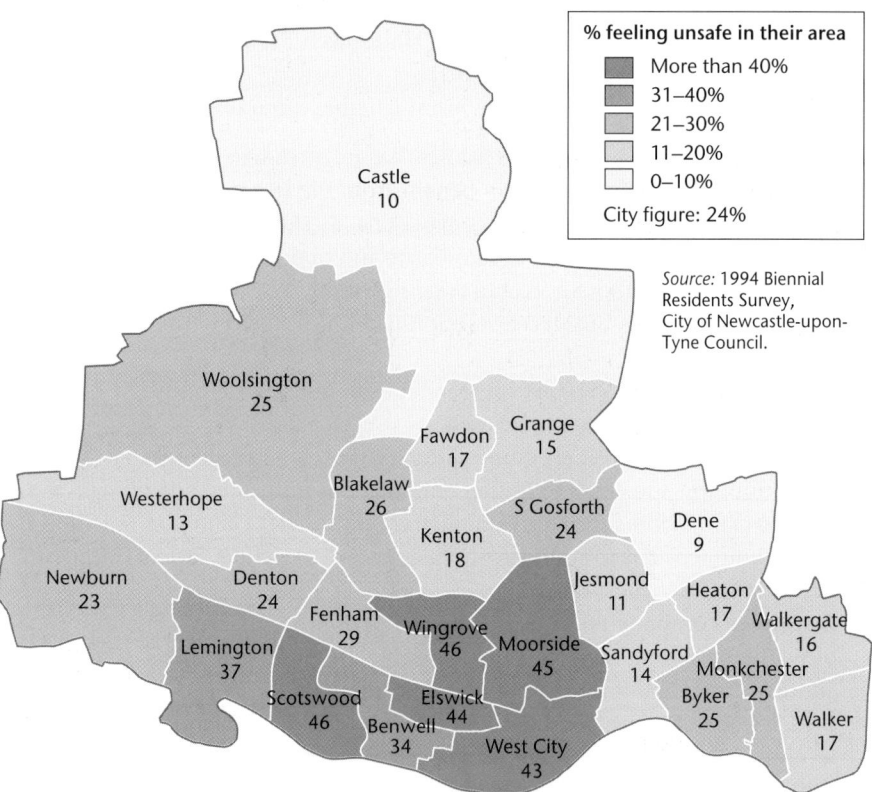

% feeling unsafe in their area
- More than 40%
- 31–40%
- 21–30%
- 11–20%
- 0–10%

City figure: 24%

Source: 1994 Biennial Residents Survey, City of Newcastle-upon-Tyne Council.

Castle 10
Woolsington 25
Fawdon 17
Grange 15
Blakelaw 26
Westerhope 13
Kenton 18
S Gosforth 24
Dene 9
Newburn 23
Denton 24
Fenham 29
Wingrove 46
Jesmond 11
Heaton 17
Walkergate 16
Lemington 37
Moorside 45
Sandyford 14
Monkchester
Scotswood 46
Elswick 44
Byker 25
Walker 17
Benwell 34
West City 43

Figure 10.16 shows fear of crime as expressed by residents of Newcastle-upon-Tyne. As expected, more residents feel unsafe in poorer inner-city areas such as Scotswood, Benwell and Elswick than in the more affluent areas of Jesmond and South Gosforth.

Management

Insurance companies '**predict**' crime in an area when they set insurance premiums.

It is difficult to distinguish between **prevention** and **protection** when dealing with burglary. Authorities often refer to 'crime prevention' in publicity and advice to householders about protecting their property. At its roots, though, prevention is not about increased home security, but entails looking at and trying to tackle the reasons *why* people commit crime. In the UK, for example, a large amount of burglary is drug related, and authorities have therefore promoted schemes to get people off drugs. In the long term, this should lead to a reduction in crime. Unemployment levels have also been connected with crime, and governments have an eye to crime figures when trying to increase employment. Many youth policies aimed at getting younger people off the streets and into work or leisure programmes have crime reduction as one objective.

People can either work individually or collectively to protect their property. **Individually**, they can protect their homes with a range of devices (alarms, walls, fences, window locks, security lighting, guard dogs) and by taking out adequate insurance to cover their losses. **Collectively**, they can organise a local Neighbourhood Watch scheme. In some areas, worried residents have formed themselves into groups that patrol residential streets at night. Such action, though, has been known to lead to violence and is not sanctioned by the police.

At an **organisational** level, various initiatives have been set up by the police and local and national government. These have included:

- community policing
- placing of CCTV cameras
- larger penalties imposed by the courts
- free or low-cost security devices
- television and newspaper campaigns
- defensible space, in which new housing developments are designed to deter burglars by offering no easy entry or quick exit. Such designs feature in many new inner-city developments

The danger with all these initiatives is that they may simply displace crime to another area. Burglary may be reduced in the particular area where schemes are in place, but the overall level of burglary remains the same.

Organisations such as Victim Support offer help to those affected by crime. This includes support (counselling) and practical advice, such as how to deal with insurance companies, emergency repairs, loss of official documents and how to obtain compensation.

Approaches to crime prevention

A variety of approaches to crime prevention has been tried in the UK. Some schemes are detailed below:

- **Hull**: putting more police on the beat. In 1999, the Humberside police were denied a Home Office grant to put more police officers on the beat. The city council agreed to put up £1 million to fund an extra 174 officers, allowing more highly visible foot patrols in inner-city areas. In 2000, it was estimated that crime in central Hull was down by 14%, with notable drops in burglaries and stolen cars.
- **Mozart Estate, Westminster**: removal of overhead walkways and other design changes on the estate brought a slight reduction in burglaries. Reports on the

situation commented on the need for economic and social regeneration before any major inroads into the problem could be made.

- **Kingsmead Estate, Hackney**: injunctions against specific individuals and repossession orders against persistent offenders, followed up by a programme of activities for young people. Crime rates on the estate dropped sharply. Burglaries were down from 340 in 1992 to 50 in 1993.
- **Possil Park Estate, Glasgow**: community security business to protect empty properties from vandalism and theft. The city council calculated that the scheme had saved £250,000 in reduced costs of vandalism and theft in 1 year.
- **Southmead, Bristol**: activity schemes for younger people, particularly during the school summer holiday. There was an estimated 64% reduction in domestic burglary during the period of the scheme.

Figure 10.17
Leicester City
Council home
security advice

How do I make my home more secure?

The threat of becoming a victim of crime, especially domestic burglary, is a worry to many people.

Recognising that there are proven steps that can be taken to reduce the risk of burglary, the city council — in partnership with Leicester Constabulary — operates a number of burglary reduction schemes across Leicester. The schemes aim to reduce domestic burglary, improve community safety and address the growing fear of crime. They offer practical advice and home security improvements to households that have been victims of burglary or that are considered vulnerable to crime.

Each community safety scheme operates in its own qualifying area and offers a service that is free of charge for people meeting eligibility criteria. The schemes are funded through the government's Single Regeneration Budget and Home Office money specifically earmarked for burglary reduction initiatives.

All city council run community safety schemes are managed by the housing department's renewal and grants team. Security surveys and the fitting of new security equipment are carried out by the team's area maintenance officers. Security improvements available include the fitting of five-lever mortice locks to front and back doors, along with window locks, door chains, spyholes and kick plates.

The following community safety schemes are running in the city:

- Belgrave Burglary Reduction Project
- CRASH project in Greater Humberstone
- Hinckley Road Burglary Reduction Initiative
- Banish Project in Highfields and St Matthews
- Braunstone Project
- Northwest Leicester Burglary Reduction Project for Beaumont Leys, Mowmacre, Abbey Rise, Stocking Farm and Anstey Heights
- Asfordby Street Project for Leicester University students living in Clarendon Park and Highfields

City residents who do not qualify or are not in project areas can make use of the Belgrave Community Safety Centre Shop, a non-profit-making service offering a range of home security equipment at low prices.

■ **Leicester**: local authorities operate schemes that give advice and practical help to households concerning burglary and other crimes. Figure 10.17 shows the general home security advice given to households in the city, together with a list of the projects they can apply to for help. The city also operates a non-profit-making service offering home security equipment at low prices.

Urban multi-hazard environments

Many major cities in the world are affected by a range of hazards, some in the physical environment and others which are human in origin. **London** has the potential to be affected by storm surges, river flooding, air pollution (smog), groundwater flooding and crime (particularly burglary). **Mexico City** has earthquakes, pollution (smog), contaminated groundwater, water shortages and crime.

Hazards in Los Angeles

Los Angeles in California is affected by a range of hazards, some directly concerned with the physical environment, some involving human modifications to the physical envrironment and some entirely within the human environment. The major potential hazards are:
■ earthquakes and tsunamis
■ heavy rainfall and river flooding
■ coastal flooding
■ drought
■ fire
■ mudslides and landslides
■ smog
■ crime (including burglary)
■ gang warfare and racial violence
■ HIV/AIDS

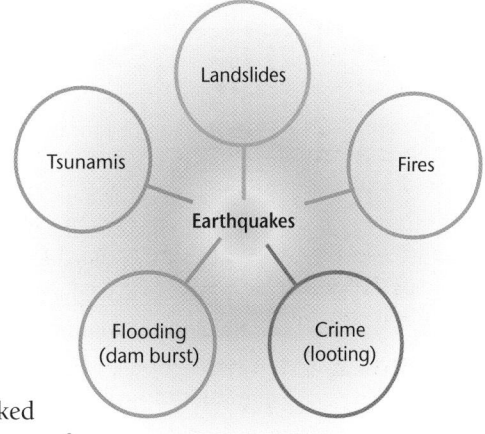

Interrelationships between hazards

Some of the hazards in an urban environment are linked and one may contribute to another. There are various forms of relationship:

Figure 10.18
Hazards associated
with earthquakes

■ **A simple direct causal relationship**: one hazard is responsible for another, for example earthquakes lead to landslides and tsunamis, drought leads to fires, earthquakes lead to crime (looting etc.), heavy rainfall leads to mudslides.
■ **One hazard responsible for several others**: earthquakes lead to many other hazards (Figure 10.18).
■ **More complex linear relationships**: drought leads to fires which lead to flooding following heavy rain. Fires remove vegetation, thereby decreasing interception, leading to more runoff and a greater flood potential.

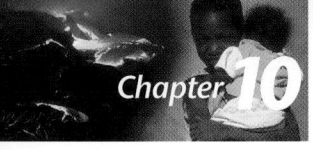

■ **Through a central factor**: there has been massive urban sprawl in Los Angeles over the last 40–50 years. This may have been a factor in several hazards:

(1) **River flooding:** as the city has sprawled, more and more land has gone under concrete and tarmac, particularly in the surrounding hills. This means less infiltration and more rapid runoff when it rains.

(2) **Fires**: the low-density urban sprawl has left vast areas of shrubland and some woodland between housing developments. There is a risk in summer of widespread fires when the vegetation becomes very dry, particularly if strong winds are blowing.

(3) **Smog**: Los Angeles covers a vast area and did not have a comprehensive public transport system until recently, so private cars are used a lot. Car exhaust fumes have been the major factor in the decline of air quality in the Los Angeles basin.

(4) **Crime/social division**: as the city has sprawled, affluent white-dominated suburbs (such as Beverly Hills and Bel Air) and impoverished Afro-American and Latin American inner areas (such as Watts) have developed. The ethnic division in the city was demonstrated in 1991 by the extensive riots and looting that followed the acquittal of white police accused of beating a black suspect (Rodney King).

A relationship in detail: drought/fire/heavy rain/flooding

Drought

Los Angeles has a Mediterranean climate which means that it experiences long, dry summers. Most of the watercourses dry up during this period.

Fire

Much of the Los Angeles basin is covered by drought-resistant chaparral vegetation. As it is too dry for tree growth, scrub dominates. At the end of the summer drought (that can last for up to 6 months), this vegetation is tinder dry. A feature of this area is the Santa Ana, a hot dry wind that descends from the mountains. This tends to increase the dryness in the vegetation to the point where a spark, a lightning strike or a carelessly discarded cigarette can start a major fire. The Santa Ana wind tends to spread the fire and make it extremely difficult to control. In late October and early November 1993, for example, a firestorm swept the area around Malibu, destroying over 1,000 homes. The event gained international coverage as the homes of several famous film and television personalities were affected.

Heavy rain

The Mediterranean climate also produces winter storms. The rain and strong winds can be particularly severe during an El Niño event. Storms that produce 50–100 mm of rain in a day are fairly common.

River flooding

Although most rivers in the Los Angeles basin are short and seasonal, they can transport huge amounts of water during times of flood. Deforestation and brush fires on steep hillsides have decreased the interception rate and this, together with rapid urbanisation, has increased surface runoff. In February 1992, for example, after 2 days of torrential rain, floods devastated an area south of Malibu, killing eight people.

Heavy rain can also trigger mudslides. In early 1994, mudslides devastated the Malibu region after over 30 mm of rain fell in 24 hours. The fires of October/ November 1993 mentioned above had removed large areas of trees and scrub in this area, leaving little vegetation to intercept the rain. Officials in Malibu had anticipated that mudslides would occur because of the bare hillsides. Sandbags had been stacked to protect multi-million dollar beachfront homes, and straw bales were used as makeshift dams.

Responses to the flood hazard

The Los Angeles basin is drained by two main river systems, the Los Angeles and the San Gabriel Rivers (Figure 10.19). Although the annual rainfall in the basin is often less than 400 mm, amounts in the surrounding upland areas, such as the San Gabriel Mountains, are usually over 1,000 mm. The first major flood on the Los Angeles River was recorded in 1815, and during the rest of the nineteenth century there were numerous floods, particularly in 1825 and 1861.

The 1914 flood caused $10 million worth of damage and brought a public out-cry for action to address the problem. Some minor attempts at channelisation were attempted, but after the destructive floods of the 1930s, money was made available from federal sources and the US Army Corps of Engineers took a major role in trying to bring the river under control. They carried out the following schemes:

- **channelising** the entire river (apart from three sections) and its main tributary, the Rio Hondo. This involved both straightening the rivers and paving the beds and banks to create completely man-made channels

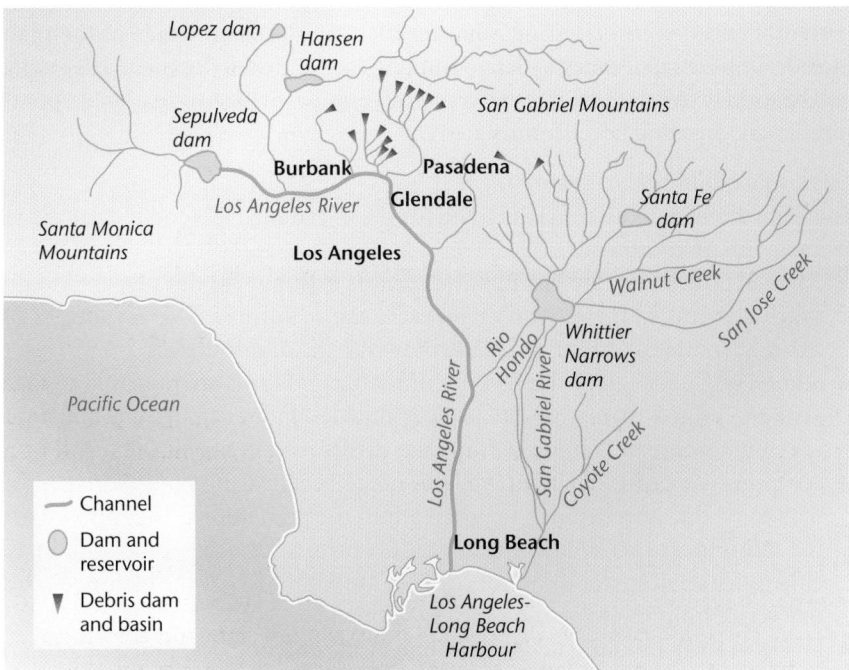

Figure 10.19
The Los Angeles and
San Gabriel River
systems

- building **small dams** on the tributaries in the mountains to contain the water near the source
- constructing **debris dams** near the sources to prevent huge amounts of river load from blocking bridges and other features downstream and causing more severe flooding
- constructing **larger dams with reservoirs**, such as the Sepulveda and Hansen dams
- developing an area that can take huge amounts of water at time of flood. A huge retaining dam was built across the valleys of the Rio Hondo and the San Gabriel River at Whittier Narrows, where they come close together. This construction is only used during the largest floods — normally the two rivers flow through it
- designating large areas as **spreading grounds** for surplus water at time of flood

In both 1992 and 1994, however, the river flooded again. As the flood waters at one time were within a few centimetres of the top of the channel wall, the Corps of Engineers suggested that they should make the wall higher by at least 0.7 m on the whole length and over 2 m in some places. The Corps insists that any scheme should be designed to get water to the ocean as quickly as possible. It points out that the price of land in LA County is so high that even a minor breach of the present channel would cause millions of dollars of damage to nearby residential and commercial property.

Many consider this policy to be a mistake, treating the symptom (channel capacity) rather than the cause (urban runoff). The Los Angeles County Departments of Public Works, Parks and Recreation, and Regional Planning are currently engaged in preparing a master plan for the river. Goals of the plan include: improving river aesthetics, promoting economic development, preserving and restoring environmental resources, providing a variety of recreational opportunities and meeting flood control needs.

Living in a multi-hazard environment

There are three main ways in which the people of Los Angeles tend to respond to living in this environment:

- Some people accept the fact that the hazards exist but are not prepared to do much about it — the 'it won't happen to me' syndrome. The advantages of living in southern California greatly outweigh the disadvantages.
- Some people recognise the severity of the hazards and are prepared to take action to protect themselves. At an individual level they can fit burglar alarms, take out insurance, attend earthquake drills, install gas cut-offs, collect a comprehensive earthquake kit, move house to be above the smog. Such people tend to be the more affluent who can afford the precautions. They also expect the authorities to act by insisting on smog control, earthquake-proof buildings, emergency services, flood control etc. They are willing to pay the increased taxes that result from implementing such measures.
- Living in such a multi-hazard environment may become too much of a strain for some people, forcing them to move away to a less hazard-prone environment.

This tends to happen after a major event, such as the Northridge earthquake of January 1994, or the riots and looting which followed the acquittal of the policemen in the Rodney King affair.

Hazards in LEDC cities

The hazards in cities in LEDCs are linked by unconstrained growth: the unplanned and haphazard development of the urban area makes it difficult for the city authorities to provide basic infrastructure and services.

Cities that are growing quickly contain large numbers of poorly constructed and badly maintained buildings that lead to unnecessary deaths when earthquakes, hurricanes and landslides strike. As a high percentage of people in LEDCs (often 20–60% of the population) live in informal settlements, solutions that might apply in MEDCs, such as building codes, zoning measures and urban planning techniques, are difficult to enforce and have had little impact in reducing the vulnerability of the urban poor.

The effects of hazards on LEDC cities could be reduced, as the technical knowledge does exist. Both structural measures (e.g. making buildings earthquake resistant) and non-structural options (e.g. limiting types of land use or providing tax incentives that direct development away from hazard-prone areas) are available. Such measures need not always be costly — building earthquake-resistant hospitals, for example, can add less than 10% to construction costs. These measures do not always have to be sophisticated either — planting trees, for example, shelters buildings from strong winds and makes hillsides less prone to erosion and landslide.

However, the amount of capital available to the government and the city authorities in LEDCs, and the high level of foreign debt, tend to limit schemes of improvement. There are also large proportions of poor people in these cities, which means the authorities do not have the capital to implement environmentally friendly policies and strategies.

Table 10.3 illustrates the effect of this lack of investment in hazard-proof infrastructure.

Table 10.3 The death toll in selected earthquakes

Location	Year	Magnitude (Richter scale)	Death toll
Loma Prieta (San Francisco area)	1989	7.1	63
Mexico City	1985	8.1	7,000 (est.)
Gujarat (northwest India)	2001	7.9	30,000 (est.)

Conflict over resource use at a local scale

There have been a number of local conflicts in the UK over large building projects that have achieved public notoriety. These include the Newbury bypass (see case study), the second runway at Manchester Airport, the new container terminal in the Southampton area, and the Trafford Centre (large out-of-town shopping

complex) on the M60. Such conflicts are resolved by market processes, planning processes or, in some cases, a combination of the two.

Market processes operate in an environment where the ability to pay the going rate takes precedence over any local or national concerns. Often, objectors cannot afford to outbid the developer and the development goes ahead with the minimum of consultation. When it does occur, consultation often takes the form of an opportunity to voice objections or propose counter-arguments, but with no right of independent arbitration or appeal.

Planning processes attempt to provide a means by which local authority planners:

- listen to the local community (more democratic)
- listen to the organisation responsible for a proposal
- have overall development control

Any refusal to grant planning permission by a local authority committee may lead to an appeal, or may result in the developer going to a higher body, for example national government in the form of the Department for the Environment, Food and Rural Affairs. Planning processes are costly, in terms of both time and money, for local authorities. Planning committees may:

- require or negotiate modifications to be made to offset the opposition
- request additional provision of facilities (such as better road access) which the authority would have to provide if the development went ahead, and which might placate some of the local opposition

Planning committees need to weigh up:

- the gains from the proposal against its negative aspects
- the conflicts between differing groups within a local community
- the wider benefits of a scheme versus the local opposition

Case study *The Newbury bypass*

Newbury, in Berkshire, lies just south of the M4 on the A34, which links the south coast of England to the Midlands. Newbury became prosperous in the 1980s, when many high-technology companies were attracted to the town, including the town's main employer Vodaphone (mobile phones), MicroFocus (computer software), Quantel (video systems) and the headquarters of Bayer (pharmaceuticals). Newbury also has a major racecourse with several meetings through the year.

Bypass proposals

The A34 (also part of Euroroute E05, linking southern Spain to Glasgow), which runs through the town, carries large numbers of vehicles, and a bypass was first suggested in the late 1970s. There was a public enquiry in 1980 over the plan, but the project was shelved. A second enquiry, 10 years later, considered 39 route options, coming to the conclusion that the western route was the 'least bad' option (Figure 10.20 overleaf).

A 1993 Berkshire council survey showed 50,600 vehicles a day using the existing A34 through the town, 15% of them heavy goods vehicles. The road through Newbury had busy intersections and traffic congestion was severe for large parts of the day. In the summer it often continued through the night as tourist traffic added to the volume. The cross-flow caused by a new superstore added further to the congestion. Traffic was also going through Newbury to reach the M4 and gain access to the west of

England and south Wales. In 1994, bowing to local pressure, the government put the bypass project on hold, but in 1995 it reversed that decision and gave the go-ahead for work to start.

Parties for the proposal

A number of groups supported the bypass proposal:

- **National government (Department of Transport and the Highways Agency)** had listened to the arguments from both sides and came down in favour of the road. The Newbury bypass was part of the government's overall transport strategy.
- **Newbury council** had conducted surveys that indicated the bypass would reduce traffic through the town by 36%, and goods vehicles by 88%. Only three houses would have to be demolished, and public transport policies would have a greater chance of success.
- Some **residents** of Newbury (including the Newbury Bypass Supporters Association) believed the bypass would lead to reduced traffic congestion and air pollution in the town, with far fewer heavy vehicles, and that it would be safer and easier to get to work, school and shops.
- The **Freight Transport Association** and the **AA** wanted the bypass because it would reduce journey times from the south coast to the north and west, thus costing less in time and money.

Parties against the proposal

Groups who voiced objections to the proposal included: Friends of the Earth (FoE), Road Alert!, some residents of Newbury including the Third Battle of Newbury group, English Nature, the National Rivers Authority (NRA), local farmers and landowners, and eco-

warriors such as Swampy and Muppet Dave. Their detailed objections concerned:

- loss of farmland to the western side of the town
- the removal of 0.6 ha from Snelsmore Common Site of Special Scientific Interest (SSSI), a heathland nature reserve with one of the few remaining populations of nightjars in England
- disturbance of several archaeological sites as well as the site of the First Battle of Newbury (1643)
- increased risk of flooding where bridging structures crossed the Rivers Lambourn and Kennet

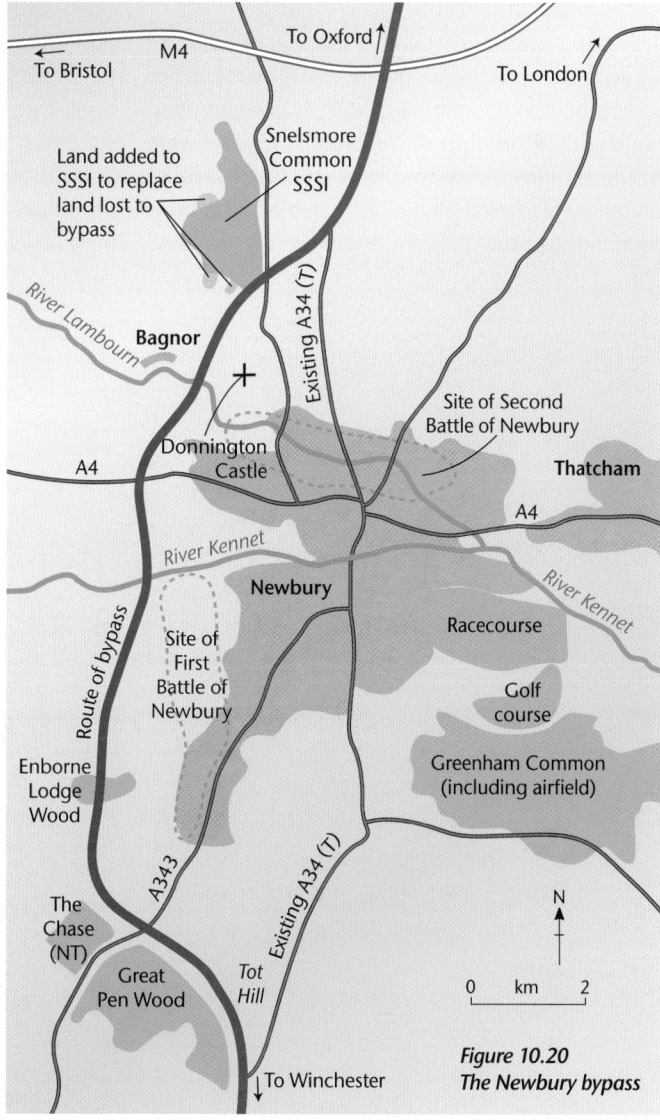

Figure 10.20 The Newbury bypass

273

- damage to the wildlife of the Rivers Lambourn and Kennet and their valleys. These are well-known chalk streams famous for brown trout as well as rare mayflies, damsel flies and an endangered species of snail, *Vertigo moulinsiana*, which is a relic from the ice ages
- the fact that the bypass route cut through existing woodland and close to a National Trust property and would require the felling of many trees

Processes used to resolve the conflict

This was a large project with a number of complex issues and several groups involved. It is therefore not surprising that the planning process allowed interested parties some input. The bypass proposal went to appeal, and a public enquiry was held in Newbury under the leadership of a judge. People and groups covering all shades of opinion were able to put their points of view, but after some time and cost, the decision to construct the bypass was upheld.

In late 1994, a document published by the Standing Advisory Committee on Trunk Road Assessment (SACTRA) stated that new roads simply generated more traffic. New road building was becoming increasingly unpopular in certain areas of the country and the government postponed six major schemes, putting the Newbury bypass on hold. In July 1995, however, the scheme was allowed to proceed and it was completed in 1998, not without opposition. The site became a target for conservationists and concerned local people and there followed several bitter battles to remove protesters from trees and tunnels. The construction sites suffered a lot of vandalism. The final cost of building the road (13.5 km) has been estimated at around £75 million with, on top of this, an additional £25 million for policing to make construction possible.

Chapter 11

Human geography
Global change in the last
30 years

Transnational corporations and the global economy

A transnational corporation (TNC) is one that operates in at least two countries. Many have their centres of production in one or more countries, while the headquarters and research and development (R&D) department are in a different part of the world. The organisation therefore tends to be hierarchical, with the headquarters and R&D in the home country (country of origin) and branch manufacturing plants overseas. As the organisation becomes more global, regional headquarters and, in some cases, regional R&D departments will develop in the manufacturing areas. TNCs take on many different forms and cover a wide range of companies involved in the following primary, secondary (manufacturing) and tertiary (service) activities (Table 11.1):

- resource extraction, particularly in the mining sector, for materials such as oil and gas
- manufacturing in three main sectors:
  (1) high-tech industries such as computers, scientific instruments, microelectronics, pharmaceuticals
  (2) large-volume consumer goods such as motor vehicles, tyres, televisions and other electrical goods
  (3) mass-produced consumer goods such as cigarettes, drinks, breakfast cereals, cosmetics, branded goods
- services such as banking/finance, advertising, freight transport, hotels and fast-food operations

Key terms

Globalisation The increasing integration of national economies through international trade, investment and capital.

Global products Recognised branded products that have been marketed worldwide, such as Coca-Cola or Big Mac. The term can also apply to products for which the components are manufactured in several countries and exported to be assembled in another. Some cars are manufactured on this basis.

Newly industrialising country (NIC) A country which has shown rapid growth in its manufacturing industry since the 1960s. Examples include Singapore, South Korea, Taiwan, Malaysia, China and Brazil.

Transnational corporation (TNC) A company that operates in more than one country. TNCs include the largest companies in the world and many have total sales larger than the gross national product (GNP) of a small country. The size of many TNCs means that they have considerable influence on the economy, and therefore the politics, of the countries in which they operate.

Table 11.1
The world's largest
companies, 2004

Rank	Company	Home country	Total revenue (US$billion)
1	Wal-Mart	USA	256.33
2	BP	UK	232.57
3	Exxon Mobil	USA	222.88
4	General Motors	USA	185.52
5	Ford	USA	164.20
6	DaimlerChrysler	Germany	157.13
7	Toyota	Japan	135.82
8	General Electric	USA	134.99
9	Royal Dutch/Shell	UK/Netherlands	133.50
10	Total Fina Elf	France	131.64
11	Chevron Texaco	USA	112.94
12	Mitsubishi	Japan	112.76
13	Mitsui	Japan	111.98
14	Carrefour Group	France	96.94
15	Allianz	Germany	96.88
16	ING Group	Netherlands	94.72
17	Citigroup	USA	94.71
18	Nippon Tel/Teleph	Japan	92.41
19	Volkswagen	Germany	91.33
20	ConocoPhillips	USA	90.49
Other UK companies that featured in the top 80			
54	Aviva	UK	52.46
56	Unilever	UK/Netherlands	50.70
63	Vodafone	UK	47.99
73	HSBC Group	UK	44.33
77	Tesco	UK	41.48

The significance of TNCs

TNCs control and coordinate economic activities in different countries and develop trade within and between units of the same corporation in different countries. Because of this, they can control the terms of trade and can reduce the effect of quota restrictions on the movement of goods.

They are able to take advantage of spatial differences in factors of production at a global scale. They can exploit differences in the availability of capital, labour costs, and land and building costs — for example they can take advantage of cheaper labour costs in LEDCs. In 2002, the household appliance manufacturer Dyson announced plans to move production from its manufacturing plant in Wiltshire to Malaysia, in order to take advantage of much lower labour rates (Figure 11.3 on page 282).

Location

TNCs can also locate to take advantage of different government policies in other countries, such as reduced tax levels, subsidies/grants or less strict environmental controls. They can get round trade barriers by locating production within the markets where they want to sell. Japanese car firms, for example, have been attracted to locations within the European Union (EU) because of quota restrictions on the import of Japanese vehicles to the EU. By producing vehicles within Europe, they are considered to be European manufacturers and gain entry to the European market.

The first Japanese motor vehicle manufacturer to establish in the UK for this reason was Nissan at Washington (near Sunderland, Photograph 11.1), followed by Toyota (at Burnaston, near Derby) and Honda. Honda's plant at Swindon employs 2,500 workers producing over 150,000 vehicles annually, mainly for the European market with some exports to the Middle East and Africa. Such companies have geographical flexibility and can shift resources and production between locations at a global scale in order to maximise profit.

The large size and scale of operations of TNCs enables them to achieve economies of scale, allowing them to reduce costs, finance new investment and compete in world markets.

Large companies also have a wider choice when locating a new plant, although governments may try to influence decisions as part of regional policy or a desire to protect home markets. Governments are often keen to attract TNCs because inward investment creates jobs and boosts exports which assist the trade balance. TNCs have the power to trade off one country against another in order to achieve the best deal.

Photograph 11.1
An aerial view of
Nissan's Washington
factory

Nissan

Case study *Nissan*

Nissan is a Japanese company mainly involved in the automobile industry. It has manufacturing plants in 16 countries and produces over 2.5 million vehicles annually. Outside of Japan, Nissan has major manufacturing plants in the USA, the UK, Spain, Mexico and Taiwan. Different models are produced for different markets, but some are manufactured in one country and exported to another. A growing trend in car manufacturing is to make components in several countries and then export them to one plant where the final vehicle is constructed. Major plants include:

- Mexico City, producing the Nissan Sunny, AD Wagon and Nissan Pickup for the north American market
- Tennessee (USA), producing the Altima, Xterra and the Frontier for the north American market
- Yokusa (Japan), producing the Maxima for the north American market and other models for the home market
- Barcelona (Spain), producing the Patrol, Terano II and the Vanette Cargo for the European market
- Washington (near Sunderland, UK), making the Almera, Micra and Primera ranges for the European market. Engine components, axles, body pressings and stampings from Nissan plants in Japan are imported, as well as other components from Italy, Belgium and Germany (Figure 11.1). This illustrates the linkages that often develop across national boundaries

Nissan operates from Japan but has regional headquarters in its main areas of operation. It has also set up local R&D establishments to develop new models within those areas. The European R&D centre, for example, is at its manufacturing facility at Washington in the UK.

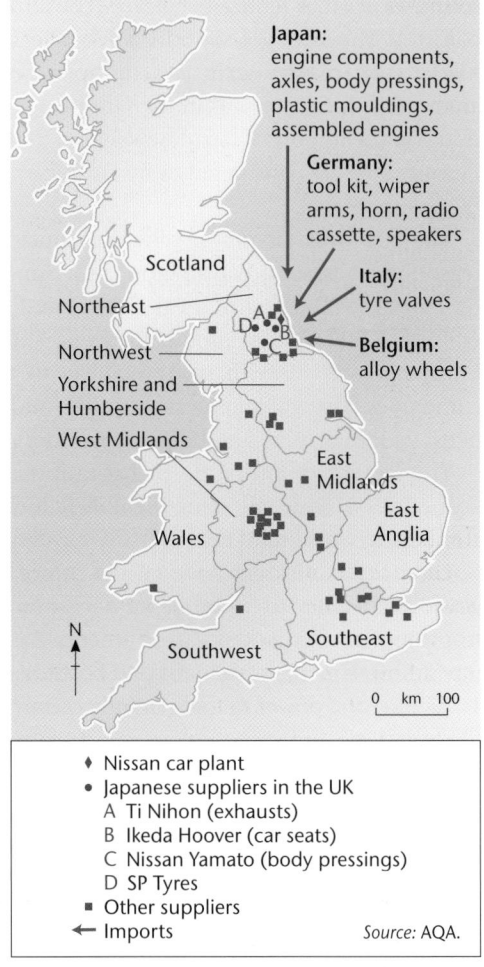

Japan: engine components, axles, body pressings, plastic mouldings, assembled engines

Germany: tool kit, wiper arms, horn, radio cassette, speakers

Italy: tyre valves

Belgium: alloy wheels

Scotland

Northeast

Northwest

Yorkshire and Humberside

West Midlands

East Midlands

Wales

East Anglia

Southwest

Southeast

N

0   km   100

◆ Nissan car plant
• Japanese suppliers in the UK
  A Ti Nihon (exhausts)
  B Ikeda Hoover (car seats)
  C Nissan Yamato (body pressings)
  D SP Tyres
■ Other suppliers
← Imports   *Source: AQA.*

Figure 11.1 Suppliers to the Nissan car plant at Washington

Within a country, TNCs have the financial resources to research several potential sites and take advantage of the best communications, access to labour, low cost of land and building and government subsidies.

Globalisation

TNCs serve a **global market** and they **globalise** their manufacturing operations in several ways:

- by producing just for the country in which the plant is situated. This is known as host market production
- by producing for a number of countries, for example specialising in goods for the EU

Case study *Nike*

Nike began in the 1960s as a shoe importer to the USA. It became a shoe manufacturer before branching out into producing a whole range of sporting goods, both clothing and equipment.

Nike still has its headquarters in the state of Oregon (USA) where it has also located its advertising and R&D centres. All of Nike's products, however, are made by contract suppliers operating outside the USA. Nike originally operated through contract manufacturers in South Korea and Taiwan, but in the 1990s it began to switch to lower-cost areas such as Indonesia, Vietnam and China. There is intense competition in the sports goods industry, particularly in the market for shoes, and companies constantly seek to drive down costs and increase profit margins.

East Asian countries are also the production base for other companies in this field such as Adidas, Puma and Reebok.

- by integrating production, where each plant performs a separate part of a process. Linkage takes place across national boundaries in a chain sequence (vertical integration) or components are moved to a final assembly plant in one country (horizontal linkage)

TNCs often locate in areas of industrial concentration, or they themselves encourage such concentrations. A major factor promoting this is **external economies of scale (agglomeration economies)**. These advantages arise from outside the company, unlike internal economies of scale that result from the large-scale operation of the production plant. External economies can be categorised into:

- **Localisation economies**, which occur when firms linked by purchase of materials and finished goods locate close together. This reduces transport costs between supplier and customer, leading to faster delivery times (which in the case of TNCs could enable just-in-time operations) and allows better communication and personal contact between firms (important in monitoring and maintaining quality).
- **Urbanisation economies**, where cost savings result from an urban location allowing linkages between manufacturing and services. Manufacturing industries depend upon a wide range of services, and although they can often provide these themselves, it may be cheaper to contract out the work to specialist companies. Savings also result from the economic and social infrastructure of the area that exists before the arrival of new companies.

Two very different examples of TNC structures are given in the case studies above.

Social and economic impacts of TNCs

Positive impacts on host countries

Inward investment by TNCs can have a significant effect on social and economic developments within a country, at both a national and a regional scale. In the UK, for example, 869 projects were initiated by foreign companies in 2000–01, generating over 71,000 jobs. This may seem relatively small compared with a total UK workforce of over 20 million, but it must be remembered that investment in such developments came largely from outside the UK. They were not financed by

the government using taxpayers' money, although some companies would have been eligible for government subsidies.

At a local level, such investment can trigger the **multiplier effect** through the process of cumulative causation, resulting in even greater levels of employment and capital injections into the local economy.

The location of a TNC in a region has the potential to increase jobs through linkage, providing business for:

- companies that supply components to the new plant
- companies that distribute goods from the new plant
- companies that supply services to the new plant, from servicing plant machinery down to supplying the canteen

The arrival of a TNC in an area may decrease unemployment and inject capital (in the form of wages) that stimulates the local economy. More disposable income in the area will create a demand for more housing, transport and services. All of this promotes an upward spiral (virtuous circle) in economic terms.

As well as creating employment, many TNCs introduce new methods of working in their own plants and those of supplier companies. These include quality management systems which monitor the standard of output in supplier firms, and **just-in-time (JIT)** component supply which requires total quality or zero defect. In this way, TNCs are influential in modifying working practices and in the **transfer of technology** to the host country, creating a more skilled workforce. There may also be greater opportunities for female employment in low-skilled manufacturing jobs, particularly in LEDCs.

Negative impacts on host countries

TNCs moving into a country may be in direct competition with local companies. If the local companies are less efficient, they will lose business and, ultimately, employees. When Russian markets were opened up after the collapse of communism, the arrival of Western chocolate manufacturers had an adverse affect upon the home confectionery industry.

In LEDCs, the introduction of TNCs can lead to increased urbanisation as younger workers migrate to cities to take advantage of employment opportunities. This will have serious effects in the rural areas which lose people of working age and are left with an ageing population.

TNCs may also bring unwelcome environmental change in the form of damage to the atmosphere, water and land. Many LEDCs have less strict pollution laws than MEDCs. Agricultural land may be lost, along with habitats for wildlife.

There have been allegations that TNCs exploit cheap, flexible, non-unionised labour forces in LEDCs. Nike, for example, has been challenged over the rates of pay offered by its subcontracting firms for shoe assembly. TNCs have also been accused of not training the local workforce but importing skilled and managerial personnel from the country of origin.

The capital generated by TNCs does not all stay in the host country, which may end up less well off than it had expected from the presence of a TNC. Critics point to decisions made outside the country and the fact that some companies

Closure of food plant is a recipe for confrontation

The closure by a multinational firm of its small factory on the coast of Cumbria might have passed almost unnoticed: another harsh but inevitable realignment of the area's declining economy.

But 5 months after the announcement, the threatened removal of Homepride from Maryport, with the loss of 123 jobs, has become a cause célèbre which its owners, the American-based Campbell Soups, may rue.

JOB AXE PLAN MADE IN USA

The decision to shut a local firm, with the loss of 95 jobs, was taken in the USA.

Now Bury South MP, Ivan Lewis, is writing to Manchester Circuits' American parent company to find out the reasons behind the shock closure.

He said, 'The managing director here says the decision to close at the end of August was taken by the company's owners in America.'

remove their investment as soon as economic conditions are less favourable (Figure 11.2).

Figure 11.2
Global decisions

Impacts on the country of origin

The impact of TNCs on their country of origin is largely negative in an economic sense. When a company takes its production overseas, there is likely to be increased unemployment, both in that industry and in component suppliers. The amount of disposable income available within a region will decrease, leading to a downward spiral (vicious circle) — in other words the multiplier effect working in reverse. This is particularly the case in traditional industrial regions within MEDCs, which might rely on only one or two industries for their economic base.

The move by Dyson of the production of its vacuum cleaners from Malmesbury (Wiltshire) to the Far East cost 800 jobs (Figure 11.3). The company claimed that moving production to Malaysia would enable it to cut about 30% from production costs. Other moves in recent years have included Kenwood's decision to move production of the Chef food mixer to China, Clarks Shoes moving most of its production to Romania, and Black & Decker moving production from its factory in Spennymoor (County Durham) to a plant in the Czech Republic.

It is not only in manufacturing that employment can be moved abroad. British Airways, Great Universal Stores, BUPA, HSBC and Barclays all have administrative centres outside the UK, many in the Far East and India. In 2004, Norwich Union announced plans to move several hundred back office jobs to India and Sri Lanka. Such countries offer a much lower cost base, as pay for staff in India is only around 20% of the UK rate (Figure 11.4). Prudential, a major UK insurance company, also announced (in October 2002) that it was considering moving 1,000 call-centre jobs from Reading to Mumbai in India.

800 jobs to go as 'sad' Dyson moves factory to Far East

James Dyson, multi-millionaire investor and champion of British engineering, is to move production of his bagless vacuum cleaners from Wiltshire to the Far East. It means the loss of 800 jobs at his Malmesbury headquarters.

Since he started production of his vacuum cleaners in 1993, Mr Dyson has become the leading evangelist of British manufacturing.

'I do not believe that the nation that was home to the Industrial Revolution can remain great if it loses its ability to make things,' he said recently. Yesterday he was forced to abandon his beliefs to the economic imperative. He said it was a 'very sad day, both for the company and personally.'

He added, 'No one could have tried harder to make it work in Britain. I feel very sad but we are minnows in comparison with our multinational competitors and we need to make substantial savings to take them on.'

Moving production to Malaysia, where Dyson Appliances already produces two of its four vacuum cleaner models, or possibly to China, would shave about 30% from production costs, Mr Dyson said.

'We need an enormous amount of cash to invest in new technology, to launch into new markets and to launch more products faster.

'Most of our suppliers are also in the Far East. And our markets are there too. We're the best-selling vacuum cleaner in Australia and New Zealand. We are doing well in Japan and we are about to open in America. It makes more sense for us to produce in the Far East.'

About 800 reseach and development staff will remain at Malmesbury where Dyson has invested £32 million in the past 2 years. Another 150 staff will continue production of the company's range of contrarotator washing machines in Wiltshire. Mr Dyson said that despite yesterday's announcement, he would still be employing twice as many people at the site as he did 4 years ago.

Figure 11.3 Dyson moves east

Adapted from the *Daily Telegraph*, February 2000

Fury at insurer's plans to double Asian workforce

Britain's biggest insurance company Norwich Union sparked a fresh row over outsourcing to India yesterday when it announced plans to nearly double its Asian workforce to 7,000.

Parent company Aviva said about 760 new jobs will be created next year in India to carry out back office work for its life and general insurance businesses.

A further 190 finance support roles will be created in Sri Lanka.

Aviva, which already has 3,700 staff in India expects to create another 2,350 positions offshore by the end of 2007, bringing the total up to 7,000.

It said next year's move will result in up to 70 compulsory job losses in Norwich, 60 in York and another 20 at other sites. Aviva employs 51,000 staff worldwide, including 30,000 in Britain.

The average pay for staff in India working for British financial firms is £3,000 a year — about 20% of the UK rate.

Mike Kirsch, Norwich Union Life operations director, said a 'small number' of the group's 8 million customers had complained about jobs being outsourced to India.

However, he added: 'We are doing this in a progressive way with minimal impact on our staff.'

Figure 11.4 Service industry also moves east

Adapted from the *Daily Telegraph*, September 2004

The higher-salary jobs, though, often stay in the country of origin. Dyson, for example, intended to retain around 800 research and development staff at Malmesbury. Many TNCs that move to cheaper locations produce increased profits which benefit their shareholders and improve the balance of payments situation for the donor country.

Newly industrialising countries and TNCs

During the last 20–30 years there have been increases in the level of industrialisation in several developing countries which have become known as newly industrialising countries (NICs). Examples include South Korea, Hong Kong, Taiwan, Singapore (known collectively as the Asian Tigers), Malaysia, China, Brazil and Mexico. This growth in manufacturing has been rapid and TNCs have played a big part.

NICs as hosts

The main attractions of NICs as host countries for TNCs were cheaper labour and land costs, and weaker environmental and planning laws. There was also a developing domestic market for consumer goods and the fact that many of the materials required by TNCs could be found locally. Rapid developments in communication and transport technology allowed TNCs to exploit these advantages. Improved telcommunications technology such as satellites and the internet made it possible to control production in NIC plants from headquarters in a developed

Case study *Malaysia*

Malaysia is a good example of an NIC which attracted a number of foreign companies. Until the 1980s, Malaysia's economy was largely based on primary products (Figure 11.5), but the government of the time set the country on a course of rapid industrialisation by encouraging TNCs to locate there. This was done through a mixture of tax incentives, liberal investment rules, cheap land and the development of infrastructure such as ports, roads and airports. Major areas that attracted investment by foreign companies included:

- the Bayan Lepas Industrial Park in Penang, which has major companies such as Hewlett Packard (4,000 employees) and Bosch (3,500 employees)
- south of Kuala Lumpur, where the government is trying to establish an Asian version of silicon valley. TNCs operating here include Microsoft and IBM

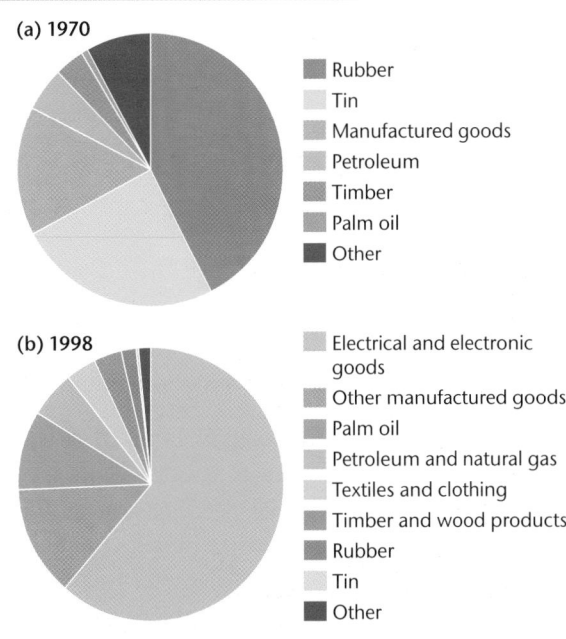

Figure 11.5 Malaysia's changing pattern of exports

Case study *South Korea*

South Korea is a good example of an NIC that has developed its own TNCs. The growth of national companies was encouraged in South Korea by the government, which owned many companies. Family-owned business conglomerates known as *chaebol* also experienced a lot of growth — their success has been at the heart of the South Korean industrial expansion.

Large companies such as Hyundai, Samsung, Daewoo and Lucky Goldstar (the growth of which is shown in Table 11.2) became established. The home market for their products was small at first, so they began to

sell in export markets. This led to the establishment of sales offices and service networks overseas. The tariff barriers imposed by the EU and other areas made it necessary to manufacture in those markets if expansion of sales was to continue. The major companies set up manufacturing operations in Europe and it was not long before many of their component suppliers followed, establishing their own operations in Europe.

Other South Korean firms, seeking cheaper labour, have set up factories in neighbouring Asian countries such as China, Indonesia and the Philippines.

1947	Lucky Chemical Co. is established, producing a range of items including toothpaste
1958	Lucky Chemicals creates GoldStar Co. Ltd to manufacture domestic electrical items
1959	Lucky Goldstar (LG) produces Korea's first transistor radio
1965	LG produces Korea's first refrigerator
1966	LG produces Korea's first television
1969	LG produces Korea's first washing machine, air conditioner and lift
1975	LG creates its first R&D centre
1978	LG establishes GoldStar Electronics International in New Jersey, USA
1980	LG creates GoldStar Deutschland in Willich, Germany
1982	LG produces Korea's first microcomputer and video camera
1983	LG opens an R&D centre in the USA
1986	LG establishes a video-cassette plant in Worms, Germany
1989	LG starts joint trading ventures with companies in Turkey, the Philippines, Egypt, Italy, Indonesia and Thailand
1993	LG starts to produce CD players, CD-ROM drives and other multimedia products
1995	LG takeover of Zenith in USA

Table 11.2 The growth of Lucky Goldstar, a Korean TNC

country, and the cost of transport over long distances was reduced by the introduction of bulk carriers and container ships.

NICs as countries of origin

The growth of NIC economies allowed a large domestic market to develop and this encouraged the growth of home-based companies. These were aided by the transfer of technology (including the development of managerial and research skills) from the incoming TNCs, and the advantages afforded by infrastructural developments. Many of the new domestic companies were joint ventures between state-owned companies and local entrepreneurs, but it was not unusual for TNCs to enter into joint ventures with local companies.

The emphasis at first was on low-technology, labour-intensive products such as clothing. Expansion later occurred into products such as machinery, electronics and motor vehicles. These companies could not be described as TNCs — they were domestic companies producing for a largely domestic market, but many were beginning to find export markets.

The development of such companies into fully-fledged TNCs came about for two reasons:

- A rise in wage levels encouraged them to look for cheaper locations in neighbouring LEDCs. These new TNCs from NICs followed the pattern established by TNCs from developed countries which moved into LEDCs in search of cheaper labour and locations. Many companies from South Korea, for example, sought cheaper locations in China.
- The growth of tariff barriers in areas like the EU meant that companies could no longer manufacture goods in NICs and then transport them to the market. They were forced to open manufacturing outlets in Europe. Many such companies have located in the UK, including LG (Lucky Goldstar), Samsung and Hyundai.

Different generations of NICs

First-generation NICs

South Korea, Taiwan, Singapore and Hong Kong formed the first wave of development. This happened as European, American, and particularly Japanese, companies sought cheaper labour sources.

Second-generation NICs

As the economies of first-generation NICs developed, the level of wages increased. This led European, American and Japanese TNCs to seek locations in a second generation of NICs, where improvements in physical and human infrastructures now satisfied their demands but wages were still low. Emerging companies within first-generation NICs also sought to move routine tasks to their cheaper labour neighbours. This second wave of NICs included such countries as Malaysia and Thailand.

Third-generation NICs

This process is now being repeated in a number of countries which are seeing significant growth rates in their economies. China is a typical example of this most recent wave (see case study, pages 309–312), as is India, although it has had a very different route to its NIC position (see case study overleaf).

Attitudes to TNCs

Positive attitudes to the presence of TNCs in a host country include the following:

- governments favour them because they invest in the country, bring increased revenue and create jobs (although some employment in home-based industries may be lost)
- component companies are able to expand and supply the incoming TNCs
- construction companies benefit by building plant, infrastructure and new housing

Case study *India*

India has achieved economic growth using a different route from existing NICs in Asia. Two main features mark India's emergence as an NIC:

- its economic growth in the 1990s was due to the expansion of its service sector rather than manufacturing — by 2000 the service sector accounted for over 45% of its GDP
- the filtering down of employment to India has been via companies operating in North America and Europe rather than those with their headquarters in Japan and other emerging Asian nations

The service sector is growing rapidly in India. The main area of growth has been in software and ICT services, but the media, advertising, retail, personal financial services, entertainment, tourism and leisure have all expanded at a significant pace in the last decade.

Software and ICT services

This sector has been at the forefront of India's economic growth in recent years. It has been estimated that by 2010 it will contribute just under 10% of India's GDP. It has benefited significantly from European and North American companies outsourcing work to Indian companies. This has occurred for the following reasons:

- Indian labour costs are much lower than those in Europe and North America
- several MEDCs have significant shortages of ICT skills
- there is a large and very able English-speaking workforce in India

Back office functions

This type of work involves call centres and medical transcription (companies converting dictation by American doctors into medical records). It has been estimated that by 2008 India will employ over 1 million people in this sector and earn $17 billion per year. (Figure 11.4 on page 282 provides an example of a UK company moving jobs to India.)

- transport and energy supply companies benefit from extra business
- unemployed people gain from the increase in job opportunities
- women may find job opportunities
- providers of local services, shopkeepers, farmers etc. benefit from the increase in local disposable income

Negative attitudes to the presence of TNCs in a host country include the following:

- competing firms face loss of sales and, ultimately, closure. They may resent foreign firms receiving subsidies, grants and other incentives that are not available to domestic companies
- people employed by domestic companies may lose jobs if they have to cut back or even close due to competition
- local councils experience increased pressure on services (although they benefit from the extra revenue created by new companies)
- local residents may complain about the increase in traffic and loss of greenfield sites when factories are built
- environmentalists are likely to object if development results in increased pollution or loss of habitat
- trade unions fear loss of influence if TNCs take an anti-union stance or adopt a single-union approach (although they welcome the increased job opportunities TNCs bring)

International migration and multicultural societies

International migration can be:

- voluntary or forced
- temporary or permanent
- legal or illegal
- of varying scale, in both volume and distance
- motivated by economic, political, social, cultural or environmental factors
- selective on the basis of age, gender and educational level, or a mass movement involving all social groups

International migration is usually subject to distance-decay, in that the number of migrants declines as the distance between their origin and destination increases. Refugees tend to move only short distances, often into a neighbouring country, whereas economic migration frequently takes place over great distances. These patterns can be distorted by:

- economic differences between neighbouring countries (differences in GDP and economic opportunities)
- colonial links (migrants move from former colonial territories to the former mother country)
- family links (family reunification)
- political regime (movement to a more democratic society)
- government policies and immigration controls

Major forms of international migration

Economic migration

One of the main reasons for international migration is the existence of greater economic opportunities in other countries, which prompts individuals to migrate in search of a 'better life' for themselves and their families. This has traditionally been movement from LEDCs in the developing South to MEDCs of the affluent North (Figure 11.6). Across the North–South divide the major movements have been:

- from Mexico and other Latin American countries to the USA
- from north Africa to Europe
- from territories that were formerly part of a European colonial empire to the former mother country, e.g. Indians, Pakistanis and west Africans to the UK

Key terms

International migration The movement of people across national frontiers involving a permanent change of residence (for at least 1 year, according to the United Nations).

Multicultural society A social grouping which contains members from a wide variety of national, linguistic, religious or cultural backgrounds.

Refugee Defined by the United Nations in 1951 as 'someone who, owing to a fear of being persecuted for reason of race, religion, membership of a particular social or political group, is outside the country of his/her nationality and is unable or, owing to such fear, is unwilling to return to that country.' In recent years, this definition has been widened to include those fleeing from civil wars, ethnic, tribal or religious violence, and environmental disasters such as earthquakes, volcanoes and famine.

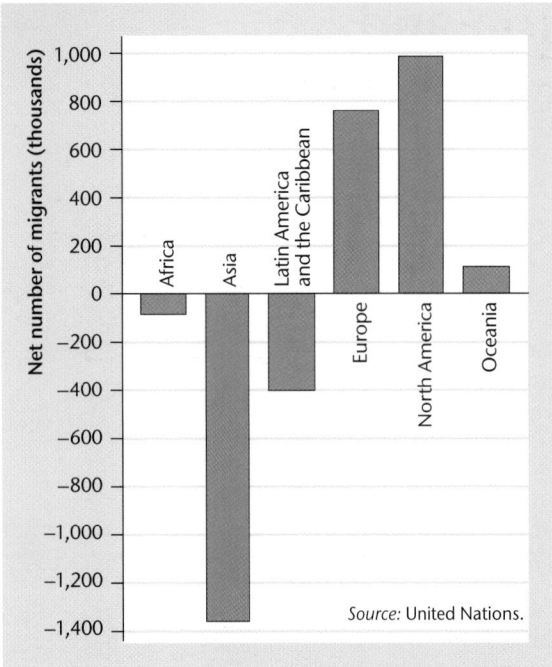

Figure 11.6 Annual net migration totals in the world's major areas, 1990–95

In the 1980s, for example, 3.5 million migrants entered the USA from Mexico. Of these, 700,000 were legal migrants, 2.3 million were amnestied or legalised (granted legal status once in the USA) and around 500,000 were illegal migrants. There is a huge gap in wealth between Mexico and the neighbouring state of California that promoted this movement, coupled with the fact that people were looking for casual employment within the farming communities of the southwest.

Migration into Europe, particularly France and Italy, has been triggered by the same economic reasons but also made easy for illegal migrants by the long Mediterranean coastlines of both countries. Much of the migration into France also reflects previous colonial ties, for example with Algeria, which mean that the potential migrant is likely to have some basic knowledge of the French language.

In recent years, the growth of certain Asian economies has prompted the migration of people within Asia, from the poorer countries to richer ones like Singapore and Hong Kong. This has also been seen in the Middle East where people have migrated from countries such as Pakistan to the oil-rich states where they can find work in the oilfields or in domestic service. Before the Gulf War in 1991, it was estimated that well over half the population of Kuwait consisted of migrant workers and their families (40% Kuwait born, 60% foreign nationals).

It can also be argued that much of the movement within the developed world in the second half of the twentieth century was prompted by economic improvement. Those who went to Australia from the UK on the 'assisted passage' scheme could be thought of as economic migrants, as could those who migrated from Europe to the USA. The US government has always encouraged entry of highly qualified or skilled workers from other countries, so the term that came to describe this movement was 'brain drain'.

Refugees

In March 2002, the United Nations High Commissioner for Refugees (UNHCR) estimated that there were some 22 million people of refugee status in the world. Many refugee movements are large-volume, non-selective migrations over short distances, often caused by war. The refugees flee across a border into a neighbouring country. Such migrations can be temporary — when the cause of the migration is ended, refugees move back to their former homes. Major refugee movements at the end of the twentieth century included:

Montserrat to Moss Side

Refugees left volcanic island to make a new life in Manchester

It has been a year since the first refugees from the volcanic island of Montserrat arrived in Manchester.

Over the last 12 months, 26 families, uprooted in one of the world's most violent natural disasters, have rebuilt a little piece of their homeland in Moss Side.

And looking back, some are glad to have been given the opportunity to make a new life in England.

Janice Skeet and her 2-year-old daughter, Caroline, were the first two Montserratians to be relocated by Mosscare Housing last August.

They fled their home town of St Peters after a series of volcanic eruptions threatened to bury it under tons of ash. Janice said, 'When I left things were getting worse and I was really scared.

'I gave up my job as a sales clerk and travelled to Manchester, even though I had no family here and nowhere to live.'

When Janice arrived on 6 August 1997, charity Mosscare Housing found her a small house in Moss Side — an area that is now home to six families of refugees from Montserrat.

Initially, Janice felt very dejected at finding herself in a cold, dark, terraced street thousands of miles from her home.

'I took one look at the weather and swore I'd return to Montserrat as soon as I could,' she said.

But a year on Janice, now aged 21, has decided to make Manchester her permanent home.

She said, 'I feel so settled that I've decided to stay for good. The street where I live is full of people I knew from Montserrat so I almost feel like I'm back home.

'The one thing that is different — apart from the weather — is I'm having to get used to climbing up and down stairs. In Montserrat everything was at ground level. Also I used to leave my house open in Montserrat. Here I have to be more security conscious. But the neighbours are very friendly and we all look out for each other. The food is OK and I can get all the things I had in Montserrat in Moss Side.'

Janice is on a college computer studies course, and was recently joined by her brother Harrison.

Her daughter Caroline is starting nursery school in September and Janice is happy that she will grow up a Mancunian.

She said, 'The main thing I miss about Montserrat now is that I used to compete at athletics for the island. I was the best at the 400 metres. Here I've been playing netball, but I'd like to take up my athletics again and I've not had the chance yet.'

Adapted from *Manchester Metro News*, July 1998

- 2 million from Ethiopia, Sudan and Somalia primarily due to famine, but also because of civil war
- 6 million from Mozambique as a result of famine, civil war and extensive flooding
- 1 million Kurds fleeing oppression in northern Iraq
- 100,000 Tamils fleeing oppression and civil war in Sri Lanka
- 7,000 residents of Montserrat (over 75% of the population) fleeing the eruption of the Soufrière Hills volcano in 1995 (see Figure 11.7).

Figure 11.7
Montserrat to Manchester — a migrant's view

The changing forms of international migration

Since the late 1980s, it has become obvious that patterns of international migration are changing. These changes have included:

- A decrease in legal, life-long economic migration, particularly from LEDCs to MEDCs. Host countries have provided fewer opportunities for migrants as the number of low-skilled jobs available has dropped. At the same time there have been growing employment opportunities in the countries of origin, through the presence of TNCs. Host countries have also tightened up entry requirements with increased legislation and the introduction of quotas. Monitoring at the point of entry has become more rigorous.

- An increase in attempts at illegal, economically motivated migration as a response to the above restrictions.

- An increase in those claiming asylum. Some people claim that many asylum seekers are economic migrants in disguise.

- An increase in movement between MEDCs, particularly the countries of the EU where restrictions have been removed to allow free movement of labour. The Schengen Agreement has allowed cross-border movement without passport control between most of the countries of the EU, but not in and out of the UK.

- An increase in short-term migration as countries increasingly place time limits on work permits.

- Increased movement between LEDCs, much of it because of a perception of better conditions in other countries.

- A decline in the number of people who migrate for life. Many newer migrants want to return 'home' at some time in the future.

- Fewer people migrating for the purpose of reuniting family members.

The pattern of migration into European countries illustrates some of these changes:

- From the 1950s to the early 1970s migration was mainly selective, with many single men from LEDCs seeking work in Europe where there was a demand for unskilled labour. Some of this migration was linked with colonial ties — migrants from the Caribbean, India and Pakistan came to the UK and those from north and west Africa to France.

- From the mid-1970s to the early 1980s migration slowed as families moved to join those who had already become established. Governments of European countries began to impose much stricter

Table 11.3
EU asylum applications, 2002

Country	Asylum applications	Applications per 1,000 inhabitants
UK	110,700	1.9
Germany	71,127	0.9
France	50,798	0.9
Austria	37,074	4.6
Sweden	33,016	3.7
Belgium	18,805	1.8
Netherlands	18,667	1.2
Ireland	11,634	3.1
Italy	7,281	0.1
Spain	6,179	0.2
Denmark	5,947	1.1
Greece	5,664	0.5
Finland	3,443	0.7
Luxembourg	1,043	2.4
Portugal	245	0.0

controls on new migrants, but allowed reunification of families as social migration.

- From the mid-1980s to the present day there has been an increase in the migration of asylum seekers — people who have no legal right of entry but claim some form of persecution in their home area. Some of these are genuine refugees, forced to leave their homes, often under desperate conditions, but others are economic migrants fleeing poverty in their home countries. There has also been an increase in the numbers attempting to enter Europe illegally. Table 11.3 shows asylum applications in EU countries in 2002. Figure 11.8 shows the origin of those applying in the UK.

Multicultural societies

Migration of various ethnic groups has led to the creation of multicultural societies. In Africa, many multicultural societies have grown up because the boundaries of the former colonies were drawn to serve commercial interests, largely without regard for the territorial claims of the indigenous peoples. Different ethnic groups, speaking different languages, with differing customs and even religions, were put into one country. Nigeria, for example,

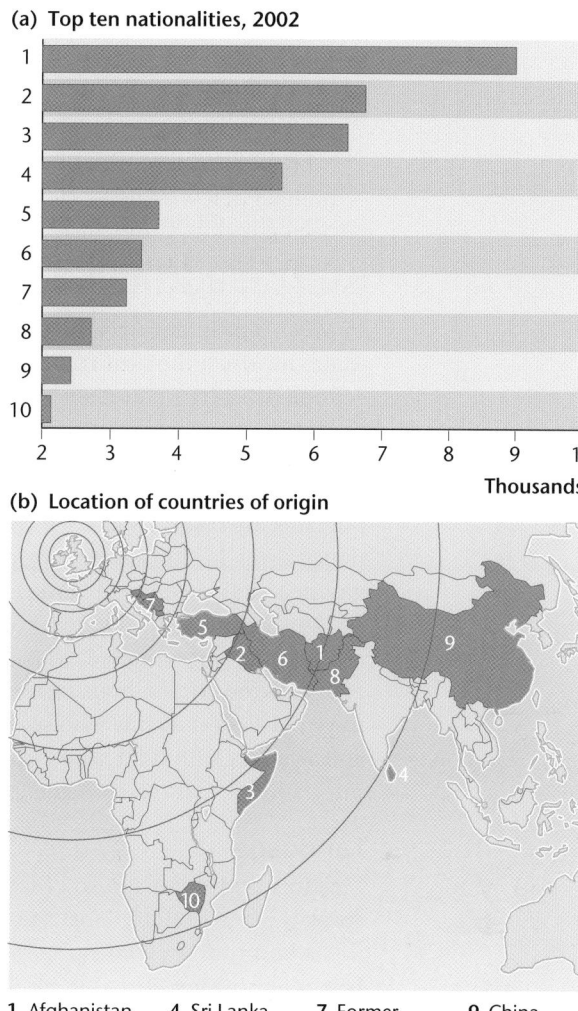

(a) **Top ten nationalities, 2002**

Thousands

(b) **Location of countries of origin**

1 Afghanistan	4 Sri Lanka	7 Former	9 China
2 Iraq	5 Turkey	Yugoslavia	10 Zimbabwe
3 Somalia	6 Iran	8 Pakistan	

Figure 11.8 Origin of those seeking asylum in the UK

consists of around 300 different ethnic groups, of which the Hausa-Fulani, the Ibo and the Yoruba are the main ones. Unlike the other two, the Hausa-Fulani are Muslim, and they have come into conflict with the federal government as they have tried to impose strict Islamic law in the northern provinces where they are in the majority.

In most countries there is at least one minority group and, while they may be able to live peacefully with the majority, it is more likely that there will be a certain amount of prejudice and discrimination leading to tensions and conflict. This is therefore an emotive and sensitive issue, particularly when cultural differences are interpreted as racial differences.

Current scientific research suggests that modern humans have descended from three main racial types, namely caucasoid, mongoloid and negroid, but the distinctions between them are now so blurred that race has little scientific status.

Skin colour remains a visible distinguishing feature but people also differ in their ethnic background which is expressed in terms of language, religion and culture.

Multicultural societies are often the product of migration, but they may also be the stimulus for it, as persecuted groups seek to escape oppression. The level of integration of minorities varies between societies. South Africa under apartheid clearly illustrates a lack of integration, whereas both Singapore and Brazil are highly multicultural countries where tolerant attitudes prevail.

Issues and attitudes

In MEDCs attitudes towards migrants and multiculturalism tend to reflect the extent to which the host country encourages and welcomes the opportunity for cultural diversity, or attempts to minimise the differences in culture. Some of the major concerns and issues to do with migrants are discussed below.

Ghettos

Migrants are often a source of cheap labour, working in low-paid construction, transport or health-service jobs. (In the UK in 2000, 27% of London Underground's staff, 26% of all workers in the health services and 14% of all workers in the catering industry were foreign born.) Migrants tend to be poor upon arrival, and this, coupled with low wages, leads to concentrations of ethnic groups in the poorest housing areas of major cities. Such concentrations are reinforced by later migrants who seek the support and security of living near friends and relatives within an ethnic community. The opportunity for assimilation or integration is reduced, and ghettos develop in these areas.

Language

Migrants find it difficult to obtain employment and to integrate if they do not speak the host-country language. Second-generation migrant children, educated in the host country, grow up speaking the language and have different aspirations from their parents. They are more likely to integrate, and this can cause tension within the ethnic group if they adopt the culture of the host country.

Religion

Migrants are likely to follow a different religion from the host population and this may cause friction with employers and authorities when migrants wish to adhere to their own religious calendars and practices.

Education

Educational opportunities may be restricted for migrants. Language difficulties and prejudice can result in problems for ethnic-minority children, who are often in poorer, lower-achieving schools. Local authorities may have to fund additional language teaching; some view this additional expenditure as an issue.

Economic issues

Migrants may not have equal opportunities in obtaining employment and may be subject to discrimination, prejudice or racism (Figure 11.13). The cost of state

Case study **South Africa**

For a long time in South Africa the minority white population was in political and economic control. The black people who made up three-quarters of the population had virtually no say in how the country was run (Figure 11.9). From its early days, first as a Dutch and then a British colony, South Africa pursued a policy of segregation. In 1946 this policy of apartheid (separate development) was legally established by the ruling National Party.

Under apartheid, white citizens were considered first class, the 'coloureds' and Indians were given some

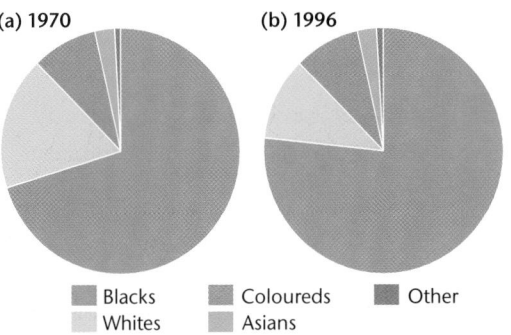

(a) 1970   **(b) 1996**

■ Blacks   ■ Coloureds   ■ Other
■ Whites   ■ Asians

rights but were distinctly second class, and blacks had virtually no rights outside of their homelands, the areas where they were forced to live. Mixed marriage between groups was illegal and the different groups had separate housing, education, employment, amenities and political rights. Apartheid was complete in that there was segregation in restaurants, places of entertainment, transport and even on beaches. Blacks were also not allowed to vote in general elections.

Under F. W. de Klerk's government, reforms paved the way for the eventual dismantling of the system. In the 1990s the African National Congress (ANC), the Communist Party and the Pan-African Congress were all recognised. Nelson Mandela, a prominent black activist, was released after 28 years in prison. The country held its first democratic elections in 1994 and Mandela became president of the new black-majority government. Apartheid may be over, but thousands of black people still live in poverty, with poor housing, high unemployment and inferior schooling.

Figure 11.9 Ethnic groups in South Africa

Case study **USA**

The USA has long been a 'melting pot' in which many ethnic groups have been assimilated. There are more Irish people in the USA than in Ireland, more Greeks than in Greece, more Jews than in Israel, and so on. Figure 11.10 shows the ethnic breakdown from the 2000 census (although the sensitivity of questions about race and ethnicity on the census form made the results difficult to interpret).

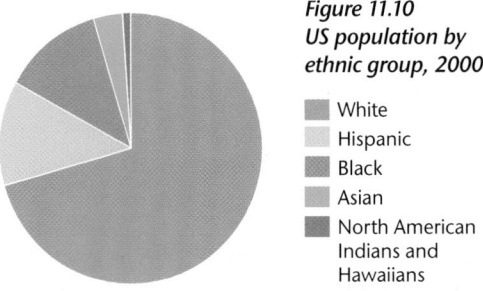

Figure 11.10 US population by ethnic group, 2000

■ White
■ Hispanic
■ Black
■ Asian
■ North American Indians and Hawaiians

The USA is a country of immigrants — the only indigenous people are the native North Americans and Hawaiians who formed only 1% of the population in 2000. Until the 1920s, immigration to the USA was dominated by Europeans, which is why whites form the largest group today. Most of the ancestors of the current black population were imported as slaves to work in the plantation system.

The Hispanic population has mainly migrated from Mexico, Puerto Rico and Cuba within the last 40 years, attracted by the high living standards in the USA compared to their home country. Many Hispanics first arrived in the USA as illegal immigrants. It was easy for them to hide from the authorities in cities in Florida and California, where many of their fellow countrymen lived. Asian Americans, who have mainly come from Japan, Taiwan, Korea and Vietnam, were also

attracted by the prospect of higher living standards and better job opportunities.

One of the most important features of this multi-racial society, like many others, is that it is unevenly distributed. Black people mainly live in the south and east and, outside of this, are almost exclusively in urban areas, often in very concentrated residential areas (for example Harlem in New York City and the South Side in Chicago). Not surprisingly, Hispanics, the fastest-growing group (Figure 11.11), are concentrated in those states which border Mexico and also in Florida, the regions which have Spanish as the most widely spoken language. Whites, as one would

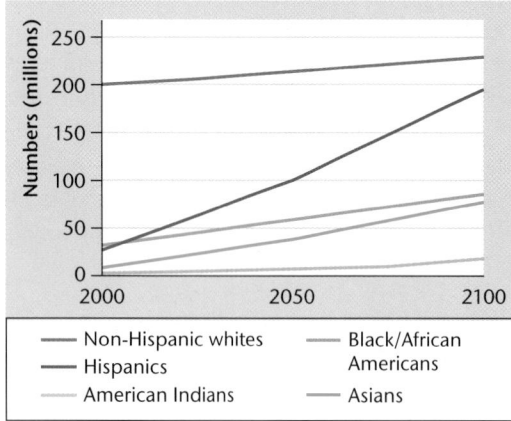

Figure 11.11 Projected populations of US ethnic groups, 2000–2100

expect, have the most widespread distribution, but as a general rule, the further north a state, the greater the proportion of whites living in it.

Although racial diversity and harmony are promoted in the USA, the reality is that the mix is often volatile. Conflict between various ethnic groups is rife, particularly in Los Angeles, where Hispanic, black and Armenian gangs fight. This was clearly demonstrated by the Los Angeles riots of 1991 which occurred after white police were acquitted of beating black driver Rodney King. In the trouble that followed, 53 people were killed and £600 million damage was done in 3 days of looting and arson. Los Angeles is also the home of the largest communities of Koreans and Iranians outside of their own nations. The 2000 census showed that whites now only make up 49.8% of California's population.

The same census showed that the white population is now in a minority in almost half of America's biggest cities, even in traditionally white cities such as Boston (whites 49.5%) and Milwaukee (whites 45%). Already the political consequences have been felt. In California, for example, the surge in the Hispanic population has led to bitter campaigning by the disappearing white majority for cuts in government services for illegal migrants and an end to 'affirmative action' (positive discrimination) on jobs and university places.

Case study *Singapore*

The population of Singapore is dominated by the Chinese who make up over three-quarters of the people (Figure 11.12). There are three main races:

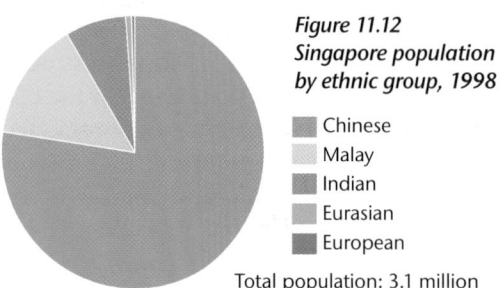

Figure 11.12 Singapore population by ethnic group, 1998

- Chinese
- Malay
- Indian
- Eurasian
- European

Total population: 3.1 million

Chinese, Malay and Indian, and each group is tolerant of the others. At one time there were ethnic residential areas in Chinatown, Arab Street and Little India, but the government has removed most of the older houses and such ethnic concentrations have been broken up . Most inhabitants of Singapore now live in modern high-rise blocks.

There is a very positive attempt to promote racial harmony, with educational, poster and media campaigns, and the government has created a National Day, an occasion for all ethnic groups to celebrate.

benefits for migrants' housing, education and unemployment may cause resentment and racial intolerance from members of the host population. The government policy of constructing centres to house asylum seekers in rural areas of the UK has caused concern and resentment among local residents.

Migration is often welcomed in periods of economic growth, but resentment occurs during economic recessions, when migrants are accused of taking jobs. This has occurred in Germany, France and the UK. It has been calculated, however, that the net tax contribution of migrants to the UK economy is £2.5 billion per year and that a 1% population increase through migration can lead to a 1.25–1.5% increase in GDP (UK Home Office statistics, May 2002).

Immigration controls

Many countries are tightening their rules on immigration and allocation of work permits. This makes it more difficult for both economic refugees and genuine asylum seekers to gain entry. There has been growing pressure for external controls to restrict immigration into Europe now that movement within the EU is easier. Some people are concerned that Europe's traditional role as a place of sanctuary is being replaced by an attitude of hostility — a 'fortress Europe' repelling migrants.

Some countries want to encourage migration of skilled workers to fill gaps in the employment structure. In March 2002, Germany introduced a scheme by which highly qualified foreigners — engineers, experts in information technology, mathematicians and managers with experience in science and research — could apply for permanent residency. Self-employed foreigners prepared to invest in Germany, and with a good chance of creating new jobs there, would also be allowed to settle in the country. The UK government announced a similar scheme

*Figure 11.13
Racist attacks on
migrants in Europe*

Thousands protest over arson attack

Thousands of people took to the streets in the German industrial town of Solingen yesterday after a fire bomb attack which killed five members of a Turkish family in their house only days after the German parliament had voted to tighten immigration controls.

Police were said to have detained a teenager linked to neo-Nazi groups after the arson attack on Saturday. The federal prosecutor said a number of right-wingers were being questioned but none was a prime suspect. The German government offered a DM100,000 (£38,000) reward for evidence leading to arrests.

About 2,000 Turks took part in two marches through Solingen, one led by Muslim leaders, the other by left-wingers and Turkish Kurds. Police kept the two groups apart. Mourners and neighbours laid flowers and lit candles outside the burnt-out house.

Last night Turkey, outraged by the racial assault, condemned Germany for failing to curb neo-Nazis. The Turkish president said he was horrified and called on Germany to tighten security. For Helmut Kohl, the German chancellor, the incident was so embarrassing that he did not visit the scene. Barely 2 weeks ago he visited Turkey and promised that the Turkish community in Germany would be better protected.

The Times, 31 May 1993

*Photograph 11.2
Asylum seekers from
Afghanistan at a
community centre
in Halifax*

Paula Solloway/Photofusion

in January 2002, under which some workers would be allowed to enter the country, depending on their 'educational qualifications, work experience and past earnings'. Initially, though, the UK was thinking of allowing such migrants to stay for only 1 year.

Migrants to LEDCs

In LEDCs, the main issue surrounding migrants is often cost, as countries cannot afford to support a large influx of migrants or to build temporary housing or shelter. Such governments may have humanitarian principles, but need help from the United Nations (UN), other governments and aid organisations such as the Red Cross and Oxfam. In August 2001, when Australia refused to allow a ship crammed with Afghan refugees into its territorial waters, Chris Janovski of the UNHCR said, 'If anybody should be speaking about being swamped, it is the Pakistanis and the Iranians, who for years have supported 3 million Afghan refugees.'

Changing attitudes

Governments have taken some positive steps to address migration issues. In the UK, there has been legislation on anti-racism, employment and equal opportunities to combat discrimination. Although children have been encouraged to integrate in schools, there has also been freedom to build Muslim and Jewish schools within the community. In social terms, cultural differences are better tolerated than they once were. Food, dress and custom have diminished as issues, and differences in religion are probably less significant because religious observance is decreasing in many host populations.

The decrease in the number of permanent migrants, in theory, should reduce pressure on housing, but second-generation migrants with higher social and economic aspirations do not want to remain trapped in low-cost inner-city areas. They resent the racism that they experience when they try to move to other areas.

Language differences become less obvious as migrant populations become established and this helps integration. In some urban areas ethnic groups are well integrated and have a high profile in the local community, with representatives on local councils or as MPs. In other areas there is tension and intolerance, and resentment and racism tend to increase during periods of economic difficulty and high unemployment.

In LEDCs, migrants are not provided with state benefits, and housing schemes tend to be more integrated. Migrants are less obviously segregated in urban areas and there is greater social tolerance.

Political geography

Within Europe today, there are two apparently contradictory processes in operation. Regions within some countries, as in other parts of the world, are seeking to break away from the control of centralised government. These are signs of **fragmentation**, with separatist pressures threatening to break up national units into smaller areas. At the same time, countries in the EU are strengthening their economic and political links. They are **integrating** and moving towards greater centralisation of power.

Separatist pressure

When the people of a region feel alienated from central government, they often seek to gain more political control. Such groups may have a different language, culture or religion from the rest of the state and are often geographically peripheral. They feel remote from centralised government and maintain that they do not receive adequate support, particularly with regard to economic development. Reasons for separatist pressure in a region include:

- an area which is economically depressed compared with a wealthier core
- a minority language or culture with a different history
- a minority religious grouping
- the perception that exploitation of local resources by national government produces little economic gain for the region
- peripheral location to the economic/political core
- collapse of the state, weakening the political power that held the regions together (e.g. the former USSR, Yugoslavia)
- the strengthening of supranational bodies such as the EU, which has led many nationalist groups to think they have a better chance of developing economically if they are independent

Key terms

Autonomy The right of self-government.

Separatism The attempt by regional groups to gain more political control from central government over the area in which they live. For some groups, the ideal would be total independence.

Supranational organisation A grouping of countries into a much larger organisation such as the European Union.

There are examples of separatist movements all over the world. Some of them have succeeded in their aims, while other struggles for independence are ongoing. In a few cases they have become bitter and violent. Some of the best known are:

- in Spain, the **Basque** area (northern Spain and southwest France, see case study) and **Catalonia** (northeast Spain), which now has the autonomy to decide many of its own affairs. The Catalan language, for example, has been taught in all schools in the region since 1983 and has become the official language in education
- the collapse of **Yugoslavia** and the formation of Croatia, Slovenia, Bosnia-Herzegovina and FYR Macedonia (see case study)
- in Canada, the question of independence for French-speaking **Quebec**, and pressure from the Inuits in the north that led to the creation of a self-governing region known as **Nunavut** in 1999
- **Scottish** nationalism. Before its union with England, Scotland was a separate kingdom and it still has its own national church (Presbyterian), separate educational and legal systems and its own language (Gaelic) which is spoken in parts of the country. The Scottish National Party (SNP) feels that the exploitation of North Sea oil and gas has done little to develop the economy of Scotland. The drive for independence was partly satisfied by the establishment in 1999 of a parliament with limited tax-raising powers
- **Welsh** nationalism. Wales has its own language and culture and its nationalist party, Plaid Cymru, has gained increasing power. The drive for independence has been partly satisfied by the creation in 1999 of a Welsh Assembly (with some devolution of decision-making powers, but not tax raising)
- the break-up of the former **Soviet Union (USSR)** into its 15 constituent republics, including Russia, Moldova, Latvia, Ukraine, Georgia, Kazakhstan
- national groups *within* former Soviet republics pushing for independence, for example **Chechnya** in the Russian republic. Rebels have been put down with extreme force by the Russian army
- **Czechoslovakia,** which separated into the Czech Republic and Slovakia in 1993
- **Belgium**, which consists of a Flemish-speaking north and a French-speaking south (Wallonia), is almost two countries
- in France, where concessions on self-government have had to be granted to **Corsica**, but there is also a movement for autonomy in **Brittany** (Breton nationalism)
- in Italy, where the **Northern League** has been agitating for autonomy for some of the northern provinces, particularly Piedmont and the Veneto (Venice-Verona region)
- **East Timor**, which sought independence after being annexed by Indonesia in 1976. After a long and bloody struggle between the separatists and militia gangs supported by the Indonesian Army, the UN took control in 1999, handing over to a new government in May 2002
- the bitter struggle against the Sinhalese majority in Sri Lanka by the **Tamils**, who want to set up their own state in the northern part of the island. The civil war began in 1983, and since then has claimed over 60,000 lives, including that of the Indian prime minister, Rajiv Ghandi

Case study *The Basque region*

A number of communities in Spain see them-selves as distinct, based upon their culture but primarily on their language differences. The main Spanish language is Castilian, but other languages are used in parts of Spain, most notably in Catalonia (Catalan), Galicia (Galician) and the Basque area (Euskara). The Basque area covers northern Spain and the very southwest of France (Figure 11.14). There is some debate over its extent, but it is generally agreed that it covers four provinces in Spain — Vizcaya, Guipuzcoa, Alava and Navarre — and three regions of France — Labourd, Basse-Navarre and Soule.

Figure 11.14
The Basque area of Spain

Persecution of the Basques

In 1936, the Spanish government granted the Basque region some autonomy, but this govern-ment was overthrown by Franco in 1937, during the Spanish Civil War. Franco, who led Spain until 1975, executed or imprisoned thousands of Basque nation-alists. The use of Euskara was banned on buildings and road signs and in publications, and the teaching of the language was declared illegal. Basque culture and language were suppressed for over 40 years.

ETA

In 1959, a Basque nationalist organisation, ETA (Euskadi ta Askatasuna), was formed, which declared war on the Spanish state in the 1960s. ETA has operated a violent campaign, targeting police, security forces, and legal and government figures and buildings. It has stated that its armed struggle will continue until it has achieved an independent Basque state comprising the seven regions listed above (in both France and Spain).

In 1979, a government-held referendum resulted in massive support for autonomy of the Basques, and a Basque parliament was created. The entire leadership of the political party that many people link to ETA, Herri Batasuna (People's Unity), was imprisoned in December 1997 for collaboration with terrorists, although these convictions were quashed 20 months later.

After the 2001 elections, the regional Basque parlia-ment became dominated by the moderate Basque National Party (Partido Nacionalista Vasco), although it did not have an overall majority. This parliament controls health, education, police and many taxes within the Basque areas and is proof, according to the Spanish government, that the Basques enjoy the highest level of autonomy of any European region. That is clearly not enough for some, and the ETA campaign of bombing resumed in the summer of 2002 with attacks along the Costa del Sol, targeting the Spanish tourist trade.

In June 2002, the Spanish parliament passed a law to ban any political party that 'promotes, justi-fies or excuses terrorism' and, following this, the Batasuna party was banned for 3 years in August 2002, being forbidden to hold meetings, call street demonstrations or stand in municipal elections.

Case study *The former Yugoslavia*

The history of Yugoslavia is very complex because it is a region that has long been prey to the ambitions of its neighbours on every side. In the seventh century an area of the Balkans was occupied by itinerant Slavic peoples from the east, the Slavs. The areas covered by modern-day Slovenia and Croatia resisted orthodox religions from Byzantium, maintaining a Roman Catholic identity. The ground for repeated future conflict was therefore based on the region's significance as a dividing line between the eastern Orthodox Church and the western Roman Catholic Church. To further complicate matters, followers of Islam migrated into the southern part of the region, now covered by Bosnia-Herzegovina and Albania.

Between 1941 and 1945, during the Second World War, the area was occupied by the Germans and Italians. During this time the Communist Party of Yugoslavia under Josip Broz (known as Tito) gained wide support for its partisan activities. Tito took power in Yugoslavia in 1945, adopting a Soviet-style constitution which gave considerable autonomy to the country's constituent republics. Internally,

therefore, ethnic divisions were suppressed, but not extinguished.

Fragmentation

Following Tito's death in 1980, the rotating collective presidency (which Tito had designed to replace him) became increasingly concerned with holding the country together. The disparity between the richer northern areas of Slovenia and Croatia and the rest of the country became a source of friction. The two wealthier republics increasingly questioned the use and distribution of central funds, to which they were the sole net contributors. Friction between the different ethnic groups, which Tito had suppressed, also came to the fore.

The disintegration of Yugoslavia began in earnest in the spring of 1990. Slovenia and Croatia held multi-party elections which returned governments committed to the pursuit of independence in both republics. In October 1990, the Serb minority in the Croatian region of Krajina declared autonomy and, as Croatian forces sought to quell the uprising, the

(a) Before secession

(b) Dates of secession

Figure 11.15 The shrinking republic of Yugoslavia

Serb-led federal army intervened on the side of the Serbs. European Community negotiators sought a political settlement. Meanwhile the other Yugoslav republics staked out their positions in the conflict: Montenegro backed Serbia; Macedonia pushed for independence; and the three ethnic groups in Bosnia-Herzegovina (Muslim, Serb, Croat) all had different aspirations. By 1991, both Slovenia and Croatia were separate countries (Figure 11.15).

By 1992 the Yugoslav Federation was falling apart and nationalism had replaced communism as the dominant force in the Balkans. A further conflict had broken out in Bosnia, which had also declared independence. The Serbs who lived there were determined to remain within Yugoslavia and to help build a Greater Serbia, and they received strong backing from extremist groups in the Serb capital, Belgrade. Muslims were driven from their homes in carefully planned operations that became known as ethnic cleansing. In 1993, the Bosnian Muslim government was besieged in the capital Sarajevo, surrounded by Bosnian Serb forces which controlled about 70% of the country. In central Bosnia, the mainly Muslim army was fighting a separate war against Bosnian Croats who wished to be part of a larger Croatia.

The Dayton Agreement

American pressure to end the war eventually led to the Dayton Agreement of November 1995, which created two self-governing units within Bosnia — the Bosnian Serb Republic and the Muslim (Bosnjak)-Croat Federation. The Muslim-Croat and Serb units have their own governments, parliaments and armies.

Croatia, meanwhile, took back most of the territory captured by Serbs when it waged lightning military campaigns in 1995, which also resulted in the mass exodus of around 200,000 Serbs from Croatia. Macedonia declared its independence and set itself up as the Former Yugoslav Republic (FYR) of Macedonia. Unfortunately, it too has been affected by strife, as the Albanian population, backed by the National Liberation Army, has attempted ethnic cleansing of Macedonians in some areas.

Photograph 11.3
Kosovar Albanians
fleeing fighting in
February 1999

Popperfoto/Reuters

Kosovo

In 1998, the Kosovo Liberation Army, supported by the majority ethnic Albanians, came out in open rebellion against Serb rule. The international community, while supporting greater autonomy, opposed the Kosovar Albanians' demand for independence. Evidence that the Serbs were conducting a ruthless campaign in Kosovo led to NATO intervention in early 1999 with air strikes on Belgrade and other parts of Yugoslavia.

Within days of the intervention, tens of thousands of Kosovan refugees were pouring out of the province (Photograph 11.3), with accounts of killings, atrocities and forced expulsions at the hand of Serb forces. Returning them home became a top priority for the NATO countries. Since 1999, Kosovo has remained within the Federal Republic of Yugoslavia but is effectively under international protection. At the end of the conflict, the UN set up a criminal tribunal to ascertain whether the leaders of the former Yugoslavia were guilty of war crimes in Kosovo. In 2001, Slobodan Milosevic, the president of Yugoslavia, was forced to stand trial for atrocities he is said to have ordered in Kosovo.

- the southern region of **Sudan**, where the population is mainly Christian and is fighting for independence against the majority Muslim population of the north
- **Western Sahara**, which has been fighting for independence since 1975 when armed forces occupied the country and incorporated it into Morocco following Spain's withdrawal

The grouping of nations: the European Union

In 1957 the signing of the Treaty of Rome created the European Economic Community (EEC). The original six members were added to over the years to give a membership of 15 by the beginning of the twenty-first century. The organisation changed its name first to the European Community (EC) and then, in November 1993, to the European Union (EU). Countries joined as follows:

- **1957:** the six initial members were France, Italy, West Germany, Belgium, Netherlands, Luxembourg
- **1973:** UK, Ireland, Denmark
- **1981:** Greece
- **1986:** Spain, Portugal
- **1995:** Austria, Sweden, Finland

The EU increased in size in 2004 when ten countries, mainly from eastern Europe, gained membership: Cyprus, Czech Republic, Estonia, Hungary, Latvia, Lithuania, Malta, Poland, Slovakia and Slovenia. It is expected that Romania, Bulgaria and Turkey will join at some future point. In 2001, a referendum in Switzerland rejected demands for the government to open membership talks with the EU.

Enlargement is controversial. Existing members are worried about:

- the cost of supporting new, poorer members
- a flood of cheap labour into western countries from the east
- the effects of applying the Common Agricultural Policy (CAP) to poorer and more highly agricultural countries in the east
- the lack of human rights in some potential members, notably Turkey
- the fact that Cyprus is still partly occupied by Turkey

Initial aims of the EEC

The 1957 treaty was established to develop closer ties between the member states, with a very definite economic emphasis. It was considered important that West Germany should be tied into an economic union that could guarantee future peace on the continent. By promoting prosperity and encouraging greater international cooperation the EEC hoped to maintain peace.

The aims of economic integration have been promoted through the years by:

- reducing tariffs and barriers to trade between members
- establishing a common external tariff for imports from outside the EU
- allowing free movement of labour, capital and other factors of production
- establishing common policies on agriculture, fishing, industry, energy and transport

Countries have developed greater interdependence and, as their economies become more integrated, all should benefit through rising living standards.

Trade encourages competition and this promotes a greater efficiency through economies of scale as each producer has access to a much larger market. The potential EU market stood at 377 million people in 2002. Less efficient producers, however, are forced out of business and this encourages countries to specialise in those industries where they are most productive. This is known as developing **comparative advantage**. Trade barriers between countries have gradually been removed, creating a customs union, but one which is protected by external tariff barriers.

Treaty of Maastricht

The Treaty of Maastricht, which was signed in December 1991, further strengthened the economic union and the political links between members. This significantly reduced the control that nations have over their own economies and policies. Although there was much to negotiate, all countries in the EU had ratified the treaty by 1993, although the UK and Denmark were able to insert substantial opt-out clauses for themselves. This treaty started the move towards monetary union and the common currency, the euro, which came into use on 1 January 2002 throughout the EU, with the exception of the UK, Denmark, Sweden and Greece.

*Figure 11.16
The EU and the eastern enlargement, 2004*

□ EU countries
□ Eastern enlargement

D Denmark   SV Slovenia
N Netherlands   BS Bosnia
B Belgium   SB Serbia
L Luxembourg   M   Macedonia
S Switzerland   AB Albania
SK Slovakia   MV Moldova

The constitution

The constitution brings together for the first time the many treaties and agreements on which the EU is based. It was agreed by the member nations in Brussels in June 2004 and signed on 29 October 2004. Among other things, the constitution sets out:

- the powers of the EU, stating where it can act. The EU already has rights to legislate over external trade and customs policy, the internal market, the monetary policy of countries in the eurozone, agriculture and fisheries and many area of domestic law, including the environment, and health and safety at work. The constitution extends these rights into new areas such as justice policy, including asylum and immigration
- the principle of voting by qualified majority, although members will retain the right of veto in foreign policy, defence and taxation. The definition of a qualified majority is 'at least 55% of the members of the Council, comprising at least 15 of them and representing states comprising at least 65% of the population of the EU'
- that the European Council, i.e. the heads of state or government of the member states, shall elect a president, by qualified majority, for a term of $2\frac{1}{2}$ years (renewable once). The new president will have to be approved by the European Parliament
- that a minister of foreign affairs is to be appointed by the European Council
- that the European Commission will consist of one national from each member state. This will be slimmed down in the future

The constitution has to be ratified by all the EU member states within 2 years of the signing date. This process varies between the member states — some countries initially favoured direct democracy (via a referendum) while others favoured parliamentary democracy. In 2005 the success of the 'no' vote in the referendums held in the Netherlands and France made the status of the constitution very uncertain. Many believed new negotiations would have to be held to determine the future structure of the organisation.

The Common Agricultural Policy

The Common Agricultural Policy (CAP) was one of the first to be set up when the EEC was formed. It was established with a number of basic aims:

- to increase agricultural productivity within member states
- to ensure a fair standard of living for farmers
- to stabilise agricultural markets within and between member states
- to ensure reasonable consumer prices
- to maintain employment in agricultural areas

These aims replaced existing national agricultural policies and often caused conflict between member states. Under the CAP, farmers were given guaranteed prices for their produce, known as the 'intervention' price. If world prices fell below this, then the intervention price was paid to farmers, encouraging them to maximise production. This created surpluses in a range of products, sometimes known as 'mountains' or 'lakes'. Over the last 40 years products in surplus have included cereals, butter, beef, apples, oranges, tobacco and wine.

Agriculture provides only 5% of the EU's total income, but at one time 70% of its budget went on supporting agriculture. The net gainers from the CAP tended to be countries such as France and those in southern Europe with smaller, less efficient farms. The net losers tended to be those countries with a smaller agricultural sector but with efficient farms, such as the UK. By the mid-1980s it was accepted that the CAP had brought great benefits, such as close to self-sufficiency in food production, but it had also caused problems:

- the surplus production detailed above
- over-intensive farming which was damaging the environment, especially the use of fertilisers
- growing tension between the EU and some of its main trading partners, such as the USA, Australia and New Zealand, over the impact of EU-subsidised produce on world markets
- large, prosperous farmers benefiting more than the medium to small farmers. This caused many smaller farmers to leave the land and migrate to urban areas

CAP reform

In 1992, radical reforms to the system were introduced, in which:

- the support for cereals, beef and sheep was reduced
- quotas were introduced, particularly in dairy farming
- there was to be an increase in set-aside policies
- environmentally sensitive farming was to be encouraged, decreasing the use of fertilisers and pesticides
- early retirement plans for farmers aged 55 and above were to be implemented

Although surpluses fell dramatically through the mid-1990s, several member governments were still not happy with the way in which the CAP operated. Germany, which was the CAP's main paymaster, was particularly anxious to reduce its net contributions. There was also the problem of accommodating the agricultural economies of the countries of central and eastern Europe that were lining up to join the EU.

Germany and the UK were aware that the CAP could not sustain this level of funding without financial problems. A move to curb open-ended production-based subsidies was inevitable, to prevent the collapse of the CAP. The other major factor driving reform was the need for the EU to comply with World Trade Organization (WTO) negotiations to work towards freer trade in food commodities. Import tariffs and export subsidies needed to be cut, and European farmers forced to rely more on world prices. Consumers in Europe should then benefit from cheaper food and the environment should benefit from a shift in emphasis from agricultural production to 'rural stewardship'.

These reforms, which were agreed in March 1999, did not go as far as many wanted. Some member states voted to reduce the level of changes, claiming that the effect on their agriculture and farming communities would be too great. In 2002 a new plan was put forward by the European Commission that will switch funds gradually from intensive production to schemes that promote rural life, safer food, animal welfare and a greener environment. Farmers will no longer be subsidised on the basis of crop area or head of livestock, ending the incentive

	Area farmed (million ha)	Number of farms (thousands)	Average farm size (ha)	Share in employed working population (%)	Share of agriculture in the GDP (%)
UK	15.7	233	67.7	1.4	0.7
Belgium	1.4	62	22.6	1.8	1.0
Denmark	2.7	58	45.7	3.2	1.8
Germany	17.0	472	36.3	2.5	0.8
Greece	3.9	817	4.4	15.8	6.5
Spain	25.3	1,287	20.3	5.9	3.4
France	29.6	664	42.0	4.1	2.1
Ireland	4.3	142	31.4	6.9	2.0
Italy	15.3	2,154	6.1	4.9	2.3
Luxembourg	0.12	3	45.3	2.0	0.6
Netherlands	1.9	102	20.0	2.9	2.0
Austria	3.4	200	17.0	5.7	1.2
Portugal	3.8	416	9.3	12.5	2.5
Finland	2.3	81	27.3	5.5	1.2
Sweden	3.0	81	37.8	2.5	0.6

Table 11.4
Agriculture in the
EU, 2002

that leads to over-production. It is intended that no farming operation will receive more than £200,000 per year, ending the anomaly in which 80% of CAP funds go to the big farmers while the smallest producers receive nothing. Table 11.4 shows the agricultural situation in the EU member states in 2002.

These plans were adopted by EU farm ministers in June 2003. The new CAP is geared towards consumers and taxpayers, but gives EU farmers the freedom to produce what the market wants. In applying the new regulations in the UK in early 2005, the Department for Environment, Food and Rural Affairs website informed farmers that:

> the CAP reform will simplify arrangements for subsidy payments by replacing ten major CAP payment schemes with one new single payment. Farmers will have greater freedom to farm to the demands of the market, as subsidies will be decoupled from production. At the same time, environmentally friendly practices will be better acknowledged and rewarded.

Attitudes to the growth of the EU

Attitudes to membership of the EU vary within and between countries. The countries of central and eastern Europe can clearly see the advantages of belonging to this powerful trading bloc, but some politicians and others within the existing membership have a very different view of the EU. Arguments in **favour** of the EU and its policies include:

- increased trading opportunities
- reduced tariff barriers between member states
- ease of movement for workers between member countries
- ease of movement for travellers following removal of border checks in the Schengen agreement
- support for agriculture through the CAP
- support for remote agricultural and declining industrial regions

- availability of urban regeneration funds
- protection against cheap imports
- a common currency, the euro, preventing currency fluctuations and simplifying transactions
- reduced risk of war between countries
- representation in world affairs for smaller nations

Those people who **oppose** the EU or some of its institutions cite:
- an increased feeling of centralised government and loss of sovereignty, as key decisions are taken in Brussels or Strasbourg by what some regard as an undemocratic bureaucracy
- loss of financial controls, particularly over the economy
- the cost of the CAP, the bureaucracy associated with it (farmers complain of being buried under paperwork), the fraud to which it is vulnerable and the environmental damage it has caused
- quota systems which affect national industries such as fishing
- pressure to adopt EU legislation, e.g. the Social Chapter, minimum wage, Bosman ruling on soccer transfers, weights and measures, food legislation
- difficulty in meeting requirements and costs of EU directives, e.g. on pollution
- the threat of a Federal Europe as the EU becomes more powerful and national governments exercise fewer powers (some European politicians see this as the real *strength* of the EU and would include it in the advantages section)
- the financial burden of the eastern enlargement of the EU
- the increasing loss of national vetoes and the move towards more qualified majority voting within EU institutions

Development issues

The core–periphery relationship

Friedmann produced a model of the economic development of a country, with particular reference to the changing economic relationships within that country (Figure 11.17). The model passes through four stages of development:
- **Stage 1** The country begins with a number of relatively independent local centres, each of which serves a small region. The country does not have a settlement hierarchy.

Figure 11.17 Friedmann's development model

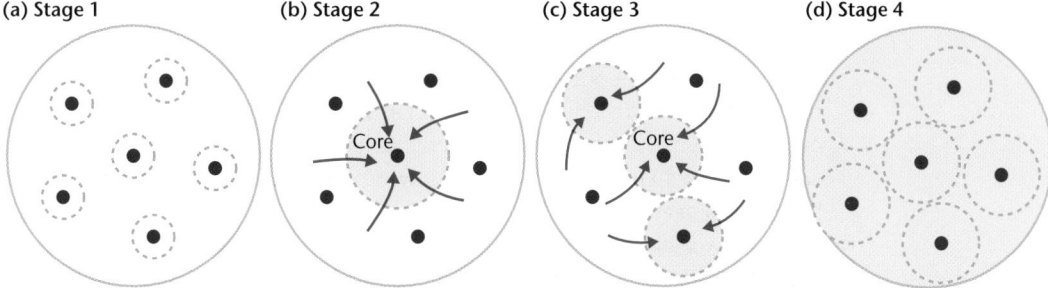

(a) Stage 1   (b) Stage 2   (c) Stage 3   (d) Stage 4

Key terms

Core The name given to that part of a country which has the highest concentration of economic development and is therefore the most prosperous.

Periphery An area of low or declining economic development within a country. Some geographers refer to everything outside the core as the periphery.

- **Stage 2** A single strong core develops during the initial phase of industrialisation, with an undeveloped **periphery** making up the remainder of the country. Economic development occurs in the core region, which has specific advantages over the rest of the country, for example a natural resource, dense population or a good location for transport (e.g. a river estuary). The initial advantage is maintained by cumulative causation (Figure 11.18) as more capital, entrepreneurs and labour move to the core from the periphery.
- **Stage 3** The core–periphery structure becomes transformed into a multi-nuclear structure with the national core and a number of **sub-cores**.
- **Stage 4** A functional interdependent system of cities develops, resulting in national integration and maximum growth potential.

Four types of region can be designated from this model:
- the **core** region — the focus of the national market and seedbed of new industry and innovations
- **upward transitional** areas which have some form of natural endowment, characterised by the inward migration of people and investment
- **downward transitional** areas (or **periphery**) which have unfavourable locations and resource bases, characterised by outward migration of people and investment. Living standards tend to be low. Decline may be irreversible, with the region locked into a downward spiral (or vicious circle) (see Figure 11.19)
- **resource frontiers** where new resources are discovered and exploited

Figure 11.18 Cumulative causation

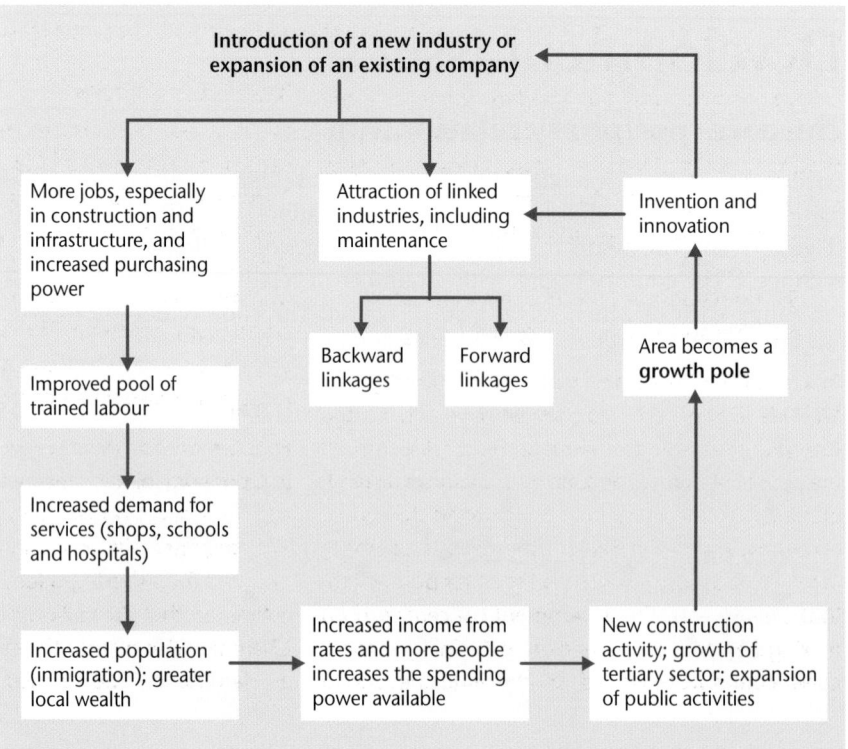

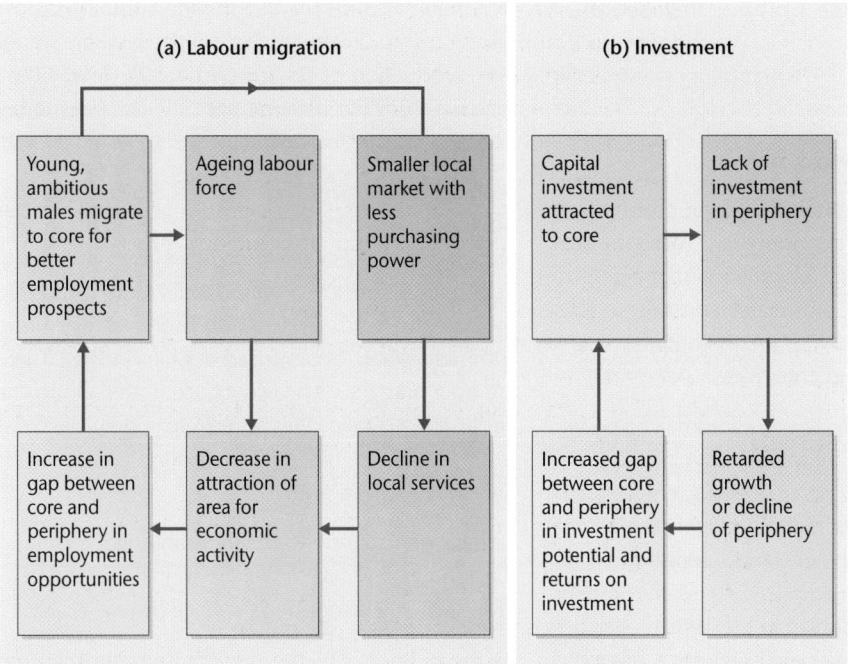

*Figure 11.19
Downward spiral
in the periphery*

Case study *The People's Republic of China*

The People's Republic of China is the world's most populous country. In 2001 it had a population of 1.295 billion. It also generates a huge GDP ($1,159 billion) but, because of the size of the population, the GDP per capita is just under $900. However, the purchasing power parity of that figure is close to $3,600. China's GDP is generated as follows:

- 51% from industry (manufacturing, mining and construction)
- 33% from transport, commerce and services
- 16% from agriculture, forestry and fishing

In 1949, China became a Communist country and Mao Ze Dong came to power. The aim of the Communists was to create a better life for all the people of China. The state took charge of all land, factories and businesses and controlled the whole economy. A series of 5-year plans was established, in which each sector was set a target for production. In the early years, the government concentrated on making steel and textiles, and on coal mining, as these commodities were needed for other industries and to

meet people's needs. The first 5-year plan ran from 1953 to 1957. The tenth 5-year plan began in 2001.

Mao's initiatives

In 1959, the government decided to increase the pace of development and launched the **Great Leap Forward**. This plan stressed the need for people to work together. Instead of having their own land, families worked the village land together as a community, or commune, and each commune was set targets for production. The government decided all wage levels. Agricultural communes were encouraged to begin to make small-scale industrial products.

Food production did increase, but bad management and poor organisation meant that transport links were not ready and food was left to rot in the communes. This, together with environmental damage, resulted in severe food shortages for 3 years, and it has been estimated that 20 million people died as a result. Chairman Mao blamed the failures on managers and academics. In 1966 he tried to control the people

even more through the **Cultural Revolution**, in which managers and academics were sent to work in the fields. Mao died in 1976 and the new leadership tried to develop new ideas.

Introducing market forces

In 1978 the Chinese leadership began moving from a **centrally-planned economy** to a more **market-orientated** system. This operated within a political framework of strict Communist control (socialist market economy), but with increasing influence of non-state managers and enterprises.

The authorities increased the decision-making powers of local officials and plant managers in industry, permitted a wide variety of small-scale enterprise in services and light manufacturing (such as electrical goods and small machines), and opened the economy to increased foreign trade and investment. Agricultural output doubled in the 1980s as farmers were encouraged to make profits for themselves and, overall, the economy grew at a rate of 10.2% annually.

Special economic zones

These ideas were taken a stage further with the creation of **special economic zones (SEZs)** and **open cities**. These areas of the country are allowed to attract foreign companies to set up plants. Foreign investors receive preferable tax, tariff and investment treatment. This means that foreign companies now have access to Chinese markets, in areas where wages and production costs are particularly low. China benefits from this arrangement by earning money from abroad and increasing the skills of its workforce.

Both SEZs and open cities were initially concentrated on the coast, facing Taiwan and the Pacific. For example, four SEZs were established near Hong Kong, at Zhuhai, Shenzhen, Shantou and Xiamen, and also on Hainan Island. Hong Kong itself, with its huge commercial hub, became part of China in 1997. There are now many SEZs heavily concentrated along the Pacific coast. Open cities also stretch down the coast from Tianjin in the north, through Shanghai to Zhanjiang in the south.

Larger open zones have been established, including the Yangtze River delta near Shanghai and the Pearl River delta in Guangdong Province (Figure 11.20).

As the economy of the eastern and coastal areas boomed it was clear that this area was developing into the economic core of the country. It accounted for about two-thirds of China's industrial production, 80% of export earnings and 90% of foreign capital. There are large numbers of foreign-built factories (Photograph 11.4) which assemble imported components into consumer goods for export, as well as many clothing manufacturers.

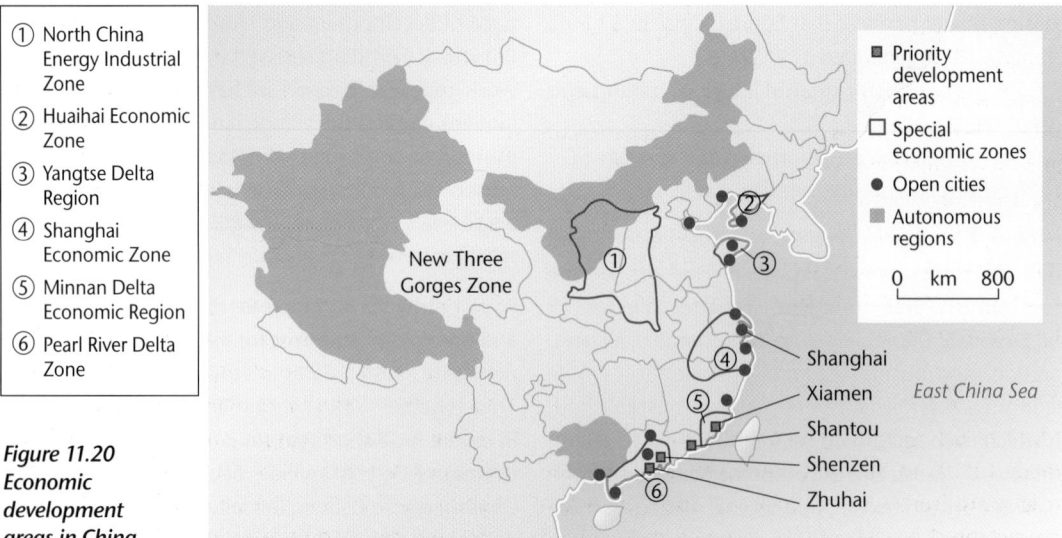

1. North China Energy Industrial Zone
2. Huaihai Economic Zone
3. Yangtse Delta Region
4. Shanghai Economic Zone
5. Minnan Delta Economic Region
6. Pearl River Delta Zone

Figure 11.20 Economic development areas in China

- ■ Priority development areas
- □ Special economic zones
- ● Open cities
- ▨ Autonomous regions

0   km   800

New Three Gorges Zone

Shanghai
Xiamen
Shantou
Shenzen
Zhuhai

East China Sea

Photograph 11.4 An evening market outside the Volkswagen car plant in Shanghai

A growing economy

The economy, as a whole, has continued to grow. The figure for the period 1990–2000 was 10.3% a year. China has authorised some foreign banks to open branches in Shanghai and allowed foreign investors to purchase shares in Chinese stocks. In 2001, over $41 billion was received as **foreign direct investment (FDI)**, the main sources being the USA, Taiwan, Japan and TNC investment in Hong Kong. FDI has taken place not only in the Pacific coastal area, but also in key locations within the body of the country. China's entry into the World Trade Organization in October 2001 encouraged further economic growth and foreign investment.

China's periphery

Much of the rural interior of China and the mountainous areas to the west and south constitute the true **periphery** of the country. Such areas are characterised by subsistence farming and pastoral activities, and people have very low living standards. This has meant much rural–urban migration to the cities, particularly those on the eastern coast, in search of work.

The government needed to promote such areas at the expense of the southeast. As a result some tax breaks in the southeast have been ended, with reduced incentives for foreign investment in the area. A great deal of attention has focused on the Chongqing region (Figure 11.21), which the government sees as the new growth pole for the west of the country. It is the hub of road, rail and water routes connecting western China with other regions in the south and west.

Chongqing is typical of many Chinese cities away from the core in having a large number of debt-ridden state-controlled enterprises. In fact, 90% of Chongqing's industry is state-owned, a lot of it defence related. Unemployment in the region is high, nearly seven times higher than the national average, and 20% of the population of the city is either unemployed or living on subsistence wages

311

Figure 11.21 China's core and periphery areas

provided by factories operating well below capacity. The establishment of the Three Gorges Zone here (Figure 11.20) will give the area greater autonomy from Beijing and should begin to attract foreign investment.

The Three Gorges Dam

One of the most contentious projects in the area is the Three Gorges Dam scheme on the River Yangtze. The estimated cost of this is around $25 billion, of which 10% is being met by foreign investment. The aims of the project are to:

- control the river (to prevent serious flooding and loss of life)
- generate HEP (up to 18% of China's needs)
- provide water for irrigation (allowing more cash-crop production)
- improve river transport (by eradicating rapids on that section of the river)

The scheme has come in for much criticism, both inside China and internationally. The main objections have focused on the following:

- over 2 million people will be displaced
- around 130 towns and cities will be submerged
- up to 7,000 factories will have to close
- many architectural, cultural and historic sites will be flooded
- thousands of hectares of valuable farmland will be submerged

- once it is in use, the reservoir will sediment up badly
- the river is currently used for a massive amount of waste disposal. Other systems are not in place to cope with all of this
- doubts have been raised as to the market for the electricity produced

Resource frontiers

There are a number of sites throughout China where **resource frontiers** exist. They include:

- the major oilfields at Daqing and Liaohe in north-eastern China, which produce nearly 70% of China's oil
- the offshore oilfields in the Bohai Sea (east of Tianjin), the Pearl River mouth and the Gulf of Tonkin
- the gas field in Xinjiang province to the west

Problems caused by growth

The recent economic growth in China has not been without problems. Some of the major concerns are:

- a dramatic increase in the income gap between rich and poor
- the relative poverty of much of the interior and the mountain areas, which has led to massive rural–urban migration. As people leave the authoritarian rural communities and move to the more liberal cities, government control over its one-child population policy is declining
- unemployment has grown and many people are moving around China looking for work
- inflation has risen
- the large state-owned industries (steel, armaments, textiles) have hardly changed and are not able to keep pace with foreign competition
- the environment is deteriorating rapidly in some areas. There have been large losses of arable land to soil erosion, and extensive air pollution. In 2000 the World Health Organization said that seven of the world's ten most polluted cities were in China. It is predicted that carbon monoxide emissions will grow substantially by 2030

Assessment exercises

This section contains questions for self-assessment based on the content of Module 4. There are two types of question: structured questions and essay questions. Each question is worth 25 marks and you should allow yourself 45 minutes when attempting to answer them. For the structured questions this time should be divided according to the mark allocation for each part.

Structured questions

Vegetation and soils

1 **a** Describe and explain the vegetation succession known as a 'lithosere'.   (7 marks)

  b Describe two ways in which vegetation successions may be modified by human activities within the British Isles.   (5 marks)

  c With the aid of a labelled diagram, describe and suggest reasons for the characteristic features of either: (i) a podzol, or (ii) a brown earth   (8 marks)

  d Explain two ways in which human activity alters the characteristics of a soil.   (5 marks)

          (25 marks)

International migration and multicultural societies

2 **a** Describe and suggest reasons for the major forms of international migration that have taken place in the last 30 years.   (8 marks)

  b Describe two pieces of evidence that suggest there are different attitudes to international migration.   (5 marks)

  c What is meant by the term 'multicultural society'? Illustrate your answer by reference to one such society.   (5 marks)

  d Discuss the issues that have arisen from the development of multicultural societies.   (7 marks)

          (25 marks)

Essay questions

Question 1

'The hazards of burglary and transmittable disease have significant impacts on people and the environment in which they live, and these impacts can be identified easily. Consequently, these hazards can be both prevented and/or managed effectively.'

  Discuss this statement. In your answer you should:

- identify the impacts of these two hazards
- describe the responses that have been made to prevent and/or manage these hazards
- discuss the effectiveness of these responses   *(25 marks)*

Question 2

Write an account of the conflict on the island of Harris, Scotland as detailed below.
In your answer you should:

- identify both the proposal that led to the conflict and the main participants in it

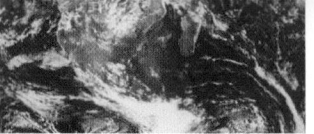
- discuss the attitudes of participants to the issues that have arisen in the conflict over the use of the resource
- describe and explain the processes that may operate to resolve the conflict *(25 marks)*

On a clear summer's day it is hard to imagine a more idyllic place than Lingarabay on the island of Harris in the Western Isles of Scotland. Meadows of ferns and lilies run down to the edge of a cove etched into the rocky shoreline. Buzzards and herons fly overhead. Rising above is the mass of Roineabhal mountain, covered in heather.

But, if an English quarry company has its way, Lingarabay will not remain a beautiful place for much longer. The company wants to develop a superquarry to excavate the massive reserves of hard rock which make up the mountain, in order to meet the demands of road builders and cement makers throughout the British Isles and Europe.

Table A
Population
of Harris

1931	4,500
1951	3,990
1971	2,880
1991	2,140

Table B
Likely economic impact of the superquarry

Years from start of scheme	Number of persons employed	Annual cash injection into local economy
0–5	210	£1.3 million
6–10	245	£2.2 million
11+	230	£3.5 million

Table C
Some socioeconomic aspects, 1991

	Harris	Scotland
Proportion of second/holiday homes (%)	13	2
Decline in school rolls 1981–91 (%)	27	4
People competing for each job vacancy	28	11
Average per capita income (£)	3,250	5,000

Question 3

'There has been debate over the origin of the vegetation of the tropical regions of Africa with wet and dry seasons. Some view it as a natural response to the seasonal changes that take place in the climate of this area, whereas others regard it as a product of human interference over the last 2,000 years.'

Discuss this statement. In your answer you should:

- identify the main features of the natural vegetation and the climate of this area
- describe the ways in which the vegetation is adapted to the seasonal climate of this area
- discuss the importance of human activity on the vegetation of this area *(25 marks)*

Question 4

'The growth of transnational corporations has had a major impact on the economic and social geography of a number of countries around the world.'

Discuss this statement with reference to the more economically developed countries (MEDCs), the less economically developed countries (LEDCs) and the newly industrialised countries (NICs).

In your answer you should include the following:

- a definition of a transnational corporation (TNC), illustrated by an example
- an explanation of the reasons for the growth in the number of such companies
- a discussion of their economic and social impact in named countries
- an examination of the attitudes of individuals and groups to their expansion *(25 marks)*

A2
Module 5

The synoptic module

Chapter 12

Answering synoptic questions

Synoptic assessment

In geography, synoptic assessment tests candidates' abilities to draw on their understanding of the connections between different aspects of the subject. The demonstration of this range of skills is called **synthesis**.

There are a number of ways in which synopticity can be assessed in this module. The following list covers the main ways in which candidates can demonstrate synopticity:

- showing an understanding of the interrelationships between different fields of geography, for example the links between physical, human and environmental geography; the links between different aspects of each of those fields; the links between places; and the ways in which these links operate at a variety of different scales
- applying facts and ideas which have been learnt and understood about one location, to gain a better understanding of another location
- using skills developed as part of the study of geography in a new situation

You will be given credit for any work that falls into the above categories, so long as it is relevant to the task given.

The examination

The synoptic unit test is an exercise in issue evaluation which may involve decision making. It requires you to use the range of geographical skills, knowledge and understanding that you have acquired throughout the A-level course. In general, you should be able to analyse evidence, establish criteria for the evaluation of an issue or for making a decision, and either evaluate the range of options concerning the management of an issue or justify the

---

Box 1
Issue evaluation

- Recognition and definition of the issue/problem.
- Selection and consideration of evidence from a variety of sources and points of view.
- Establishing the criteria for the evaluation of the relevant evidence.
- Evaluating a range of options concerning the management of the issue.
- Identifying and analysing potential areas of conflict.
- Considering ways of resolving and/or reducing conflict.
- Recommending ways of managing the issue, which may involve making a decision.
- Justifying the recommendations.
- Review of the process, including identifying possible outcomes.

---

decision that you make on the basis of that evidence and those criteria. Box 1 is an overall summary of the process of issue evaluation.

The issue evaluation exercise is always based on material that is sent out to candidates in advance, usually 4 weeks before the written examination. This is called the Advance Information Booklet (AIB).

Succeeding in the examination

Doing well in this type of examination involves two main phases:

- preparation for the examination using the pre-released Advance Information Booklet
- performing well in the examination itself

It is important to note that the assessment is conceived as a single exercise. Although it may be divided into a number of separate questions, these are written in a clear order of progression. The initial questions are to enable candidates to set the scene for the context within which the evaluation of the issue will have to take place.

Questions in the earlier part of the assessment are intended to direct you towards certain issues or problems that must be considered when answering the later questions. Cross-references back to earlier points can be used to illustrate good intellectual and organisational skills. It is these skills that such exercises seek to assess.

Working with the Advance Information Booklet

Before the exam you are expected to make yourself familiar with the information in the AIB. You should undertake several hours of preparation — as a rough guide 5–8 hours in total. The following are guidelines about the kind of work that should be done:

- Read and re-read the booklet to become familiar with its content and layout.
- Look up the meaning of any unfamiliar words and phrases used in the booklet. There is nothing to prevent you asking your teacher for help with this.
- Use an atlas and other sources to become reasonably familiar with the geographical area in which the exercise is located.
- List the concepts, theories, processes etc. that are referred to in the booklet, and check that they are completely familiar to you, by referring to course notes and textbooks.
- If necessary, carry out further research into the issue or the area referred to in the booklet.
- Consider some of the possible questions that might be asked, and think about how answers to such questions could be planned.

The internet

It is understandable that many candidates want to make use of the internet to support their background research into the geographical area and issue. However, bear in mind the following reservations:

■ Information on some websites may be biased towards a particular outcome and may not be based on sound geographical evidence. In other words, the material may be unreliable.

■ Information on the internet may be out-of-date, or indeed more up-to-date than the material that is provided in the AIB. It is dangerous to make use of such material because the decision you make has to be based on the information provided *within* the AIB, and nothing else. This is to ensure that all candidates can be assessed fairly on the same contextual material.

■ Information on the internet may not be relevant to the question set. Examiners may well choose to use the context of a real-world issue for a slightly different purpose — one that makes a more effective assessment exercise. Again you are reminded that the material in the AIB is of most importance to you.

Summarising the information

A number of techniques can be used to summarise the information presented in issue evaluation tasks:

■ **Tabulation:** a visual display of summarised information highlighting the key issues. This technique tends towards brevity and simplicity, and is not a good method for evaluation, only for summary.

■ **Ranking:** ordering options according to criteria, for example low priority to high priority, or most expensive to least expensive. Such a method is easier to apply to quantitative data than qualitative data.

■ **Scaling:** subjective ranking of data/information on a personal basis to allow some comparison and analysis. This may be based on relative advantages or disadvantages, or on relative impacts such as positive or negative outcomes. This technique introduces an element of personal bias, but does involve a degree of evaluation. The weighting of some factors may also feature here; for example, if the issue concerns flood control, then rainfall may have a greater influence than, say, vegetation type, and should be weighted accordingly.

■ **Matrices:** a more complex form of diagram illustrating relationships between different aspects of the information. Matrices can be used to identify possible areas of conflict, or environmental impacts. Such matrices should not be used in the examination itself, unless specifically requested. All answers require continuous prose, and even matrices with explanations of codes or numerical scores do *not* fit this requirement.

Thinking about the question

It is not a good idea to second guess too confidently what the issue evaluation tasks will be. Sometimes the tasks are easier to predict and, where this is the case, the examiners will be aware that the task appears straightforward. They are then more likely to introduce a more unfamiliar context or sub-context into the final question.

The nature of the resources tells you exactly what topic the exercise is going to focus on. Clearly if all the data relate to flooding, you should spend time reading through your notes on this topic to make sure you understand all its concepts, for example hydrology and hydrographs. If the context appears to be industrial location, then relevant general material should be read, and so on.

In many cases you will be given one or a number of opinions about an issue or problem in a particular area. It is useful to look at the issue or problem from the 'standing' of the 'owner' of the opinion. Is it based on professional research, fact or sheer emotion? Do the resources support the opinion in any way? Do the opinions conflict with each other?

While it is obviously not a good idea to prejudge the question, it is likely that you will have to compare the options in one way or another. This might, as explained above, involve weighing up advantages and disadvantages, socio-economic benefits and negative impacts, or cost–benefit analysis. It is therefore a good idea to spend time looking at the alternative options and preparing a table of some sort to summarise the information.

The best answers usually demonstrate an intimate knowledge of the advance information. References to different items within it are brought together in interesting and relevant ways, which allow the candidates to show good insights. These answers are also often supported by useful information from within the booklet. Geographical theories and ideas are well integrated into the answers, and candidates demonstrate their knowledge and understanding of their previous studies. In simple terms, these candidates are well-prepared.

The worst answers show little familiarity with the advance information. They are often brief and unstructured. They may include large sections copied out from the booklet, or with only small changes to the original wording. There are few developments of the ideas and concepts within the booklet, and little awareness is shown of the relevance of the candidate's previous studies within geography. In short, these candidates do not think geographically.

Answering the examination questions

You are not allowed to take your *annotated* pre-release material into the examination room. You may be provided with either a 'clean' set of material (for example a clean photocopy) or additional copies of the sheets that you need to use in the examination. You will also be given the task itself once you are in the exam.

You will be given a context within which you should answer the question. For example, you may be asked to imagine that you are a particular person, or have a particular role. Alternatively, you might be given a set of guidelines that should form the basis for all or some of your answers. In some cases, you may even be told that you represent a certain body with a specific 'mission statement'.

It is important that you empathise with this role, whatever it may be.

Remember:
- treat the paper as a whole exercise — all the questions are linked
- demonstrate familiarity with the data or information
- where there are maps and/or photographs, make sure you refer to them, and be accurate in your use of them (most maps and photographs will have grid squares on them — make sure you use these accurately)

The task itself will be varied, and divided into sub-tasks. There is a large number of possible sub-tasks. For example, you may be asked to:
- summarise a given set of geographical information
- identify problems/issues/conflicts
- consider a variety of proposals
- evaluate each of these proposals
- place the proposals in a rank order according to the context of the whole exercise
- justify that rank order
- recommend the best way to manage the proposals in the light of possible problems/conflicts
- examine the difficulties that may arise in their implementation
- make a decision
- justify that decision

As in other forms of examination it is important that you answer the question(s) asked, and keep a firm eye on the time. Appropriate use of the mark allocation will give an indication of the amount of time to be spent on each aspect of the exercise. By their nature the issue evaluation and/or decision-making elements and subsequent justification come at the end, so it is vital that you leave enough time to complete the task thoroughly.

The final question or section

When answering the final question or section, make sure you consider all or some of the following points:
- the short-term and long-term effects of each proposal
- the local, regional and national considerations
- the spatial (geographical) impacts of each proposal
- the social, economic and environmental impacts
- the costs and benefits of each proposal
- the effects of each proposal on different groups of people in the area

In addition, to access the highest mark ranges — the Level 4 marks in most cases — you need to demonstrate the highest level of geographical skills, including:
- critical understanding of the issue(s) being considered, often showing an awareness of the complexity of the real-world situation
- maturity of understanding of the issue(s), which is identified through a coherent, well-reasoned and perceptive argument

- evidence of high levels of synopticity (see earlier)
- insight into the issue(s), possibly including elements of creativity and/or flair

Finally, you may be asked to justify your answer. The command word **justify** is one of the most demanding that candidates face. At its most simplistic, a response to this command must include a strong piece of writing in favour of the chosen option(s), and an explanation of why the other options were rejected.

However, issue evaluation is not straightforward. All the options in an issue-evaluation scenario have positive and negative aspects. The options that are likely to be rejected will have some good elements to them, and equally, the chosen option will not be perfect in all respects. The key to good issue evaluation is to balance up the pros and cons of each option and to opt for the most appropriate, based on the evidence available. A good answer to the command 'justify' should therefore provide the following:

- for each of the options that are rejected: an outline of their positive and negative points, but with an overall statement of why the negatives outweigh the positives
- for the chosen option: an outline of the negative and the positive points, but with an overall statement of why the positives outweigh the negatives

Try to avoid repetition, but make brief references back to the previous sections of the paper. Develop the points made earlier in the context of the final task and in the context of the role you have assumed. Be logical, use the evidence accumulated from the rest of the exercise, and always remember that it was conceived to be a complete task in its own right.

Guernsey issue evaluation exercise

Dear student

Welcome to your work experience placement here on Guernsey and particularly to the Planning and Finance Department which manages the island's budget. Please take time to assimilate the background material given below and the details of three proposed development schemes for the island.

Each of these proposals will need to be put before the parish councils and the island council. No doubt these proposals will generate some heated debate. We need to be ready with a response to all the points which may be raised. Forewarned is forearmed.

So, I would like you to prepare a report by answering each of the questions on page 326.

Background information on Guernsey

Figure 12.1
Map of Guernsey

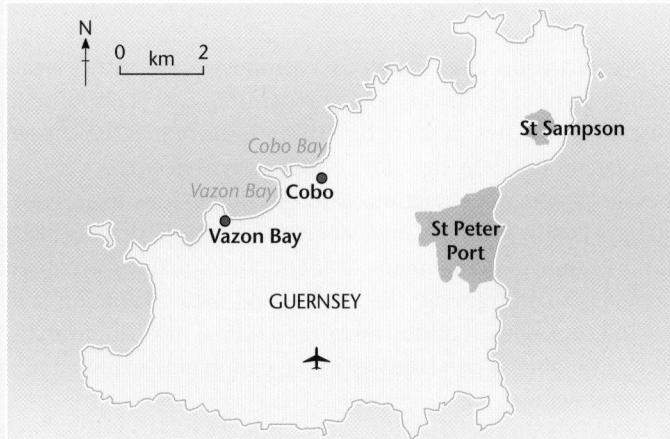

Figure 12.2
Population pyramid for Guernsey, 2001

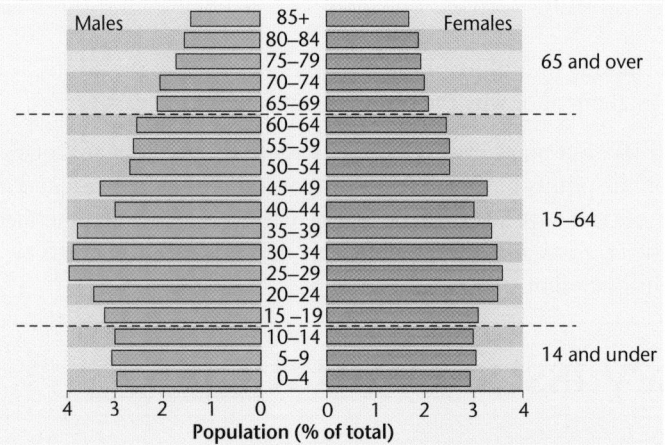

Figure 12.3
Guernsey population density, 2001, and population change by parish, 1991–2001

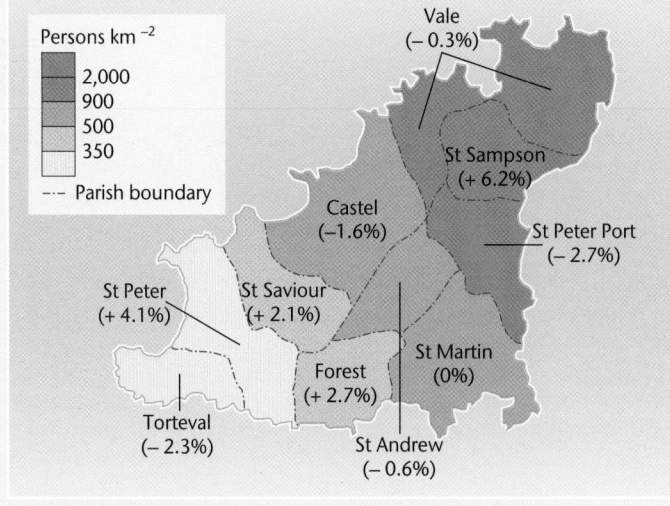

Age group (% of population)				
Year	0–14	15–64	65+	Total
2001	18	63.2	18.8	60,542
2011 (est.)	15	66	19	62,021
2021 (est.)	14	65	21	63,808

Table 12.1 Guernsey population totals and per cent in age groups, 2001–21 (estimated)

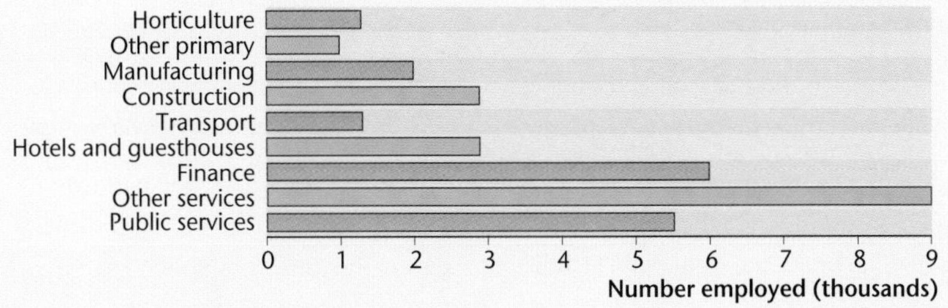

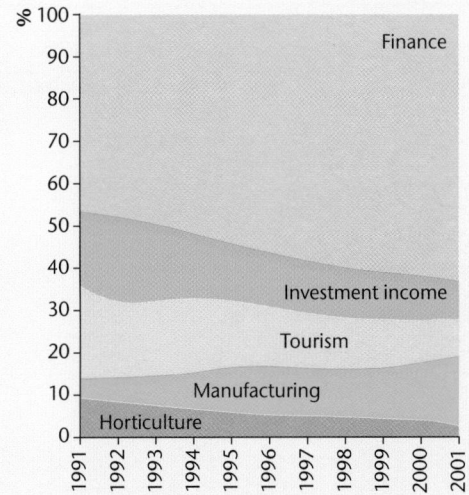

Figure 12.4 Economically active persons by sector in Guernsey, 2001

Figure 12.5 Contribution to Guernsey's exports by industrial sector, 1991–2001

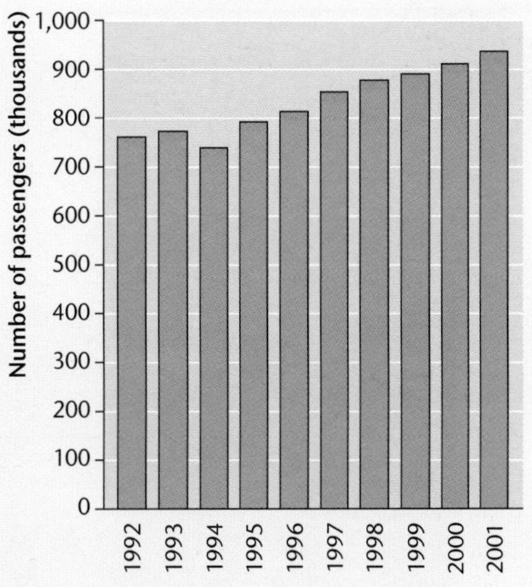

Figure 12.6 Number of air passengers (arrivals and departures) to Guernsey airport, 1992–2001

Type of visitor						
	Leisure	Business	Leisure (day trip)	Business (day trip)	Sailing	Total
1995	246	42	36	23	61	408
1998	316	52	31	16	66	481
2001	300	46	30	22	51	449
Average stay	7 days	2 days	1 day	1 day	3 days	

Table 12.2 Number of visitors to Guernsey, 1995–2001 (thousands)

Proposal 1 To extend the runway at Guernsey Airport

A new terminal building has recently been built at the airport. This proposal is that the runway should also be extended. Here are some views on the proposed expansion:

> Unless the current runway is extended, the move to bigger aircraft will result in fewer airlines being able to operate here in future. The airlines that serve Guernsey will not be able to operate their larger aircraft. The island will become an insignificant backwater.
> *An aviation journalist*

> Our airline has invested heavily in short take off and landing aircraft to ensure we can and will continue to operate in Guernsey with the runway at its present length.
> *Channel Islands Airline manager*

> We had a cargo plane crash on approach in 2000 hitting a house to the east of the airport and narrowly missing a petrol station. Any extension to the runway with more people passing through on larger planes will increase the risks of living under the flight path. People to the west of the runway extension will see property prices crash and the quality of life decline. Some closer to the runway may even lose their homes.
> *Councillor for St Peter Port living in the vicinity of the airport*

> Why should we, living in the rural parish of St Peter, have to lose land of landscape value and suffer more noise to satisfy the bankers' and hoteliers' desire for more money? Hotels are never full and the bankers do not need larger planes.
> *Small farmer close to the airport*

Photograph 12.1
The new terminal building at Guernsey Airport

Proposal 2 To develop a waterfront area close to the centre of St Peter Port for a hotel and exclusive housing

This proposal is to demolish some of the existing buildings in the St Peter Port area and to build a new hotel development with exclusive residential property nearby. The following criteria must be satisfied:

- The development should be well designed, in keeping with the setting and should bring about significant improvements to the area.
- It should not occupy open space or block an important view.
- Adequate car parking should be provided.

Photograph 12.2 The sea front in St Peter Port

Here are some views on the proposed scheme:

A top quality hotel close to the financial district of St Peter Port is needed for business people and corporate entertaining. It could also provide quality accommodation in the summer months when business demand is lower.   *Overseas investment manager of a major offshore bank*

We don't need hotels; we need affordable housing, especially for the elderly Guernsey-born population. Financial services have created wealth for the island. The wealth should be distributed to make the island a good place to live for everyone and not just the bankers.   *Retired former glasshouse worker living in St Sampson*

House prices have doubled in the past 10 years and an average house costs £200,000. How on earth can a young person like me begin to afford one on this island where I was born?   *Recently qualified nurse*

The provision for future accommodation on the island should be made predominantly in the urban areas. A range of development possibilities, including the use of brownfield sites, should be investigated. Housing provision for those of modest means is also specified in a current policy.   *Planning department representative*

Proposal 3 *To raise the sea wall along the western coast, starting with Vazon and Cobo Bays, to protect the coastal areas from rising sea levels*

Vazon Bay is a popular place for tourists who want a beach holiday. It is important that the beach is maintained and that the land behind the existing sea wall is protected. Here are some views on the proposed scheme:

> We need to make sure that the sea defences are strong in order to protect the properties lying behind and below the beach ridge at Vazon. Our timeshare hotel and golf course are threatened by storms attacking the sea wall. Salt water has penetrated into the drained land on which the golf course is sited.
>
> *Manager of Vazon Bay sea front hotel*

> Without the sea wall raised at the head of the Cobo Bay a storm surge will overtop the defences and sea water will flow inland threatening our houses. It is all the fault of global warming. Planners seem more interested in new housing or airport extensions.
>
> *Cobo Residents' Association*

> Once we start raising the defences here, we have to do it all along the coast. It is better to let nature take its course. Those who live at or near high water mark have made their choice. You cannot stop natural processes!
>
> *A retired geomorphology professor living on the island*

> When winter storms rage, my cottage is already washed by spray. If the sea wall is not raised and strengthened many coastal dwellers will have to move out. Defend what the island has, its existing cottages and homes.
>
> *Castel parish councillor*

Questions

Guidelines

In order to demonstrate your synoptic ability, you should, wherever possible, refer to a range of information, ideas and examples from the resources provided and from other modules you have studied. Use this method to show your understanding of the connections between different aspects of your course and the topic featured in the questions.

Report to the Guernsey Planning and Finance Department

1 Using only the information provided, summarise the main demographic and economic characteristics of the island of Guernsey. (15 marks)

2 Suggest possible conflicts that may arise in each case if each of the proposed schemes is approved separately. (15 marks)

3 Guernsey Planning and Finance Department has three strategic aims which guide the island's planning mechanisms:
- to improve the quality of life for all people on the island
- to manage the island in a cost effective and sustainable way
- to maintain and improve the quality of the environment

In the light of these strategic aims, conclude your report by ranking the three proposals in order for implementation, and justify your decisions. (20 marks)

Mark schemes

Question 1

Level 1 The answer simply restates the information given in the data in written form. There is no interpretation of the data and no recognition of trends or of significance beyond the obvious. For example, the candidate may include figures of individual age groups from the population pyramid or numbers employed in particular industries. There is no evidence of cross-reference between two or more sets of data.

(0–5 marks)

Level 2 The answer demonstrates some recognition of trends from the information in the data. These trends may take the form of analysis over time or involve projections into the future; alternatively, they may group sets of data that demonstrate similar or related factors. The answer provides evidence that the candidate can see beyond the obvious, for example, recognising the possible consequences if trends do continue. The answer may also demonstrate some evidence of cross-referencing between two or more sets of data. (6–10 marks)

Level 3 The answer must refer to both demographic and economic characteristics at the Level 2 standard to reach this level. Achieving Level 2 in both of these aspects would trigger the lowest mark within Level 3. (This is an example of a mechanical means by which Level 3 can be awarded in an examination answer.) The answer demonstrates wide-ranging synoptic ability. It refers to most if not all of the data provided and illustrates the degree to which the different elements within the data are interdependent — one impacts upon another, and not always in a straightforward manner. Short- and long-term implications of the trends in the information are stated clearly and effectively. The answer is written clearly and logically with a good command of the English language. (11–15 marks)

Question 2

Level 1 The answer contains simplistic statements regarding possible conflicts that may arise for any of the proposed schemes. It is not clear from the answer that the candidate has recognised the location of the possible conflict — it could apply to any similar area. Participants in the conflict are identified in very general terms and are usually grouped as being either in favour of or against the schemes. Attitudes contributing to the conflict are also stated in very general terms with no clear sense of attribution — for example, sentences begin with the phrase 'Some people would think...'. Answers that examine only one proposed scheme cannot go beyond Level 1.

(0–5 marks)

Level 2 The answer contains detailed statements about the possible causes of the conflicts. There is an understanding of the factors leading to the possible conflicts, together with a recognition that there will be some variation of attitude within groups of participants. The answer is clearly linked to the named and located areas. Participants in the possible conflicts are clearly identified and their attitudes are clearly attributed. However, material copied from the resources provided will not be credited at this level. The answer recognises that attitudes may vary over time, depending on the outcome of the decision-making process. (6–10 marks)

Level 3 The answer must contain Level 2 material on at least two of the proposed schemes to reach this level. The answer contains a thorough account of the possible causes of the conflicts, including background material on the area(s) and/or participants. There is some recognition of the variation in the basis of attitudes within each set of participants, depending on whether they would lose or benefit from the outcome of the conflict. There is also some consideration of the short- and long-term implications of the proposed schemes. There is evidence of critical understanding of geographical concepts and principles. The answer is written clearly and logically with a good command of the English language. (11–15 marks)

Question 3

Level 1 A basic ranking is made, but any reasons given for this are little more than unsubstantiated assertions of pros and/or cons. The candidate does not make direct reference to the strategic aims given in the question and it is impossible to see any indirect recognition of their importance. (0–5 marks)

Level 2 A clear ranking is made, together with some appropriate justification supported by clear arguments. The justification tends to be constructed simplistically. For example, the reasons for the number one proposal outweigh the reasons for the other two, or only the negative aspects of the other two schemes are given. There is clear use of and reference to the strategic aims given in the question. However, there is no recognition that these may be contradictory or may also be ranked. (6–10 marks)

Level 3 The answer is detailed and developed. The ranking is justified thoroughly. The positives and negatives of each of the three proposals are examined in depth and compared with each other. There is clear use of and reference to the strategic aims throughout the answer. There is recognition that the decision-making process in such cases would be very complex. A variety of considerations is dealt with in detail. Issues are considered at different scales and from differing viewpoints. The answer is developing a high level of synopticity — the connections within and between the information given both at the outset, and for each of the three proposals, are developed. The answer is written clearly and logically with a good command of the English language. (11–17 marks)

Level 4 The candidate has completed the task thoroughly. The answer critically evaluates the task that has been given in relation to the context of the exercise. There is evidence of critical understanding of geographical concepts and principles. There is a clear acknowledgement that there is going to be disagreement when the outcome of the ranking process is announced/published. There may be recognition that each of the different proposals would satisfy the three strategic aims but in different ways and that it is the relative ranking of these that could influence the decision-making process to a greater extent. The answer is carefully structured, shows real geographical insight, a clear sense of place and an understanding of a variety of different needs. Synopticity is shown throughout the answer. (18–20 marks)

Candidate answer

A sample answer is given, followed by the examiner's comments.

Question 1

The demographic characteristics are shown by means of the population pyramid, the table below it and the population distribution map. The population pyramid indicates that there is a falling birth rate as the percentage of people at each age group gets less towards its base. There is a bulge in the middle section of the pyramid, in the 25–39 age groups. This could mean either that there was an increase in birth rate at that time or that there has been an in-migration of young adult males to the island of Guernsey. There is almost an equal proportion of over 65s to those under 15 on the island, with more females than males in the older age groups. This illustrates that women have a longer life expectancy than males.

The table shows that the proportion of people under the age of 15 is likely to fall, and the proportion of elderly people will rise. This could mean that more retired people will be coming to the island, which is also supported by the fact that the total population of the island is likely to increase by 2021. The island's population is going to get bigger and older.

The population distribution map illustrates that most people live in St Peter Port on the east of the island, with high densities in the northeast at St Sampson and Vale. Lowest densities are in the southwest of the island.

Over 30,000 people are economically active, in other words in employment, on the island of Guernsey. If you compare this with the data in the table, where approximately 38,000 people are between the ages of 15 and 64, you can see this is a very high proportion of people who are in work. Unemployment is likely to be low. Over 20,000 of these people work in some sort of services — finance, public services and others. This is high at 66% of the workforce. Tourism is also important as shown by the fact that 3,000 people work in hotels and guest houses. The others sectors are low, with more people employed in construction than in manufacturing. Primary industries are low.

The information gives confusing data regarding tourism. The number of passengers to the airport shows signs of an increase and yet the number of people visiting the island shows a decrease between 1998 and 2001. However, it is clear that the contribution to exports that tourism has made is falling in proportion and is now only 10% overall. Although manufacturing employment is low the proportion of the island's wealth that comes from manufacturing is increasing. By far the biggest proportion of the island's wealth comes from financial affairs — over 70% if you combine finance and investment income. This is a massive proportion.

Examiner comments

The candidate has constructed an answer by carefully examining each of the resources in turn. Although this is an acceptable approach, its drawback is that some interrelationships can be missed. However, this candidate does manage to identify most of the characteristic trends in the data.

In the first paragraph the candidate identifies some of the major demographic trends and even seeks to explain them in general terms (for example, women having longer life expectancies). However, the candidate does not appreciate the scale of the estimated reduction in the number of under 15 year olds — a reduction of

nearly a quarter in 20 years. Low Level 2 would be awarded for this paragraph. The second paragraph shows a good degree of insight and moves the response well into Level 2, but the third paragraph is poor. It only makes simplistic comments and the information relating to population change is not referred to at all: St Peter Port is declining in population, with a growth in the nearby parish of St Sampson; three of the four parishes in the southwest of the island are increasing their populations.

The paragraphs on economic characteristics are also of a Level 2 standard. The candidate has attempted to make detailed use of the data with varying degrees of success. There are 32,500 people who are economically active, and it is true that unemployment would be extremely low on the island. The comment about the importance of services would have been even more accurate if the candidate had recognised that hotels, guest houses and transport are also service industries — making the proportion employed in the service sector nearly 80%. Perhaps it would have been better to make the direct link to export contributions here, where again 80% of export earnings come from some form of service activity — finance, investments and tourism. No comment is made regarding horticulture — this is surprising bearing in mind the Channel Islands' reputation for producing flowers and early vegetables.

Some interrelationships between the resources are established — for example, the comment regarding manufacturing — and both the demographic and economic characteristics are referred to at Level 2 standard. Consequently, a low Level 3 can be awarded and this answer would score 12 marks.

Candidate answer

Question 2

The proposed extension to the airport runway will cause conflict between those who want it to go ahead and those who do not. The idea of runway extension has caused conflict in a variety of locations in the UK — the most recent high-profile case was at Manchester airport. Here the argument was between industrialists who wanted to increase the amount of aircraft traffic through Manchester in order to encourage more investment in the area, together with the owners of the airport itself who obviously would make more money, and the environmentalists who wanted to save farmland, woodland areas and rare habitats. Guernsey earns a lot of its export income from service industries and tourism. An efficient airport is essential for both of these — flying tourists in and out, and flying business executives in and out.

Airlines with bigger aircraft will want a full length runway, otherwise they will not be able to use the airport, or may have to use smaller planes. The aviation journalist suggests therefore that the airline companies would want the expansion. He also suggests that it is in the interests of the island economically to have a longer runway — he says, if not, 'the island will become an insignificant backwater'.

However, both of these views are challenged by two of the other quotes. The airline manager states that an extension is not necessary due to investment in STOL aircraft. The councillor who lives near the airport, but represents St Peter Port 4 km away, who one may have thought would have supported the proposed extension, is against it on safety grounds. Perhaps he lives to the west of the airport and is worried that the value of his house may fall. If this is the case, he is not really representing the views of his constituents. The small farmers in the area are most likely to be against the expansion — they will lose land and still have to live under the noisy flight path. However, they are also most likely to gain compensation, so their views are likely to vary over time. They could even appear to be against the proposal, while actually being in favour of it.

The proposed development of the waterfront area in St Peter Port is likely to upset a lot of local people. They will be concerned that the appearance of the area will be greatly changed. You can see in the photograph that many of the properties are quite old — some of these will be demolished to make way for this new development. The people who own these properties and live there will be angry if the development goes ahead. However, the people who own the properties but do not live there will be delighted as they will make money out of the sale. The views offered also highlight some other opinions. Some people will say that this is not the kind of housing that is needed — housing should be provided for public service workers like nurses and teachers, and the elderly who will increase in number in the years to come. But then again, those supporting the development point to the fact that as Guernsey depends a huge amount on financial dealings then it is right that there is somewhere for these people to stay and to be entertained.

The proposed development at Vazon Bay is very different from the other two. Most people would seem to be in favour of it — indeed it is difficult to contemplate much opposition, apart from those who want the other schemes to go ahead first. All sorts of people would be in favour: the people who have houses on the coastline, like the councillor; the people whose businesses depend on tourists coming to the beach, such as the hotel and golf course owner; even the tourists because then they are guaranteed a nice place to have a holiday. I don't really envisage much of a conflict here — the only dissenting voice comes from a geography professor who says that you can't stop nature, and that if you raise the sea wall here, there will be a knock-on effect somewhere else along the coast. This is true, as many places in the UK, such as on the Yorkshire coast, have seen the consequences of work on one part of a coastline affecting other parts in a negative way. But, I don't think many people would listen to that view.

Examiner comments

This is a good answer. The candidate has studied the background information and has weighed up the pros and cons of each of the proposed schemes. In answering this type of question, it is very tempting to lift material directly from the quotes provided and to reuse these to illustrate the possible causes of conflict. In this answer, the candidate has only done this once, and then has used quotation marks and has used the quote to highlight a key point made in the same sentence. Use of material in this way is perfectly acceptable — but it is *not* appropriate to construct an answer using the same words and statements found in the stimulus material. Candidates should demonstrate that they understand the material fully.

The section dealing with the runway extension demonstrates a good understanding of the issue, and the complexity of the arguments that may be put forward. It recognises that some of the participants may not have stereotypical views, namely the airline manager and the local councillor. It also suggests that other participants (for example, farmers) may have varying views as the issue progresses.

The candidate refers to other similar conflicts — at Manchester Airport. This is a valid exercise, as it is highly relevant in this case. It also demonstrates a degree of synoptic ability. As the answer is clearly related to the area identified and is detailed it is well within Level 2.

Similarly, the examination of the possible conflict in St Peter Port is handled well. There is a clear sense of place, good use is made of the given quotations, and there is evidence of good geographical insight. The candidate relates the issue to the data provided for Question 1 — the importance of the financial sector for the island of Guernsey. There is also a recognition that views on the proposed development will depend on whether individuals will lose directly or gain directly. As is the case in many such conflicts, the views of the participants are not always clear cut. One concern is that the candidate has chosen, deliberately or by accident, to ignore the final quotation. However, this section also reaches a Level 2 range of credit in its own right. According to the mark scheme, this would mean that a cumulative low Level 3 could now be awarded.

The final paragraph is also good. The candidate actually challenges the basis of the question here, by stating that conflict is unlikely. This is evidence of critical understanding and/or critical thinking — a high-level skill. The answer therefore moves up higher into the Level 3 range of credit. As elsewhere, the participants in any possible disagreements are identified and their views are summarised. The reference to the Yorkshire coast is again evidence of synoptic knowledge and understanding. The final sentence also introduces an element of insight, demonstrating that the candidate is well in control of this task. Overall this is an excellent answer which would be awarded a high Level 3 and would gain all 15 available marks.

Candidate answer

Question 3

This particular task is not an easy one, but I will give my ranking as: first, the extension of the runway; second, the development of the waterfront area; and third, the raising of the sea wall.

There are a variety of ways in which this task can be examined. I have looked at the three strategic aims and in my opinion the best way to satisfy them is the rank order I have given them. I have also based my ranking on which schemes would harm least, but help most. I will now justify my ranking by taking each scheme in turn, in preference order.

I think the proposed extension to the airport will harm the least. The only people who will be affected by it are those directly in the flight path of the airport, and they will already be in that flight path, so what major effect will there be? They are selfishly thinking that the value of their houses will lower, but they would have known that when they first moved into that area. The airport has been there for many years. The farmers will not object really — they have the potential to make a lot of money out of the sale of the land. The population density of the airport area is low — so not many people will be directly affected. Against these objections are the possible benefits. There will be more tourists coming to the island, and as tourism is showing signs of declining it needs some form of stimulus. Business people will come in greater numbers on the bigger planes, increasing the potential for more business. We must also remember that Guernsey is an island. If the people want to go on holiday themselves, or visit relations in the UK or France, then they have to fly. A bigger airport will make that all the more easy, and may even lower costs if a budget airline gets involved. On consideration of the three strategic aims, the quality of life for more people will be improved, there will be little damage to the environment (a few extra metres of runway will not be noticed) and cost benefits are likely to outweigh any economic losses. This has to be the best scheme.

The second scheme I would implement is the development of the waterfront. Again this is connected to the economic growth of the island. Business, finance and tourism are vital to the island's economy. People who are involved in these need to be supported, and all the other important financial centres of the world like London and Frankfurt have high-quality hotels. The photograph of St Peter Port shows that there are no modern exclusive hotels there — it needs some modernising. The criteria for the development say that it must be well-designed, and in keeping with the area. It is a brownfield site — it is not building on new land. The planning department has acknowledged that there should be housing provision for those of modest means, but this does not mean that prestige developments should not go ahead. The reason I ranked it in second place is that it does not really satisfy the first of the strategic aims — it will not improve the quality of life for all people on the island, only a few — those who get jobs there.

In last place, I have identified the raising of the sea wall in Vazon and Cobo Bays. As a geography student I have learnt from my studies that trying to prevent nature from doing what it does is not easy. There are consequences of any such scheme, some of which can be predicted, but others cannot. It would be a very expensive scheme to implement, but the costs of it would not suddenly end. There will be on-going costs, in the bays themselves but also elsewhere on the coastline. This is not being cost effective, and certainly is not being sustainable. But who would the scheme benefit? The only beneficiaries are the owners of property and land in that area — the council tax payers of the island would be paying to protect a small minority of people living in those bays, who again should have known that living in a coastal area will mean that you get flooded every now and again. The representative of the Cobo Residents' Association states 'Planners seem more interested in new housing or airport extensions', but that is exactly what they should be doing. I think this scheme satisfies the strategic aims the least, indeed it could be argued that it does not meet any of them.

In summary, extend the airport runway first, build the hotel second, and personally, I would not even consider the raising of the sea wall.

Examiner comments

The candidate has been very clear in the ranking process at the start of the answer. This is exactly what is required by the question and is to be encouraged. Some candidates assemble their argument first, and then give the final placings at the end of their answer. Although this approach is acceptable, it should be noted that it does not assist examiners in their task. It should also be noted that there is no definitive answer to this question — any ranking would be correct — it is the quality of the justification that is important.

Taken as a whole, the answer clearly accesses the Level 3 range of credit. It is detailed and thorough. Good use is made of the stimulus material, both at the start of the exercise (in Question 1) and in the evaluation of each of the proposals. This demonstrates a degree of synoptic ability. The ranking process is justified thoroughly. There is even some evidence of empathy with the people who

will have to brief the various councils involved in the decision-making process. There are a number of statements that identify those who will benefit from each of the schemes in turn and those who will lose, and there is some limited comparison between the proposals.

The candidate recognises that the decision-making process is not straightforward and will depend on the standpoint of the people making the decision. Indeed, the answer to Question 2 has already highlighted a possible conflict of interest for one councillor on the island.

There is evidence of synoptic insight — the references to world financial centres and to previous work on coasts illustrate this. There are also frequent references to the strategic aims which are the key to this question, and to the context of the exercise. Note that the student is working within the Planning and Finance Department of the island — its aims must be paramount.

The answer is well written and logical. All the criteria for Level 3 are satisfied so 17 marks would be awarded.

Improving the answer

Examine the mark scheme which appears with the question and work out for yourself how the answer to Question 3 could have accessed Level 4. To help you in this process, think about the following two points:

- Are the three strategic aims of equal importance and relevance for each of the three proposals? Should the candidate have given some consideration to this, and if so, how?
- Has the context of the exercise been fully appreciated by the candidate? Has the task been critically evaluated? Read again some of the views expressed by the candidate — how well are these likely to be received by the people involved as well as those in the decision-making process?

A2

Module 6

The practical paper

Chapter 13

The written alternative examination

At A2 you have a choice between taking a written examination testing fieldwork and practical skills (Module 6) or completing an investigation and submitting it as a complete enquiry (Module 7). This chapter deals with the Module 6 option.

You have to answer a number of compulsory structured questions. The first assesses a piece of fieldwork that you have completed (the fieldwork-based question) and the others assess practical skills and techniques involved in the conduct of a geographical enquiry (the practical question). In this chapter some examples of questions are provided and sample answers are discussed.

The fieldwork-based question

This question asks you about a piece of fieldwork you completed during your course. Only selected parts of the fieldwork can be dealt with in the time available. Questions can therefore be based on any of the following aspects of the fieldwork process:

- understanding how to develop an overall aim and how to identify a clear research question, hypothesis or problem from this
- identifying, selecting and collecting data using a range of techniques. These must be appropriate and either quantitative or qualitative. They must also be derived from primary sources, including fieldwork, and secondary sources
- presenting data in an appropriate manner
- analysing data
- identifying and evaluating results

Chapter 14 describes all the skills needed. The difference between Units 6 and 7 lies in the way the work is assessed.

Two examples of fieldwork-based questions are given here. Try to answer these based on a piece of fieldwork you have completed. After each question, you will find a brief examiner commentary on what is required to answer it well.

Question 1

Answer the following questions in relation to a geographical enquiry you have carried out involving the collection of primary data in the field. State the aim(s) of your enquiry.

a State two hypotheses or research questions that you established as part of the enquiry and explain why you chose them. (5 marks)

b Describe the methods you used to collect the data required to test the two hypotheses/research questions. (6 marks)

c Analyse the limitations of the methods described in part (b), and discuss the precautions you took to ensure that the data were as accurate as possible. (7 marks)

d Discuss the extent to which the results of your enquiry supported your hypotheses/research questions. (7 marks)

(25 marks)

Examiner comments

a A Level 1 response to this question would simply state the two hypotheses or research questions and the concepts or ideas behind them. To access Level 2, more contextual detail would need to be offered, in other words some background would have to be given about the purpose of the enquiry. Was it to test out a theory in a textbook, or did it grow out of a real-world issue?

b A brief description of methods would access Level 1 here, but to gain more credit some detail of the methodology, clearly linked to the hypotheses and location, would have to be provided. Level 2 credit is therefore for the clarity of linkage(s) between data needed, method of collection, original hypothesis and field location.

c Here limitations must be explained in detail to achieve higher marks, and it must be clear how they applied to the situation described by the candidate. Precautions to ensure accuracy could relate to methods used to ensure accurate instrument readings, or the need for the averaging of several readings. Accuracy could be improved by better design of the data collection methods, say sampling techniques or questionnaire design. You may now begin to appreciate how the precise assessment of the answer depends very much on the quality of material offered by the candidate.

d This marks the logical conclusion of the enquiry. This answer needs to show some appreciation of the geographical significance of the results of the enquiry, and of how they add to the understanding of the environment studied. References to results have to be provided, together with statements of conclusions that were developed. References to anomalous results should also feature here, together with some evaluation of the overall success or otherwise of the enquiry. The best candidates are brutally honest here and do not try to bluff the examiner. They give accounts of accurate results and findings, and state when and why things went awry. It is very rare for a geographical enquiry to go completely to plan.

Question 2

For an investigation that you have carried out in either physical geography or human geography, or as part of a physical/human interrelationship study:

a identify the aim of your investigation and state the hypothesis or hypotheses that you established as part of the investigation (5 marks)

b for one hypothesis describe one method of analysis that was used
to test the validity of your hypothesis and explain why that method
was suitable for your investigation (8 marks)

c (i) state the results of your investigation (4 marks)
(ii) discuss how your results helped you to understand the
environment or topic you studied (8 marks)

(25 marks)

Examiner comments

a The answer should establish the general aims or purpose of the investigation.
This could be to test out textbook or classroom-based theory in a real-world
environment; for example to investigate the changes in characteristics or
processes along the course of a river channel. Hypotheses should be clearly
stated in the correct form as a hypothesis or a null hypothesis, and should
reflect the aims of the study.

b Any method of analysis relevant to the testing of the hypothesis is valid.
This could be a scattergraph, bar chart or kite diagram but it must be clear how
the method was used for *analysis* rather than simple presentation. Spearman
rank correlation or chi-squared test could also be referred to. Description of
the method is required as well as the means of analysis — for example
candidates should include some comparison with critical values or levels
of significance. It must be made clear why the method was appropriate or
suitable for the analysis of the data collected. This will clearly depend on the
nature of the investigation and what was being tested. Level 2 will be awarded
for clarity of any or all of the following: the method, the application of that
method to the stated hypothesis and the suitability of the method for the
chosen study.

c (i) Specific results must be given here. It is worth repeating that fieldwork,
which must include primary data collection as well as secondary data collection,
has to have taken place. The more specific and real the results given, for
example correlation results, or summary statistics such as mean calculations
and standard deviation, the better.

(ii) This section requires some appreciation of the geographical significance of
the results of the investigation and how these relate to established theories.
There should also be some attempt at an evaluation of the results in the context
of the study and of the environment in which it was set. There could be some
assessment of the ways in which the study contributed to the candidate's
improved understanding — it may have reinforced or confirmed ideas only
previously studied in theory. It is worth identifying and commenting on any
anomalous results that may have occurred.

General advice

Do not write at length about the weather, or your teacher's failings, or how lazy
and inefficient members of the group let you down. Do show an awareness of
any shortcomings in your fieldwork, particularly if you can suggest how those
shortcomings could be overcome in the future.

Finally, as stated earlier, the more specific your results, and the more clearly they are yours, the better. One simple point — you should write your answer in the first person singular ('I did this…' and so on).

The practical questions

The rest of the questions on the Unit 6 paper can be based on any of the following investigative skills.

Identification, collection, selection and use of salient information:
- sampling procedures
- Ordnance Survey maps, morphological maps, land-use maps, weather maps
- census returns
- questionnaires and interviews
- measurements of physical characteristics in a physical environment
- the use of texts and published data
- the use of photographs
- the use of remotely sensed imagery and geographical information systems

Organisation, recording and presentation of evidence:
- graphs: histograms, line graphs (including logarithmic graphs), proportional divided circles, dispersion graphs, bar graphs, long and cross sections, scatter-graphs, Lorenz curves and triangular graphs
- maps: isopleth maps, choropleth maps, proportional symbols, flow and desire lines, sketch maps
- tables and written text

Description, analysis, evaluation and interpretation of evidence:
- map and photograph interpretation
- descriptive statistical techniques: measures of central tendency (mean, mode and median), measures of dispersion (standard deviation, range and inter-quartile range), the normal distribution, standard error of the mean and confidence limits
- the analysis of trends, cycles and other forms of change: index numbers, location quotients and various forms of graphs
- geographical relationships and changes in them: Spearman rank correlation
- the analysis of text: summary, hypothesis formulation and decision making
- the limitations of evidence, and limitations of conclusions drawn

Approximately half the marks available in these questions will be for practising one or more of the techniques given above. Allow plenty of time for this — do not rush. Take care when using the technique, as marks can be lost as easily as they can be gained. Make sure you have the necessary equipment (ruler, calculator, protractor and a pair of compasses) to do yourself justice.

One of the tasks may be to compile a report based on an exercise and the data provided. A word limit will be given, but it is there only as a guide. You will be given instructions on what should be included in the report. These will include some analysis of both the data and the results of the practical exercise you have completed. It is likely that some assessment of the usefulness of the technique and its shortcomings will be required.

Two examples of practical questions are given here, with answers and an examiner commentary. A third question is provided, with a mark scheme, for you to attempt.

Question 3

Look at Tables 13.1 and 13.2.

The data presented in Table 13.1 were collected from secondary sources as part of an enquiry relating to population and resource issues in countries in various states of development.

Table 13.1
Economic and social indicators for selected countries, 1990

Country	GNP per capita (US$)	Energy consumption per capita (kg oil equivalent)	Infant mortality rate	Life expectancy at birth (years)
1 Kuwait	13,680	9,200	15.6	73
2 Norway	20,020	5,263	8.4	76
3 Saudi Arabia	6,170	1,900	71	63
4 Belgium	14,550	6,049	9.2	74
5 Canada	16,760	9,950	7.3	77
6 Japan	21,040	3,680	4.8	79
7 UK	12,600	6,369	9.5	75
8 Argentina	2,640	1,804	32	71
9 Mexico	1,820	1,227	50	66
10 Colombia	1,240	685	46	66
11 Bolivia	570	367	110	53
12 Sri Lanka	420	106	22.5	70
13 Zaire (now DRC)	170	62	108	53
14 Malawi	160	56	130	49
15 Mali	230	27	117	45

n	0.05 +/−	0.01 +/−
10	0.56	0.75
11	0.54	0.73
12	0.51	0.71
13	0.49	0.68
14	0.46	0.65
15	0.45	0.63
16	0.43	0.60
17	0.40	0.58
18	0.40	0.56

Table 13.2
Critical values for R_s at 0.05 and 0.01 significance levels

a Use the data for GNP and energy consumption (Table 13.1) to draw a scattergraph on logarithmic graph paper, showing the relationship between these two indicators.

Identify each country by its reference number and insert a best-fit trend line. (5 marks)

b Calculate a Spearman rank correlation coefficient for GNP and infant mortality rate. Comment on the statistical significance of the relationship.

Critical values for R_s are shown in Table 13.2. (5 marks)

c Assess the extent to which data on life expectancy are useful indicators of the level of development in a country. (4 marks)

d State two indicators, other than those shown in Table 13.1, which you would use to assess the extent to which a country was overpopulated. (1 mark)

e Write a brief report (no more than 300 words) to summarise the findings for this part of the enquiry. In the report you must:
(i) discuss the nature of the relationships between the indicators
(ii) evaluate the extent to which these indicators are appropriate in assessing the level of development

(10 marks)

(25 marks)

Sample answer and examiner comments

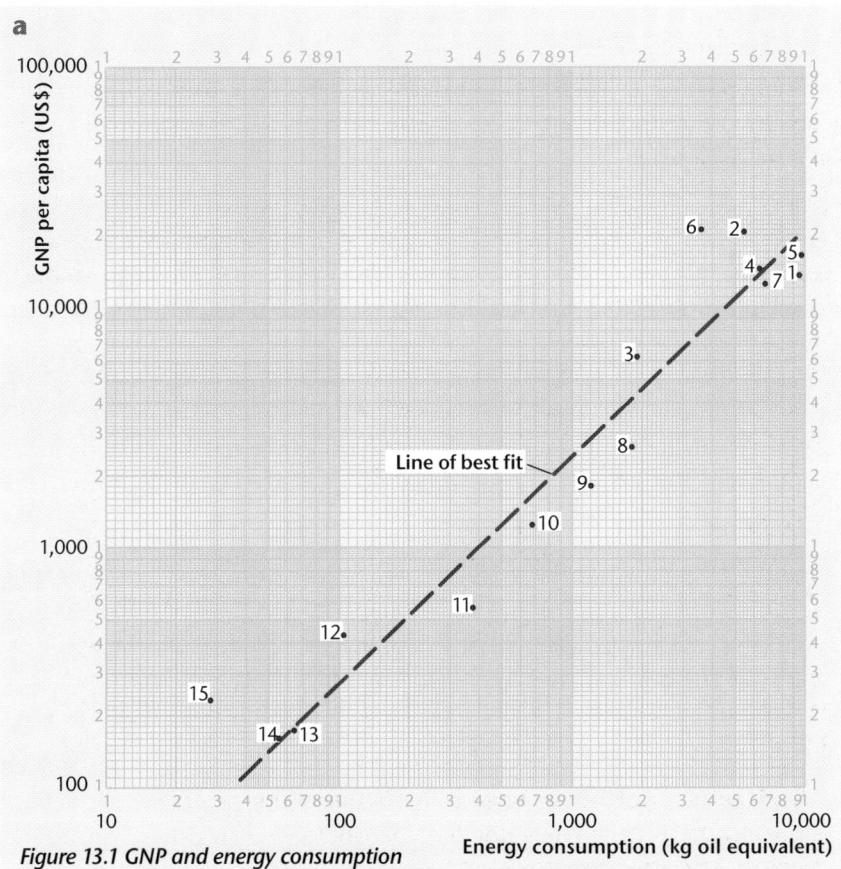

Figure 13.1 GNP and energy consumption

As the graph paper provided is logarithmic on both axes, it does not matter which variable is placed on which axis. The 5 marks are awarded as follows:
- 1 mark for correct labelling of the axes
- 1 mark for correct use of the scales on these axes
- 2 marks for the plotting of the points
- 1 mark for the insertion of the best-fit trend line.

b The Spearman rank correlation coefficient is –0.90. This is significant at the 0.01 significance level. This means that the likelihood of this negative correlation occurring by chance is 1 in 100.

Three marks are reserved for the calculation process, and 2 marks for the commentary on the statistical significance.

> **c** Life expectancy reflects the level of medical care in a country. The longer people live tends to be a measure of how rich or developed the country is because the country can afford to have good medical facilities. Life expectancy increases as there are more medicines, more hospitals and more opportunities for long-term care as people grow old. Life expectancy is also a response to the quality of diet in a country — the better food people eat, the longer they will live. Life expectancy is therefore quite a good indicator of the level of development of a country.

This is a generalised answer which makes some valid points regarding medical care, diets and long-term care. The candidate could have referred to the data to identify possible anomalies in the overall trend. For example, people in Saudi Arabia have relatively low life expectancies, whereas those in Sri Lanka have longer ones. Clearly other factors come into play, for example wealth distribution, incidence of preventable diseases and lifestyles. This answer gains 3 of the 4 marks available.

> **d** The number of people who are starving, or who cannot read or write.

Any two indicators that suggest there are too many people for the resources available are acceptable. Other examples include food intake, number of patients per doctor, number of school places per 1,000 people and so on. This answer gains the 1 mark.

> **e** (i) The relationships between the indicators in the early part of this question demonstrate clear links. The relationship between GNP and energy consumption is positive and strong. There are some anomalies, such as Japan, where there is a lower use of energy than would be expected, and Kuwait, where the GNP is bigger than the trend. This is due to Kuwait being an oil exporter. The relationship between GNP and infant mortality is negative, but equally strong at −0.9, and this is highly significant. Saudi Arabia is an anomaly here as its infant mortality rate is nine times greater than that of Norway, yet its GNP is only one third of that country. It also has a greater IMR than Mexico, even though it is substantially richer than that country. A relationship between life expectancy and GNP has not been attempted, and I do not have the time to do it, but it looks as though the richer countries have the longer life expectancies, which is what you would expect.
> (ii) All of these indicators are appropriate for assessing the level of development. GNP is obviously used by many agencies to measure development across the world. Energy consumption indicates the amount of power used by the country, in industry, in houses and in transport, and so is an indication of wealth. However, colder countries will use more power than warmer countries. Infant mortality gives an indication of the health-care provision in a country and the amount of national wealth given to such things. Thus it indicates wealth and social considerations in a country.

Life expectancy also reflects medical care, as well as the amount of medi-
cines and quality of diet in a country. A high level of economic development is needed to pay for these.

290 words

This answer gains maximum credit. In (i) the candidate has discussed the nature
of each of the relationships between the indicators by both stating what the relationship is and pointing out anomalies. The last relationship is described briefly,
but this does not detract from the quality of discussion earlier in the answer.

In (ii) the candidate makes the more obvious points, but also qualifies the
answer by pointing out possible reservations in using these indicators. For
example, the candidate states that temperature will influence energy consumption,
and health-care provision is not just related to wealth but is also linked to
perceptions of social welfare in a country. This sophistication of argument is what
takes this answer to the highest level of response.

Finally, note that the word total is given by the candidate. Had the answer
gone over the suggested limit, marks would not have been deducted. The word
limit is only a guide — examiners believe that longer answers are self-penalising,
so any additional penalty is not necessary.

Question 4

The data in Table 13.3 (page 344) were collected as part of a survey of the
variations in social and economic conditions in an urban area in England.

Five indicators were used to assess social and economic well-being or
deprivation. The values for each indicator were ranked from low (rank 1) to
high (rank 18) to produce an indicator score. For example, for unemployment
the ward with the lowest level of unemployment, ward 14, was ranked 1 and
the ward with the highest level of unemployment, ward 5, was ranked 18.

The rank scores for each indicator were added together to produce a total
score as a social and economic index. The maximum score for the index is 90,
if one ward scores rank 18 for each of the five indicators.

a Divide the data for the social and economic index for the wards into
   four classes. Using these four classes complete a choropleth map to
   show the variation in social and economic conditions in this area,
   as indicated by the social and economic index.   (8 marks)

b Write a brief report (approximately 250 words) on the pattern of
   social and economic well-being/deprivation in this urban area.
      In your report you should:
   ■ describe the degree of segregation on the basis of wealth and
     ethnicity as shown on your choropleth map
   ■ discuss the relative usefulness of the choropleth map and these indicators
     in identifying variations in levels of well being/
     deprivation   (12 marks)

                                                                (20 marks)

Table 13.3
Variations in social and economic conditions in an urban area in England

Indicator	A		B		C		D		E		Total score
Ward	Per cent households with NCWP-born head*	Rank	Per cent unskilled workers	Rank	Per cent unemployed	Rank	Per cent households with >1 person per room	Rank	Per cent households without a car	Rank	Social and economic index
1	1.8	3	5.5	6	9.4	3	2.3	3	31	3	18
2	3.6	7	6.6	9	15.8	13	5.8	11	42	10	50
3	5.7	10	6.0	8	11.7	6	4.0	7	35	6	37
4	3.4	5	2.2	2	8.9	2	1.3	2	29	2	13
5	50.2	18	15.6	18	23.0	18	15.8	18	62	17	89
6	3.9	9	8.4	13	18.9	16	8.2	16	47	11	65
7	11.0	14	7.3	11	13.9	9	8.0	15	41	9	58
8	10.2	13	8.5	14	15.1	12	6.7	13	48	13	65
9	7.5	12	8.9	15	14.2	10	5.7	10	49	14	61
10	12.3	16	11.6	16	17.5	15	7.8	14	54	16	77
11	29.9	17	12.5	17	22.0	17	10.0	17	68	18	86
12	3.4	5	5.6	7	12.4	8	3.6	6	40	8	34
13	12.0	15	8.1	12	15.4	11	6.4	12	50	15	65
14	2.9	4	1.8	1	6.2	1	1.2	1	16	1	8
15	1.5	2	7.2	10	16.7	14	4.7	9	47	11	46
16	3.7	8	4.3	4	12.3	7	3.4	5	39	7	31
17	1.2	1	4.2	3	11.6	5	3.0	4	32	4	17
18	6.3	11	5.1	5	9.4	3	4.6	8	34	5	32

*Percentage of households where the head of the household was born in the New Commonwealth or Pakistan (NCWP).

Source: Census data.

Mark scheme

a Two marks for appropriate classes.

Choropleth map:

- 1 mark for key
- 1 mark for shading gradation
- 4 marks for accuracy of plotting of classes

b **Level 1** The answer presents comments at a very general level; it describes the map without commenting on the pattern or degree of segregation; it only presents the inner to outer change. The answer presents superficial comments on the technique or the indicators. (0–4 marks)

Level 2 Clear description of the pattern with some attempt to identify the degree of segregation; the candidate recognises that the pattern is more complex; some variation within the inner and outer areas. Good comments on the strengths and weaknesses of the choropleth. Clear assessment of the extent to which these particular indicators are useful measures of deprivation/well-being. (5–8 marks)

Level 3 Any two of the above elements done well raises the answer to the bottom of Level 3. Top of level requires good comment on the degree of segregation/ usefulness of the choropleth and some assessment of the indicators as shown in Level 2. (9–12 marks)

Sample answer and examiner comments

a My classes are: 8–28, 29–49, 50–70, 71–91

I have drawn the choropleth map as shown (Figure 13.2).

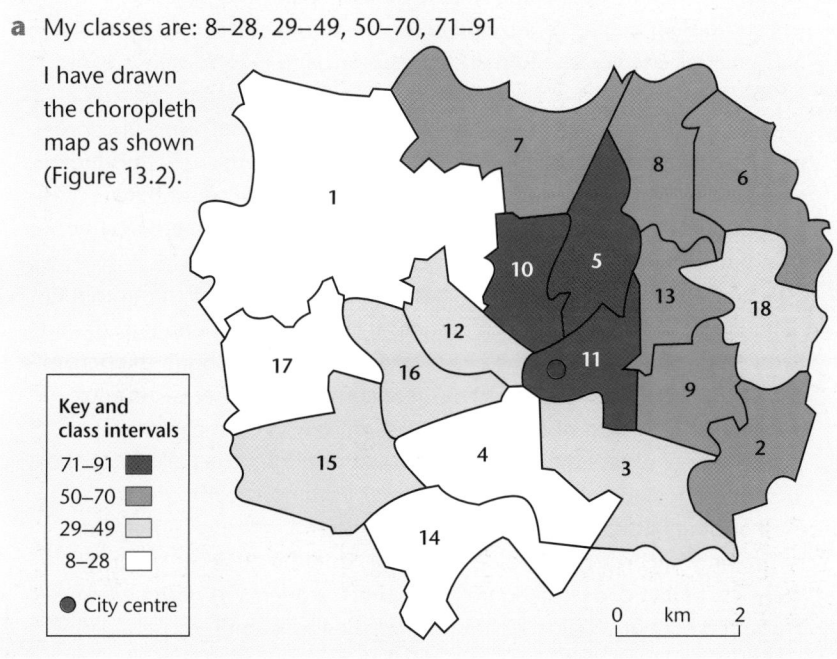

Figure 13.2 A choropleth map to show the levels of urban deprivation using a social and economic index for an urban area in England

Key and class intervals

- 71–91 ■
- 50–70 ■
- 29–49 □
- 8–28 □
- ● City centre

Although the classes provided by the candidate are uneven, they are appropriate. The choropleth map has also been drawn correctly. Maximum credit is awarded at this stage.

b The choropleth shows that the highest levels of urban deprivation are found near the centre of the urban area. These areas are the inner city, especially areas 5, 10 and 11. There is also a high concentration of deprivation totaling 50–70 to the east of the city, especially amongst the eastern periphery.

One can identify a synergy between the levels of deprivation shown on the choropleth and the percentage of NCWP (New Commonwealth or Pakistani) in the households. Areas 5, 10 and 11 have the highest levels of deprivation and also the highest levels of ethnic minorities at 50.2, 29.9 and 12.3% respectively. Furthermore, the two other areas with the fourth and fifth highest levels of ethnicity (areas 7 and 13) are also found near the centre. Moreover, the area with the least percentage of NCWP per household is area 17 on the western peripheral extremity. One also finds that it is area 17 and similar areas like 14 that, although having the least percentage of ethnicity, paradoxically have the highest levels of affluence, characterised by low unemployment, households without a car and so on.

Therefore one can conclude that this urban area is very segregated on the basis of wealth and ethnicity. Not only does there appear to be a racialisation of residential space, but also a division of affluence with the least deprivation found to the southern and northwestern extremities. Area 18, however, is an anomaly as although it has relatively low deprivation, the five surrounding areas have much higher deprivation. This may be attributed to area 18 being regenerated or receiving investment.

A choropleth map is very useful as it provides an efficient visual spatial pattern. This enables one to make conclusions and identify trends easily. Furthermore, it is able to show how, in this case, deprivation changes gradually from one area to another. Nevertheless, the criticisms of choropleth maps range from them inferring that change is rapid over boundaries when in reality it occurs gradually, to not taking into account that, although one area like area 17 may have low deprivation, there may be one road within that area which is particularly deprived.

It can be argued that although these indicators are useful in our modern materialistic lives (affluence is measured by households without a car and by unemployment), some data have failed to adapt. Although the percentage of NCWP is useful as it is often these communities that experience deprivation for a host of reasons, it does not, however, encompass the percentage of other ethnic groups such as those from the Balkans, as they are not embodied under the scope of New Commonwealth or Pakistan. Furthermore, although unemployment is a useful indicator, it may be misleading. An area that has high unemployment may occur because there is a higher concentration of pensioners, students or house workers. The criteria of households without a car may not be entirely useful either as, although homeowners may be able to afford one, they may choose not to due to the costs and more vitally due to environmental concerns of contributing to global warming.

This answer is logically presented. The first few paragraphs deal with the first bullet point, and the subsequent ones examine the second demand of the task. Presenting your answer in such a logical manner assists the examiner in the assessment of your work and this is to be encouraged.

The first paragraph makes a couple of good Level 1 points in that the pattern of deprivation is described. The second paragraph takes the response into Level 2. The candidate gives a more precise description of the pattern of segregation, in terms of both high levels of deprivation and lower levels of deprivation. The third paragraph is even more sophisticated, with some analysis of anomalies to the pattern and commentary on possible reasons for this.

The candidate now turns to examine the second part of the task. Once again the relative usefulness of a choropleth map is well discussed, and the answer moves into Level 3. The final paragraph is again excellent and, according to the mark scheme above, this has to achieve maximum credit. The criticism of the indicators is particularly good for ethnicity and unemployment, although that of households without a car is possible if a little unlikely.

The candidate gains 12 marks for this section, 20 out of 20 in total.

Question 5

Look at Table 13.4 which provides a set of summary statistics for the discharge (cumecs) of a river in northern England over a period of 19 years from 1981 to 1999.

The mean, maximum and minimum flow are shown for each month within this time period.

Table 13.5 provides further statistical information for four of these months.

Table 13.4

	Jan	Feb	Mar	Apr	May	June	July	Aug	Sep	Oct	Nov	Dec
Mean	58.90	46.30	45.88	24.37	15.15	14.29	15.68	22.96	26.27	44.75	50.58	58.95
Min.	10.00	9.30	10.04	6.34	3.24	3.60	1.88	2.23	5.16	10.20	11.22	9.98
Max.	99.70	113.97	113.81	50.62	40.70	33.02	42.80	71.33	67.01	80.60	86.03	108.94

Table 13.5

	January	April	July	October
Mean	58.90	24.37	15.68	44.75
Upper quartile	81.19	32.60	31.00	67.91
Lower quartile	31.69	13.95	6.03	30.57
IQR	49.50	18.65	24.97	37.34
Standard deviation	27.57	12.59	13.10	19.96

a Using the values for mean, maximum and minimum discharge in Table 13.4, draw a graph on which you should plot the following:
  ■ a line to show the variation in mean discharge
  ■ a line to show the variation in maximum level of discharge
  ■ a line to show the variation in minimum level of discharge
  Plot discharge on the vertical axis and months on the horizontal axis. Make sure you use an appropriate scale. (10 marks)

b These data have been collated as part of the process of setting up flood control schemes. With reference to the graphs you have drawn in (a) and Table 13.5, write a brief report (approximately 300 words) to summarise the variations in discharge and to assess the reliability of mean discharge values as a basis for river management. *(15 marks)*

(25 marks)

Mark scheme

a On the graph, the following will receive credit:
- 1 mark for the title
- 1 mark for an appropriate vertical scale
- 1 mark for correct units
- 1 mark for line placement
- 6 marks for the plots (2 per line) *(10 marks)*

b

Level 1 (0–4 marks)	Level 2 (5–9 marks)	Level 3 (10–15 marks)
Answer presents comments at a very general level; describes the graphs at a superficial level; general seasonal variations only tend to be given OR	More accurate description of variations on the graphs; considers differences between lines; extent of variation from the mean Plus either (i) or (ii) below:	Clear appreciation of the reliability issue; understands limitations of the mean. Uses figures and data to illustrate arguments And either (i) or (ii) below to get to mid-Level 3:
Answer quotes figures on measures of central tendency without attempting to make any statistical sense of the data OR	(i) More detailed comments on the mean in relation to other measures; attempts to assess the reliability of the mean as shown by reference to SD or IQR; shows some understanding of the limitations of the mean	(i) Clear appreciation of SD as indication of spread of data about mean; some understanding of variance
Very general comments of usefulness of data	(ii) Attempts to assess the extent to which this type of data is useful for prediction	(ii) Appreciates the limitations of the data given that there are other unknown factors, such as problems of planning/ recurrence intervals of floods: 1 in 50 year events and the like. Understands how SD can be used to 'predict' likelihood of levels of discharge
		(i) and (ii) done effectively take answer to top of level

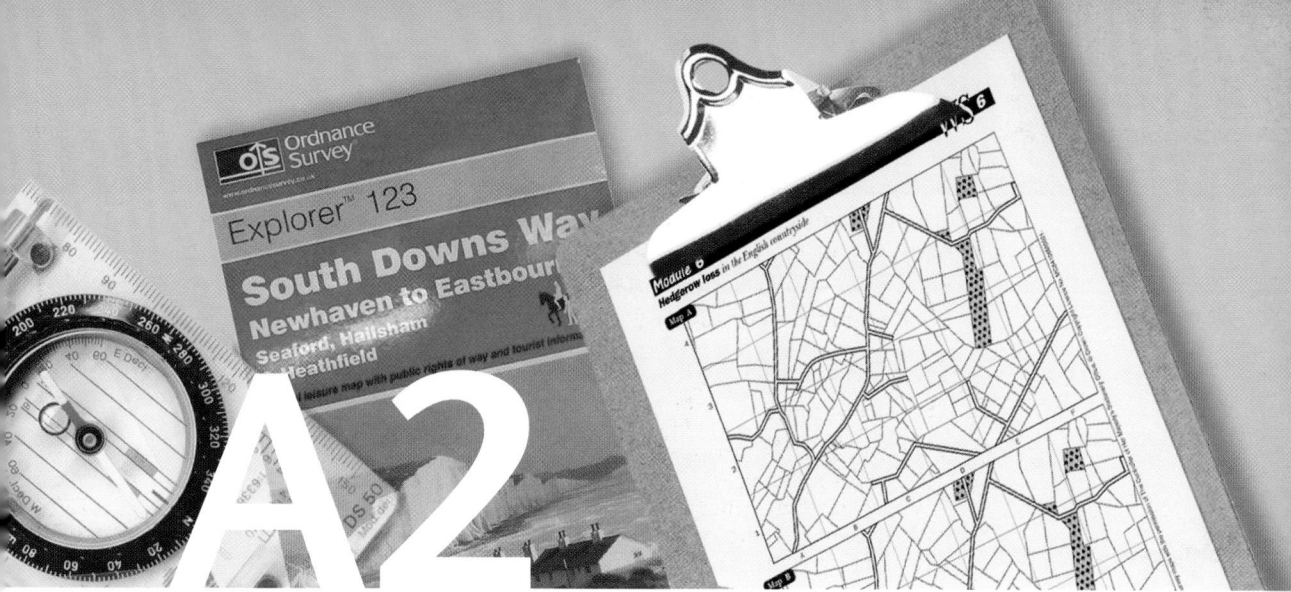

A2

Module 7

The investigation

The geographical investigation

At A2 you have a choice between taking a written examination testing fieldwork and practical skills (Module 6) or completing an investigation and submitting it as a complete enquiry (Module 7). This chapter deals with the Module 7 option. If you choose this option you are required to submit a written report of 3,500–4,000 words (excluding tables, figures and appendices) based on an enquiry into a geographical issue linked to the AS and/or the A2 specification. The report is assessed by your school, but the marks are moderated by AQA.

What is a geographical investigation?

Your investigative work must be based on evidence from primary sources, including fieldwork, and secondary sources — in most cases a combination of the two. In simple terms **primary data** are those collected by you in the field, or material from other sources which needs to be processed. **Secondary data** are derived from published documentary sources and have already been processed.

Primary research is essential — you must have some direct contact with the area of study and the subject of investigation. Your study could, for example, involve a specific economic activity, an identified group of people or a local issue. You need to visit the area concerned and talk to the people there, or record data there.

All geographical investigations should follow the same stages:

- identifying the aim of the enquiry, often in the form of testing a hypothesis or establishing research questions
- selecting and collecting data using a range of techniques, for example measuring, mapping, observations, questionnaires, interviews
- organising and presenting the data using cartographic, graphic or tabular forms, possibly with the help of ICT
- analysing and evaluating the data, noting any limitations
- drawing together the findings and formulating a conclusion
- suggesting further extension to the investigation, including additional research questions which may have been stimulated by the findings
- a statement of success or otherwise of the investigation, with some commentary on the significance of the investigation for others

The geographical investigation therefore provides an opportunity for you to demonstrate what you can do beyond the examination room.

Establishing a title

Hypotheses and research questions

The best type of investigation involves testing a **hypothesis**, or setting out one or more **research questions** that can be investigated and evaluated.

A hypothesis is a statement based on a question. For example, the question 'do the characteristics of a soil change down a slope?' could form the basis for any of the following hypotheses:

(1) *The depth of a soil varies downslope*
(2) *The clarity of the soil horizons varies down the slope*
(3) *The texture of a soil varies according to the position on the slope*
(4) *The acidity of a soil varies downslope*

and so on.

Once established, the hypothesis can be tested by collecting data. Following evaluation of the data, the results of the enquiry can be used to state whether or not the hypothesis was correct, and thus to answer the original question. The hypothesis or research question therefore forms the title of the investigation.

- A poor project title would be *Changes in the retail geography of a town over the last 30 years*. Here there is no question or hypothesis.
- A better title would be *What have been the major changes in shopping patterns in the area around Norwich within the last 30 years?* This establishes a question to which an answer can be developed.
- An alternative title, in the form of a hypothesis, would be *The CBD of Norwich has seen major decline in activity within the last 30 years caused by the growth of out-of-town retail outlets*.

Either of the last two titles are acceptable ways in which to proceed.

Aims and objectives

The purpose of the hypothesis or research question is to enable you to clarify your aims and objectives.

Aims are statements of what you are hoping to achieve. In the above example you will want to identify areas of change — evidence of decline in some shopping areas and growth in others, with possibly some reasoning for these changes.

Objectives are statements of how you will achieve your aims. What data will be needed to identify areas of decline or growth? How can these data be collected, analysed and presented? How will you obtain evidence of reasons for the changes? Do you need to write to anyone, interview anyone or devise a questionnaire? What precise form of fieldwork needs to be undertaken?

What is a null hypothesis?

Some investigations are best approached by establishing a **null hypothesis**.

This is a negative assertion, which states that there is *no relationship* between two chosen sets of variables. For example, a null hypothesis could state that *there is no relationship between air temperatures and distance from a city centre*. An alternative hypothesis can then also be established, namely that *air temperatures decrease with distance from a city centre*.

The null hypothesis assumes there is a high probability that any observed links between the two sets of data, in this case temperature and distance from a city centre, are due to unpredictable factors. If temperatures are seen to decrease with distance from the city centre in the chosen city, it is a result of chance. If the null hypothesis can be rejected statistically, then the alternative hypothesis can be accepted.

One benefit of this seemingly 'reverse' approach to an enquiry is that if the null hypothesis *cannot* be rejected, it does not mean that a relationship does not exist — it may simply mean that not enough data have been collected to reject it. In short, the investigation was not worthless, but was too limited in scope. Another benefit of this approach is that it allows the use of statistical tests on the significance of the results to be carried out.

The null hypothesis is a sophisticated way in which to approach an enquiry, and should only be used when fully understood.

Checking the feasibility of your investigation

Before embarking on this major piece of work which will, if properly undertaken, take a great deal of time to complete, you should check whether or not the tasks involved are achievable, and that an overall conclusion will be forthcoming.

Here are some questions you should ask yourself to see if your idea can really work:

Is my topic area within the requirements of the specification?

You need to check the specification yourself, or ask your teacher, or you could seek prior approval from a coursework moderator appointed by the examination board.

Is the subject matter narrow enough?

*Table 14.1
Primary data
sources*

In general it is better to study one aspect in detail than several aspects sketchily, for example to study one town rather than three. Many investigations are best suited to a local area of study — one that is large enough to give meaningful results, but not so large it becomes unmanageable. Can you visit the area of study — you may have to on more than one occasion? Again, seek guidance from your teacher on the scale of study for your chosen topic.

Quantitative	Qualitative
Land-use transects	Questionnaires
Housing surveys	Interviews
Environmental impact assessments	Field sketches
Traffic counts	Photographs (taken by you)
Climate surveys	
River measurements	
Soil surveys	

Will I be able to collect the data I need?

Remember, the use of primary data is paramount, and there is a whole range of sources of primary data, both quantitative and qualitative (Table 14.1).

As the examination board places great emphasis on data collection, it is wise to use a variety of data sources for your investigation. Do not just use questionnaires.

Will I be able to complete the investigation within the time period allowed?

One type of investigation that needs to be carefully monitored is that which deals with changes over time — in other words, one that needs a 'before' and 'after' element. Examples of this include the impact of a new bypass around a settlement or the impact of a new retail park. Check whether or not you will have time to collect data before and after the development is built.

What equipment will I need?

The amount and nature of the equipment will obviously depend on the nature of the data to be collected. However, you do need to check whether the equipment is available and in working order.

What else do I need to do?

Here is a final checklist of some of the main requirements of any fieldwork or data collection:

- **Always ask for permission if it is needed**. Any investigation that involves you going on to someone else's land, or into a building, will require permission from the owner. In most cases a letter or telephone call explaining the purpose of your visit will suffice. However, it is often better to ask your teacher to write a short note on school or college headed notepaper in support of your work.
- **Check that you can get to your area, and get back.** Is there public transport to the area or will you need help? Note that much fieldwork is undertaken at weekends and Sunday timetables are often different.
- **Wear appropriate clothing**. If you are going into remote and difficult areas, then you must wear clothing that will keep you warm and dry. Wear comfortable, appropriate footwear, even in urban areas.
- **Be safety conscious.** Never work alone and always tell someone where you are working. If possible, give details of methods of travel and times of return. Be particularly careful in coastal areas. Check on tide levels, never work underneath crumbling cliffs, and stay on coastal paths.

Collecting primary data

Once you have established the aims and hypotheses of your investigation, work out what data you need to collect and what methods you will use. Investigations at this level are essentially based upon your own observations, which means using techniques such as questionnaires, interviews, river measurements, pedestrian surveys and urban transects. The whole of your investigation can be based upon such material but, depending on the scope of your work, it should be possible to include some secondary data from published sources.

Sampling

Sampling is used when it is impossible, or simply not necessary, to collect large amounts of data. Collecting small amounts of carefully selected data will enable you to obtain a representative view of the feature as a whole. You cannot, for example, interview all the shoppers in a market town or all the inhabitants of a village, but you can look at a fraction of those populations and from that evidence indicate how the whole is likely to behave.

When you have established the need for a sample survey, you will have to decide on a method that will collect a large enough representative body of evidence. If, for example, you are interviewing the inhabitants of a village, you must ensure that your interviews cover all age ranges in the population.

Types of sampling

The main types of sampling technique you need to consider are random, systematic and stratified (quota).

A **random** sampling is one that shows no bias and in which every member of the population has an equal chance of being selected. The method usually involves the use of random number tables.

In **systematic** sampling, samples are taken at regular fixed intervals, for example every tenth person or house. On a beach you could decide to sample sites at 100 m intervals, and select pebbles at each location using the intersection points on the grid in a quadrat.

Stratified sampling is based on knowing something in advance about the population or area in question. For example, if you are surveying a population and you know its age distribution, your sample must reflect that distribution. If you are surveying an area in which you know the distribution of soil types, samples should be taken in proportion to the area covered by each type of soil. For instance, if a particular soil covers 40% of the area, 40% of the total sample points should be within the area covered by that soil.

Bias in sampling

It is possible, through poor choice of sampling method or insufficient evidence, to achieve a result that is unrepresentative of the population in question. Taking all samples on the same day of the week or outside the same shop could lead to a distortion in a shopping survey, for example.

Sample size

The size of sample usually depends upon the complexity of the survey being used. When using a questionnaire it is usually necessary to sample sufficient people to take into account the considerable variety introduced by the range of questions. Sample size can be restricted by practical difficulties and this may affect the reliability of results. Your aim should be to keep the sampling error as small as possible. You are not a professional sampler and cannot be expected to conduct hundreds of interviews, but on the other hand, sampling only 20–30 people in a market survey is not representative of the population as a whole.

Point sampling

Point sampling is carried out in surveys which involve, for example, studies of land use, vegetation coverage, soil sites and selection of such items as pebbles in longshore-drift studies. Point sampling involves the use of a grid. That produced by Ordnance Survey is ideal, but for field surveys a quadrat can be used. A **quadrat** is a frame enclosing an area of known size (often 1 m^2), and may be subdivided by a grid made of wire or string. Both random and systematic sampling can be carried out within this framework.

Questionnaires

Many investigations at this level use questionnaires. Writing questionnaires can be one of the most time-consuming and difficult aspects of individual investigations.

In designing a questionnaire it is important to have a balance between specific questions and open-ended questions. Examiners often criticise questionnaires for the use of too many open questions, because these produce results that are less quantifiable. However, they can provide additional information that is vital when trying to explain the behaviour revealed in the answers to the more direct questions.

Questionnaire surveys, for example supermarket studies, often involve a lot of effort but reveal little more than very basic information about the behaviour of shoppers. Oddities in the pattern of behaviour may not be accounted for unless open questions are included to help understand shoppers' motives.

Unless questionnaires concentrate on the spatial aspects of the sample being studied, they can become far too sociological and therefore of limited geographical value. Questionnaires that represent little more than a social survey should be avoided as they will give little scope for mapping and analysis of patterns.

At GCSE, you may have undertaken a questionnaire survey in order to demonstrate your ability to complete it. At AS/A-level the aim is completely different — the quality, quantity and reliability of the data produced are as important as the means by which they were obtained.

Guidelines

The following list summarises the guidelines for conducting a questionnaire survey:

- keep it as simple as possible — busy people do not like answering a lot of questions
- try to write a mix of closed questions (yes/no answers or multiple response) and open questions (choice of answers or free statements)
- decide on an adequate sample size
- try to introduce your questionnaire in the same way to each person sampled (write a brief introduction)
- put the questions in a logical sequence
- ask questions that will produce data that can be analysed
- think carefully about sensitive questions and use tick boxes for such information if possible

- when asking for respondents' ages and other personal information it is often better to offer categories than to insist upon an exact figure
- try to ask questions about people's behaviour, not how they perceive their behaviour
- pilot your questionnaire by testing it out to see if it will produce the material you want
- always seek approval from your teacher or tutor before proceeding in order to avoid insensitive questions and also to prevent harassment of local people by swamping the area with too many questionnaires
- obtain a document from your school/college that states exactly what you are doing
- always be polite, look smart and smile, and do not get upset if people refuse to answer
- never work alone
- if you intend to stand outside a specific service or in a shopping centre it is a good idea to ask permission

Interviews

In the course of your work it may be necessary to conduct interviews to see how some people stand on an issue or how they would act in certain circumstances. In an investigation of rural land use, for example, it might be a good idea to interview farmers — they may give you information that helps explain some of the changes you have observed. If you are investigating a conflict as part of your work, it is essential that you try to obtain the views of the parties involved. This will certainly take you beyond the scope of a questionnaire, because you will need to put different questions to each respondent or group involved. It is not always possible to meet the people involved, so you might have to write to a person or group for their opinions.

Interviews give you the opportunity to explore more detailed areas and to follow up points that are raised. They can be time consuming, so think carefully before deciding upon such a course of action, and if you decide it is a good idea, use the technique sparingly.

Always send a well-written letter requesting the interview, and prepare for it properly — people do not like wasting their time. Being prepared means you should have a list of direct questions you need to ask, making sure that the person you are interviewing has the information you require.

Do not ask questions when it is possible for you to obtain the information for yourself. Take notes at the interview or immediately afterwards. You may wish to record the interview, but always obtain permission before you start. Your final presentation can be very much enhanced by quotations from interviews, but always check accuracy and ask permission from the interviewee.

Sampling attitudes

Collecting information on people's attitudes is often difficult. You can ask about the attitude of an individual during the course of an interview, but if you are using a questionnaire survey to obtain a number of responses, you must phrase questions so that bias does not enter into your results. There are three main ways of collecting information on attitudes:

- **Bi-polar tests** involve establishing a rating scale based on two extremes of attitude (i.e. poles apart). For example, in a CBD survey you can ask shoppers if they find the shopping area attractive or ugly and offer them a sliding scale of response from (1) ugly to (7) attractive. You can put a number of such points to them and use the results to give the area an overall rating. This is useful when comparing shopping centres.
- A **point-score scale** is used to give respondents the chance to identify factors they consider to be important. You can ask respondents to place each factor on a scale from 0 to 4 in terms of importance, for example in attracting shoppers to a retailing centre.
- Using a **rating scale** allows respondents to agree or disagree with a statement. Statements can be put to them and they can be asked to place their responses in categories ranging from 'strongly agree' to 'strongly disagree'.

Types of survey

So far this chapter has mainly shown you how to collect information by asking people questions. There are other types of survey, not involving people, that you will find useful in obtaining information. Some of them are described below, with details, in many cases, of how to proceed.

Surveys in urban areas

A large and varied number of surveys can be carried out within the context of urban fieldwork.

Land-use surveys

These can be used in both rural and urban areas and must always be carried out with a clear purpose in mind. Divide land use into categories of similar types. Sub-groups within categories can be established, depending upon the detail required. Housing, for example, can be divided into terraces, detached, bungalows, semi-detached and flats.

Land-use transects

When the urban area is too large to survey as a whole, a sample has to be taken. The transect is essentially a slice through the urban area to see how land use varies. It usually starts in the middle of the urban area and runs along a radial road to the urban fringe. Such surveys are often used to show how the land use on one side of an urban area varies from another. They can also be used to show where CBD functions cease and others take over, and thus to delineate that central area.

As well as land use, other information collected on a transect can include building height, number of storeys, upper-floor use, building condition (using an index of decay) and age of buildings. It is also possible to collect information other than urban geography using this method. Such projects could include noise and pollution surveys, and temperature or humidity readings across an urban area.

Environmental surveys

These are a form of appraisal or assessment where the use of a point-score scale is recommended. Both positive and negative observations can be made. Environmental surveys can be carried out in urban areas, on beaches and within river channels and valleys. First-hand observations can be used, as well as interviews with local residents or, in the case of recreation and tourist areas, visitors. Noise, water and air pollution can also be studied.

Instruments and chemical testing kits in such surveys can be used, but simple tests or observations are also valid. Noise pollution can be estimated using a simple table, as long as the same person makes all the observations. Air pollution can be assessed using pieces of sticky tape to take samples from different surfaces in various parts of a town. Water pollution can be visually assessed (or assessed by smell!), but it is possible to use a chemical kit or a secchi disc. The latter is a black and white disc that can be lowered into the water. The depth is recorded at which it is no longer visible and at which it comes into view again when lifted.

Shop location/shopping quality surveys

Shop location studies usually take place in the central area with the aim of finding distinct patterns of land use in the CBD. Selected shop categories can then be analysed using the technique of nearest neighbour analysis which will indicate the degree to which the category is clustered. You can also calculate an index of dispersion. Surveys of service outlets and offices within the urban area can also be carried out. Shopping quality surveys can involve observational data or the use of a questionnaire.

Land value/house price surveys

These are often carried out to identify the peak land value point (PLVP) or peak value intersection (PVI) in an urban area. Taking the values of properties from estate agents and newspapers is much easier than trying to find the values used for local tax assessment (these can be obtained from the local authority's valuation office where records are open to inspection). Values should be converted into a unit per square metre of ground-floor space.

Traffic flow surveys

These involve measuring the traffic flow past several survey points within the urban area.

Pedestrian flow surveys

These are one of the recognised ways of indicating commercial activity within a CBD. There are a number of points to remember when contemplating such a survey:

- you will not be able to carry it out by yourself
- mornings and afternoons are best, avoiding the movement of office workers that takes place in the middle of the day
- do more than one survey in order to contrast different times of day or different days of the week
- at busy points use two counters, operating, if possible, back-to-back in the middle of the pedestrian thoroughway

■ shopping centres are usually private property, so it is advisable to ask for permission

Surveys in physical geography

Some of the more popular types and areas of physical geography study are outlined below.

River surveys

Measurements can be taken and used to calculate fluvial features such as discharge, load, friction and efficiency. The most popular — calculating the discharge of a river — involves finding the cross-sectional area at certain points and multiplying it by the speed of flow (usually obtained using a flow meter) to give a figure expressed in cumecs ($m^3 s^{-1}$). The calculation of the cross section will also identify the wetted perimeter. The stream gradient and bedload shape and size can also be calculated. A final measurement to consider is the extent to which the river meanders — its index of sinuosity.

Slope surveys

Surveys on slopes usually involve calculating the steepness by means of a clinometer, measuring tape and ranging poles. One feature of slopes that can be surveyed is the amount and type of vegetation present, information that can be displayed on a kite diagram.

Soil surveys

These can also be carried out on slopes, choosing sites in exactly the same way as one would for slope vegetation surveys. Various tests that can be carried out include those for soil acidity (using a chemical soil testing kit), soil texture (feeling the texture with your fingers to find the sand, silt and clay elements) and moisture and organic content (taking a sample, drying it overnight, burning off the organic matter to leave the inorganic matter, the mass of which can then be compared with that of the original sample).

Coastal surveys

These can involve measuring the direction and amount of longshore drift or examining the structure of sand dunes, including the environment for plants and animals.

Glaciation surveys

In the field, glaciation surveys usually involve a study of glacial deposits called **till fabric analysis**. This is based on the idea that stones within the ice become orientated in a direction that presents minimal resistance. This means they should be found in glacial deposits with their long axis parallel to the direction of ice movement. From this you should be able to establish the direction of ice movement and possibly its source.

Microclimate surveys

These can be carried out in your local area on a transect from rural to urban or across an urban area. Alternatively day and night conditions can be compared or, more ambitiously, seasons.

Field sketches and photographs

Sketches and photographs are both excellent ways of recording observations you have made as part of your investigation. Field sketches enable you to pick out those features within the landscape you consider important. Investigations in physical geography lend themselves nicely to this technique, particularly those on coastlines and glaciation. It does not matter if you cannot draw particularly well; it is far more important to produce a clear sketch with useful annotations.

If you are not confident about drawing, try photography as an alternative, but remember that far too many photographs find their way into fieldwork reports in the belief that they will make the work 'look good'. Only photographs which convey relevant information should be included and you need to select them carefully. As with sketches, annotation is vital, pointing out the features you have observed. Photographs should be included at the relevant points in the text and not in a large block, which sometimes makes it difficult for the reader to see their purpose.

Collecting secondary data

Secondary data collection involves the gathering of data that have already been put into written, statistical or mapped form.

If you are involved in an investigation that has a temporal context it is almost certain that you will have to access data from previous surveys. Secondary material can also provide a context in the early stages of an investigation, and can be used when explaining and discussing primary material. You may find it useful to combine field data with figures obtained from newspapers, maps, census returns, local authority and other secondary sources. This should give you a much wider database for analysis and comparison.

It is important to check the accuracy of secondary data, particularly if the information could be biased. Details of sources, including author, title, publication and date, must be given in your report, either in a reference list or as footnotes.

Remember that there is a distinction between plagiarism and the acquisition of material by research. The distinction lies in the use made by you, the candidate, of the information you have obtained, and its acknowledgement. Copying out of material from other sources and using it as if it is your own is plagiarism. If you make a direct quotation from secondary material, you must use quotation marks, and you should also ensure that it is properly referenced. This is why the examination board insists that you sign a declaration of authenticity.

Before you start an investigation, check that the sources you are intending to access include the information you need in a form you can use. It will make life very difficult if, at an advanced stage of your enquiry, you find your secondary sources do not match up with your primary research. One good example of this occurs in crime surveys. You may have carried out your primary research on an individual *street* basis, but when you come to access the secondary data, in this case urban crime figures, you will find that they are only available for *districts* within the urban area. It will therefore be impossible to match them up with your more detailed figures.

Sources of secondary material

National government material covers a wide range of data on the economy, employment, population and crime. Material is published by the Office of National Statistics and is available through Her Majesty's Stationery Office (HMSO), but can be expensive to buy and is best accessed through your local library or the internet (**http://www.statistics.gov.uk**). The most helpful publications are:

- *Annual Abstract of Statistics* covering a wide range of data and with material available for previous years
- *Population Trends*
- *Social Trends*
- *Economic Trends*
- *Monthly Digest of Statistics*, the best source of up-to-date information

You can also contact government agencies for information.
    Other national sources include:

- the media (newspapers, magazines and television/radio programmes)
- charities
- national organisations and action groups such as Shelter, English Heritage, National Trust, Countryside Commission
- environmental pressure groups such as Greenpeace and Friends of the Earth
- national company publications
- the Meteorological Office

Local data can be obtained from sources such as:

- your local authority
- the electoral register
- your local library, which will have population statistics for areas as small as electoral wards (small area statistics), census data going back into the nineteenth century, and back copies of local newspapers, as well as photographic archives and photocopying facilities
- the local chamber of commerce
- estate agents
- local newspapers
- *Yellow Pages/Thomson Local Directory*
- the local health authority (information on births, deaths, mortality rates and living conditions such as persons per room in households)
- local action groups

Geographical material comes from traditional sources including:

- geographical magazines and journals such as *Geography/Teaching Geography, Geographical, Geofile* and *Geography Review*
- maps from Ordnance Survey, the Geological Survey, local authorities and Charles Goad, whose maps show the ownership of CBD property

The internet offers an increasing number of helpful sites but it is important that you do not waste your time searching in the hope of finding something useful. Have a definite purpose in mind before accessing this source.

Presenting your results

Selecting the right technique

When you come to present your results, there is a wide range of graphical, carto-graphic and tabular techniques available to you. It is important that you select techniques appropriate to the purpose of your investigation. They should be applied to the data to enable you to describe any changes that are present, establish any differences and identify relationships. You should never be tempted to use all the techniques available — this can lead to data being presented in several different ways for no other reason than to show that you know how to construct various forms of maps and diagrams.

At this level, mark schemes award credit to those candidates who use a suitable range of techniques to provide the potential for analysis. Most investigations will choose techniques that fulfil the following functions (Table 14.2):

- identifying or describing differences
- describing spatial patterns
- identifying relationships
- classifying data according to characteristics

Whatever technique you eventually choose, make sure that it is easy to under-stand, that it is as simple as possible, and that it helps you convey information to the person reading your report.

Table 14.2 Main methods of presentation

Use	Graphical	Cartographic
A Identifying differences	Line graphs (arithmetic and logarithmic) Cumulative frequency curves (including Lorenz curves) Pie graphs and bar graphs Proportional symbols Histograms Long sections and cross sections	Pie graphs, bar graphs and proportional symbols can be placed on a base map to show spatial variations
B Describing spatial patterns		Isopleths Choropleths Flow diagrams and desire lines
C Identifying relationships	Scattergraphs	
D Classifying data	Triangular graphs	

Arithmetic graphs

Arithmetic graphs are appropriate when you want to show absolute changes in data. You will already be familiar with such graphs, but there are a number of points of which you should be aware when using them at this level:

- It is usual to plot the independent variable on the horizontal axis and the dependent variable on the vertical axis. On temporal graphs, time should always be considered the independent variable and plotted horizontally.
- Try to avoid awkward scales and remember that the scale you choose should enable you to plot the full range of data for each variable.
- Axes should always be clearly labelled.
- If you are plotting more than one line, it is a good idea to use different symbols for the plots.
- You can put two sets of data on the same graph, using the two vertical axes to show different scales.

Logarithmic graphs

A logarithmic graph is drawn in the same way as an arithmetic line graph except that the scales are divided into a number of cycles, each representing a tenfold increase in the range of values. If the first cycle ranges from 1 to 10, the second will extend from 10 to 100, the third from 100 to 1,000 and so on.

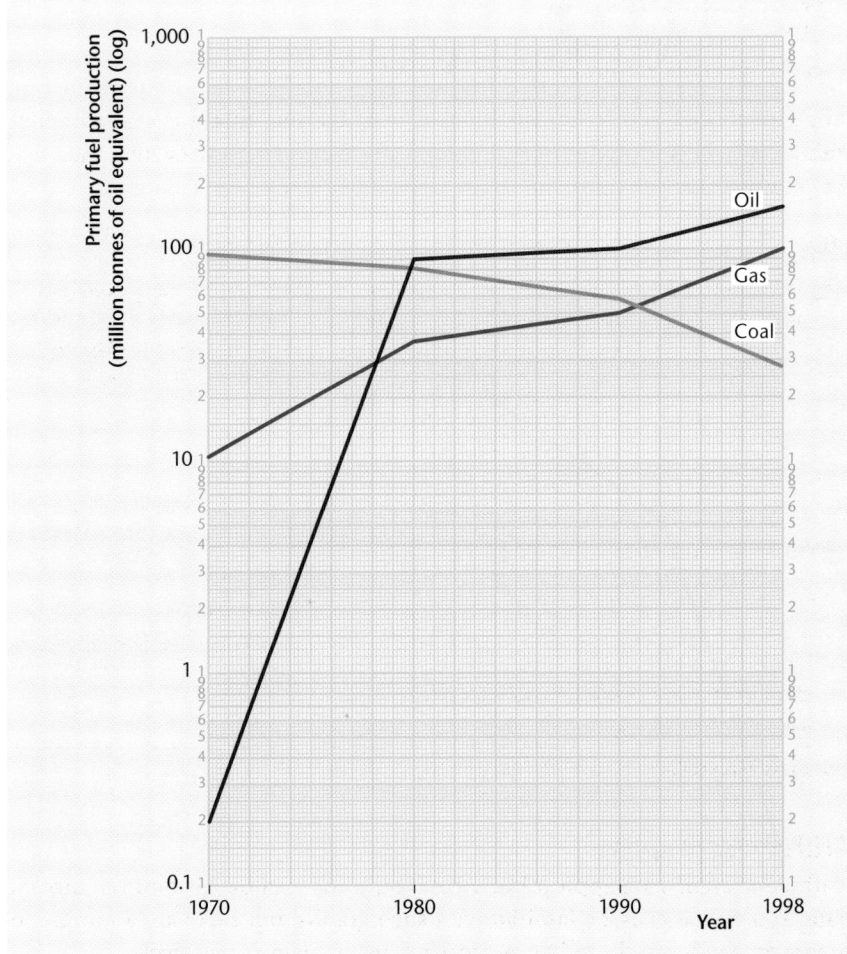

Figure 14.1 Semi-logarithmic graph showing UK production of primary fuels over time

You may start the scale at any exponent of 10, from as low as 0.0001 to as high as 1 million. The starting point depends on the range of data to be plotted.

Graph paper can be either fully logarithmic or semi-logarithmic (where one axis is on a log scale and the other is linear or arithmetic). Semi-logarithmic graphs are useful for plotting rates of change through time, where time appears on the linear axis (Figure 14.1). If the rate of change is increasing at a constant proportional rate (e.g. doubling each time period), it will appear as a straight line.

Logarithmic graphs are good for showing rates of change — the steeper the line, the faster the rate. They also allow a wider range of data to be displayed.

Remember that you cannot plot positive and negative values on the same graph and that the base line of the graph is never zero, as this is impossible to plot on such a scale.

Logarithmic and semi-logarithmic graph paper for you to photocopy is provided at the end of this chapter.

Lorenz curves

Lorenz curves are a form of cumulative frequency curve. They can be drawn on both arithmetic and logarithmic axes. Data are converted into percentages. The largest percentage is plotted first and each consecutive number added on cumulatively and plotted, until 100% is reached. For example, if the highest number is 50%, it is plotted at that point. If the second highest is 20%, then that is plotted at 70% (50 + 20) and so on, until 100% is reached (Figure 14.2).

Figure 14.2
Lorenz curve showing the cumulative percentage of service workers in advertising in the standard regions of Britain in relation to the cumulative percentage of employment in all service occupations

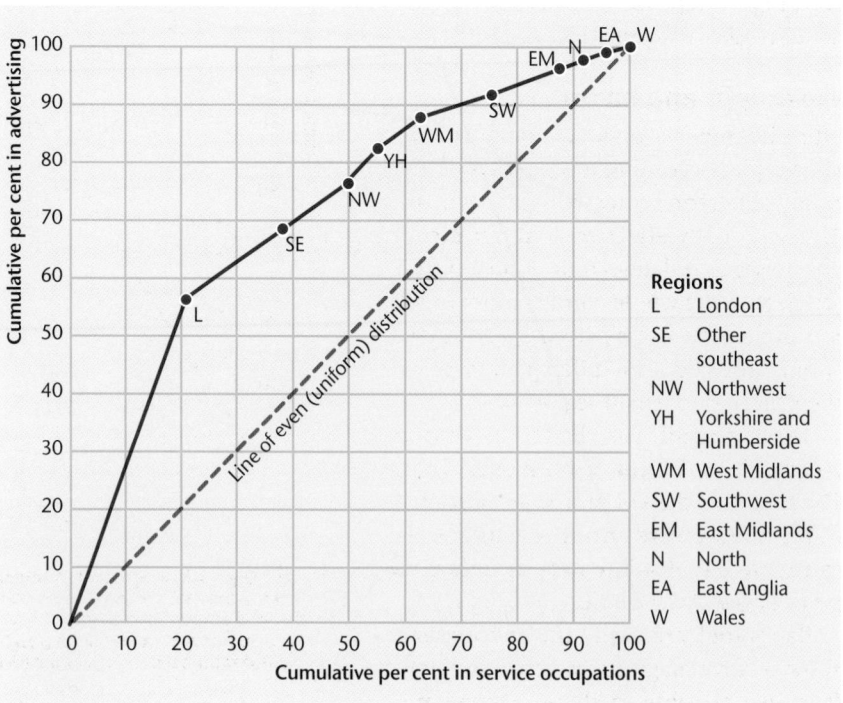

A Lorenz curve can be used to measure or illustrate the extent to which a geographical distribution is even or concentrated. In this case, the categories are ranked in order of size, with the largest first. The percentage value of the highest category is plotted first. The highest and second highest categories are then added to give the next figure to plot. This goes on until all the categories have been plotted and 100% has been reached. This method could be used, for example, if you wanted to plot the distribution of employment in an industry in relation to the workforce nationally (e.g. the distribution of all medical employment in a country against the distribution of all service employment in that country).

The location quotient, which is dealt with later in this chapter, is a method that represents such concentrations as a numerical value.

If you do not need to compare a distribution with a national or regional one, then the cumulative percentage for your data set can simply be plotted in rank order. The vertical axis is labelled *cumulative percentage* (scale 1–100%), and rank order goes on the horizontal axis (scaled to cover the total number of items in the set, e.g. eight categories will produce eight rank orders). This can allow comparisons with other distributions and a line of perfect regularity can be drawn where there is the same percentage in each category. This method is often used for showing the division of employment between particular occupations, or for showing the dependence of a country upon certain types of energy (Figure 14.3).

Figure 14.3
Lorenz curve showing energy sources used to generate electricity in Italy, 1995

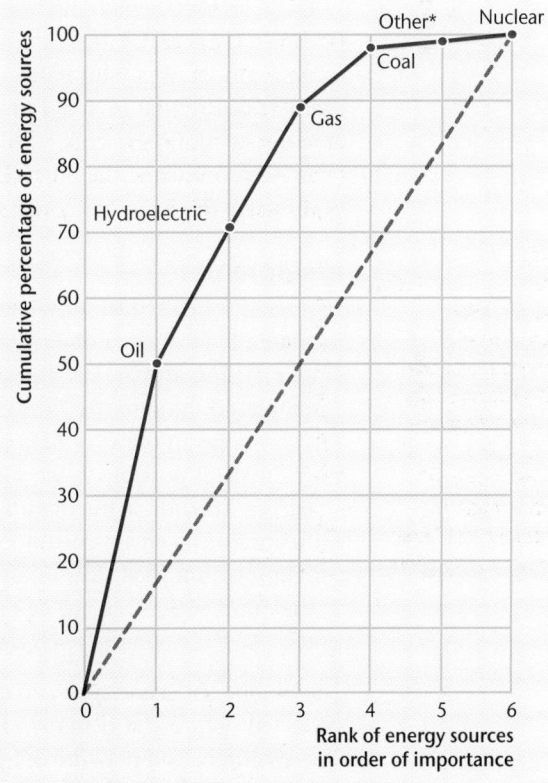

The diagonal line represents a situation in which the various energy sources make an equal contribution to the total amount of electricity
* Including wind power, solar power and energy derived from heated rocks in the Earth's crust

Pie graphs and bar graphs

The pie graph is divided into segments (like a pie) according to the share of the total value represented by each segment. This is visually effective — the reader is able to see the relative contribution of each segment to the whole. On the other hand, it is difficult to assess percentages or make comparisons between different pie charts if there are a lot of small segments.

The bar graph (or chart) has vertical columns rising from a horizontal base. The height of the column is proportional to the value it represents. The vertical scale can represent absolute data or figures as percentages of the whole.

Bar graphs are easily understood and show relative magnitudes very effectively. It is also possible to show positive and

negative values on the same graph using an appropriate vertical scale so that, for example, profit and loss can be shown on the same graph.

Values can easily be read from bar graphs by comparing the height of the bar with the vertical scale. This makes them much more useful, in most cases, than pie graphs. Many students, in their coursework, produce bar graphs that are too complicated (for example, constructing too many multiple bars or using two different scales on the vertical axis). Such complexity undermines the greatest assets of this method — simplicity and clarity of presentation.

Proportional symbols

Symbols can be used which are proportional in area or volume to the value they represent.

We have already seen that the essential element in the construction of a bar graph is that the length of the column is proportional to the value it represents. This idea can be extended to pie graphs, drawing circles proportional to the total value of the data included in each graph.

Other symbols which can be used include squares, cubes and spheres. They can be drawn independently or placed on a map to show spatial differences. If you are using symbols on a map it is very important that you take great care in placing them. It is essential to avoid too much overlap, but it must also be clear which area the symbol represents.

In A2 coursework, the most commonly used symbols are the bar and the circle as these are the easiest to construct.

Histograms

Histograms are used to show the frequency distribution of data. They use bars to indicate the frequency of each class of data, but do not confuse them with bar graphs. Histograms are used to simplify and clarify data that are easier to analyse when placed into groups, or classes, than when presented as individual data. In this way large amounts of data can be reduced to more manageable proportions which make it possible to see some of the trends present.

Before drawing a histogram you need to group the data, and this can be difficult. You need to illustrate differences between classes while keeping the variation within each class to an absolute minimum. You will need to establish:

- the number of classes to be used
- the range of values in each class — the class interval

The number of classes you use must depend upon the amount of data you have collected. Choose too many classes and you will have insufficient variation between them and may finish up with too many 'empty' classes; choose too few and you will have difficulty in recognising trends within the data.

One way of deciding on the number of classes is to use the formula:

number of classes = 5 × log of the total number of items in the set

If, for example, you have data about the size of 120 pebbles on a beach, the maximum number of classes will be:

$5 \times \log 120 = 5 \times 2.08 = 10.4$

You would therefore select ten classes.

The range of values is influenced by the number of classes that you have decided to use. This is shown by the formula:

$$\text{class interval} = \frac{\text{range of values (highest to lowest)}}{\text{number of classes}}$$

If, for example, you had data ranging from 96 to 5 and you required four classes, the class interval would be:

$$\frac{96 - 5}{4} = 22.75$$

It is important that class boundaries are clearly defined so that all individual pieces of data can be assigned without difficulty. Class intervals of 0–25, 25–50, 50–75, 75–100 should therefore be replaced with 0–24.9, 25–49.9, 50–74.9, 75–100, which do not overlap.

The number of classes and the interval will be influenced by the type of data with which you are dealing and the purpose to which they are being put. You will need to decide exactly what you are trying to illustrate or analyse. The distribution that you finish up with will fit into one of three categories:

- If your distribution has a modal class in the middle with progressively smaller bars to each side, then it is similar to the **normal** distribution (Figure 14.4a).
- If the modal class lies in the lower classes, then the distribution is said to show **positive skew** (Figure 14.4b).
- If the modal class lies to the upper end, the distribution is said to show **negative skew** (Figure 14.4c).

Figure 14.4 Histograms showing different distributions

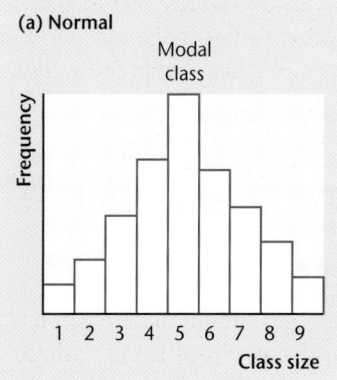
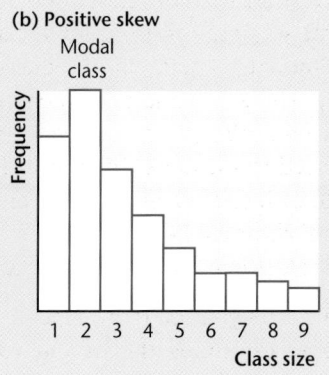
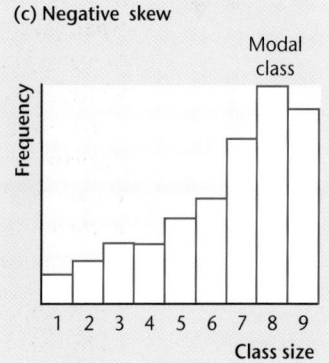

Triangular graphs

Triangular graphs are plotted on special paper in the form of an equilateral triangle (Figure 14.5). Although this looks, on the surface, to be a method that has widespread application, it is only possible to use it for a whole figure that can be broken down into three components expressed as percentages. The triangular

367

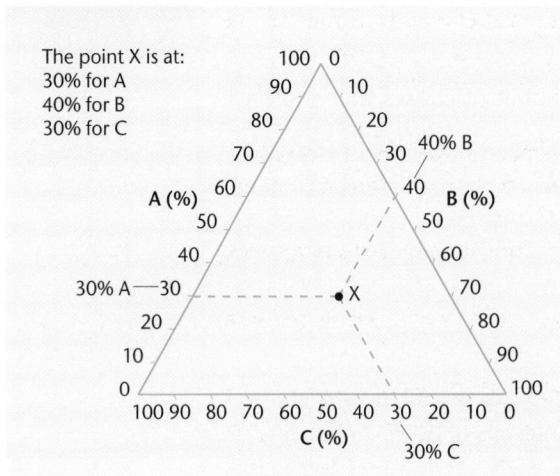

The point X is at:
30% for A
40% for B
30% for C

Figure 14.5
A triangular graph

graph cannot therefore be used for absolute data or for any figures that cannot be broken down into three components.

The advantage of using this type of graph is that the varying proportions and their relative importance can be seen. It is also possible to see the dominant variable of the three. After plotting, clusters will sometimes emerge, enabling a classification of the items involved (Figure 14.6). The items plotted could be, for example, MEDCs/LEDCs or types of soil.

Triangular graph paper for you to photocopy is provided at the end of this chapter.

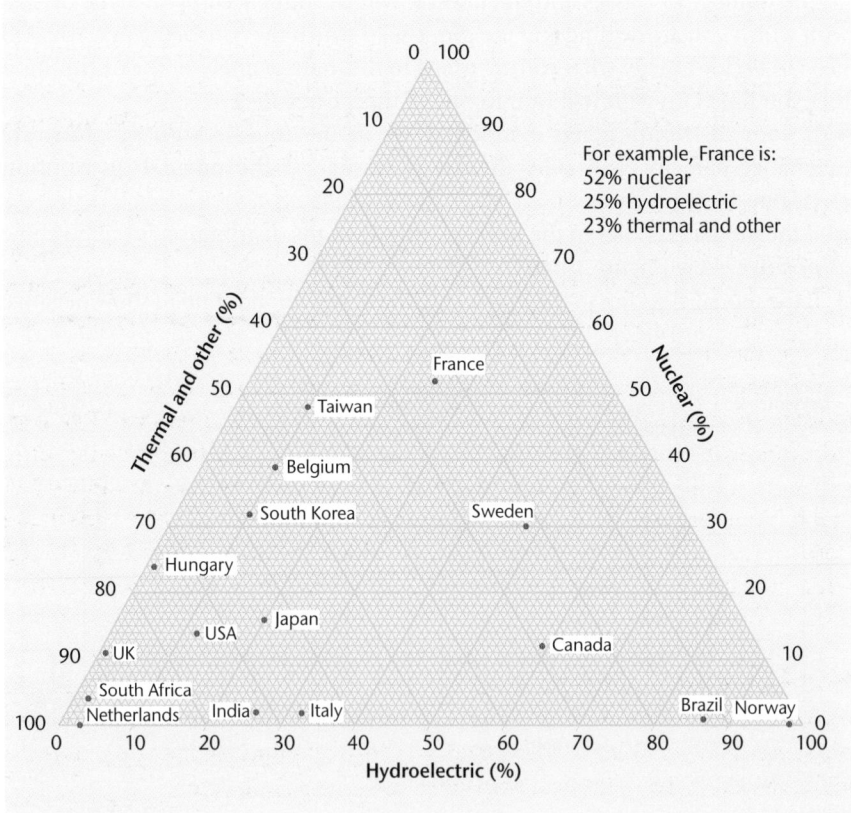

For example, France is:
52% nuclear
25% hydroelectric
23% thermal and other

Figure 14.6
Triangular graph showing percentage of electricity produced by generating source for selected countries

Scattergraphs

Scattergraphs are used to investigate the relationship between two sets of data. Here they are included as a form of presentation, but they are equally useful in identifying patterns and trends which might lead to further inquiry. If you use this method, there are several points to bear in mind:

- Scattergraphs can be plotted on arithmetic, logarithmic or semi-logarithmic graph paper.
- Only plot when you feel that there is a real relationship to be investigated. It is possible for a correlation to emerge even when a relationship is only coincidental.
- One variable usually has an effect on the other and this enables us to identify the independent and dependent variables.
- The independent variable goes on the horizontal axis and the dependent variable on the vertical axis.
- After the points have been plotted, a trend line can be drawn (best fit). If this runs from bottom left to top right, it indicates a positive relationship. If it runs top left to bottom right, it indicates a negative relationship.
- The closer the points to the trend line, the greater is the relationship, but this should be assessed with a statistical test (see Spearman rank correlation coefficient).
- Points lying some distance from the trend line are classed as **residuals** (anomalies). These can be referred to as either negative or positive. Identification of residuals may enable you to make further investigations into other factors that could influence your selected two variables.

Long sections and cross sections

These methods are useful for describing and comparing the shape of the land. Long sections are mostly used in river studies, while cross sections can be drawn for a number of landscape features. This method essentially consists of constructing a graph showing height against a horizontal scale of distance.

It is usual to take the horizontal scale from the map you are using, but to use a different vertical scale from this. If you are working with a standard OS map, the scale would reduce your section, in most cases, to a line showing little variation. It is therefore necessary to adopt a larger scale, but care must be taken that it is not massively exaggerated, changing the gentlest of valley slopes into the north wall of the Eiger! The degree of exaggeration can be calculated and presented with your finished work. The same principles apply to both long- and cross-section construction.

Isopleth or isoline maps

If you have collected data from different places, they can be represented by points on a map. It is possible to draw a map on which all points of the same value are joined by a line. This allows patterns in a distribution to be seen.

The best known example of such isolines (also called isopleths) is on Ordnance Survey maps where contour lines join places of the same height. This technique can be applied to a number of other physical factors, such as rainfall (**isohyets**), temperature (**isotherms**) and pressure (**isobars**), as well as human factors, such as travel times (**isochrones**) for commuters and shoppers.

*Figure 14.7
Choropleth map
showing population
density in a metro-
politan borough in
northern England,
1991*

Choropleth maps

A choropleth is a map on which data values are represented by the density of
shading within areas. The data are usually in a form that can be expressed in terms
of area, such as population density per square kilometre. To produce such a map
certain stages have to be followed:

- The material has to be grouped into classes. Before you can do this you have to
  decide on the number and range of classes required to display your data clearly.
- A range of shadings has to be devised to cover the range of the data. Darkest
  shades should represent the highest figures and vice-versa. It is good practice
  not to use the two extremes of black and white because black suggests a
  maximum value, while white implies that there is nothing within the area. A
  suitable method of shading is shown in Figure 14.7.

Choropleth maps are fairly easy to construct and are visually effective as they give
the reader a chance to see general patterns in an areal distribution. There are,
however, a few limitations to the method:

- It assumes that the whole area under one form of shading has the same density,
  with no variations. For example, on maps of the UK the whole of Scotland
  may be covered by one category, when it is obvious that there could be large
  variations between the central populated areas and the Highlands.
- The method implies abrupt changes at the drawn boundaries which will not be
  present in reality.

Flow lines and desire lines

Flow diagrams are drawn to show the quantity of movement, for example traffic,
that follows an existing route. Lines are drawn to widths proportional to the

quantity of movement. Straight lines linking an origin and a destination can sometimes be used to show, for example, shopping movements. Flow lines are also known as desire lines.

Analysing and interpreting your data

Statistical analysis is often used in geographical investigations. Such objective analysis of data can be used to support the conclusions suggested by a subjective view of the results.

Statistical analysis should not be used just for show. It should form an integral part of the coursework report. You should think carefully about the most effective form of statistical analysis, and why that technique is appropriate. The best statistical analysis will help in evaluating the significance of the results.

It is obvious that any statistical technique should be used correctly, and that all calculations should be performed accurately. It is perhaps less obvious that the result of any calculation should be supported by statements explaining what it means. Many calculator and computer functions will complete the mathematical process for you, so it is paramount that you understand the relevance of the values produced by them. In short, never use a statistical technique you do not understand or in which you are not confident.

Table 14.3 summarises some of the major statistical techniques for use in coursework and the written alternative examination.

Reason for using statistics	Statistical technique(s)
Summarising and comparing data	Measures of central tendency: mean, mode, median
The dispersion and variability of data	Range Inter-quartile range Standard deviation
Correlating two sets of data	Spearman rank correlation coefficient, and tests of significance
The degree of concentration of geographical phenomena	Location quotient
Measuring patterns in a distribution	Nearest neighbour statistic
Extent of differences between observed and expected data, and their statistical significance	Chi-squared test

Table 14.3
Statistical techniques

These techniques, and the rationale for their use, are described below.

Measures of central tendency

There are three such measures.

Arithmetic mean

The mean is calculated by adding up all the values in a data set and dividing the total sum by the number of values in the data set. So,

$$\bar{x} = \frac{\sum x}{n}$$

The arithmetic mean is of little value on its own and should be supported by reference to the standard deviation of the data set.

Mode

This is the value which occurs most frequently in a data set. It can only be identified if all the individual values are known.

Median

This is the middle value in a data set when the figures are arranged in rank order. There should be an equal number of values both above and below the median value. If the number of values in a data set is odd, then the median will be the

$$\frac{(n + 1)}{2}$$

item in the data set.

So, for example, if the total number of items in a data set is 27, the median will be the fourteenth value in the rank order of the data.

If the number of values in the data set is even, the median value is the mean of the middle two values. Any calculation of the median is best supported by a statement of the inter-quartile range of the data (see page 373).

Distribution of the data set

It is possible that each of these measures of central tendency could give the same result, but they are more likely to give different results. For them each to give the same result the distribution of a data set would have to be perfectly 'normal', and this is extremely unlikely when using real data. It is more likely that the distribution of the data set will be skewed (see page 367). The more it is skewed, the greater the variation in the three measures of central tendency.

None of these measures gives a reliable picture of the distribution of the data set. It is possible for two different sets of data to have the same values for mean, mode and median. Measures of the dispersion or variability of the data should therefore also be provided.

Measures of dispersion or variability

There are three measures of dispersion or variability: range, inter-quartile range and standard deviation.

Range

This is the difference between the highest value and the lowest value in a data set. It gives a simple indication of the spread of the data.

Inter-quartile range

The inter-quartile range is calculated by ranking the data in order of size and dividing them into four equal groups or quartiles. The boundary between the first and second quartiles is known as the upper quartile and the boundary between the third and fourth quartiles is the lower quartile. They can be calculated as follows:

- the upper quartile (UQ) is the value that occurs at $\dfrac{(n+1)}{4}$ in the data set when arranged in rank order (from highest to lowest)

- the lower quartile (LQ) is the value that occurs at $\dfrac{3(n+1)}{4}$ in the data set

The difference between the upper and lower values is the inter-quartile range:

IQR = UQ − LQ

The IQR indicates the spread of the middle 50% of the data set about the *median* value, and thus gives a better indication of the degree to which the data are spread, or dispersed, on either side of the middle value.

Standard deviation

Standard deviation measures the degree of dispersion about the *mean* value of a data set. It is calculated as follows:

- the difference between each value in the data set and the mean value is worked out
- each difference is squared, to eliminate negative values
- these squared differences are totalled
- the total is divided by the number of values in the data set, to provide the variance of the data
- the square root of the variance is calculated

$$\text{standard deviation} = \sqrt{\dfrac{\Sigma(\bar{x} - x)^2}{n}}$$

The standard deviation is statistically important as it links the data set to the normal distribution. In a normal distribution:

- 68% of the values in a data set lie within ±1 standard deviation of the mean
- 95% of the values in a data set lie within ±2 standard deviations of the mean
- 99% of the values in a data set lie within ±3 standard deviations of the mean

A low standard deviation indicates that the data are clustered around the mean value and that dispersion is narrow. A high standard deviation indicates that the data are more widely spread and that dispersion is large. The standard deviation also allows comparison of the distribution of the values in a data set with a theoretical norm and is therefore of greater use than just the measures of central tendency.

Measuring correlation: the Spearman rank correlation coefficient

Comparisons are made between two sets of data to see whether there is a relationship between them. Note that even if there is a relationship between two variables, this does not prove a causal link. In other words, the relationship does not prove that a change in one variable is responsible for a change in the other. For example, there may be a direct relationship between altitude and the amount of precipitation in a country such as the UK. These two variables (altitude and precipitation) are clearly linked, but a decrease in one does not automatically cause a decrease in the other — they are simply related to each other.

There are two main ways in which relationships can be shown:
- using scattergraphs (see page 368)
- measuring correlation using the Spearman rank correlation coefficient

The Spearman rank correlation coefficient is used to measure the degree to which there is correlation between two sets of data (or variables). It provides a numerical value which summarises the degree of correlation, so it is an example of an objective indicator. Once it has been calculated, the numerical value has to be tested statistically to see how significant the result is.

The test can be used with any data set consisting of raw figures, percentages or indices which can be ranked. The formula for the calculation of the correlation coefficient is:

$$R_s = 1 - \frac{6 \sum d^2}{n^3 - n}$$

where d is the difference in ranking between the two sets of paired data and n is the number of sets of paired data.

The method of calculation is as follows:
- rank one set of data from highest to lowest (highest value ranked 1, second highest 2 and so on)
- rank the other set of data in the same way
- beware of tied ranks. In order to allocate a rank order for such values, calculate the 'average' rank they occupy. For example, if there are three values which should all be placed at rank 5, add together the ranks 5, 6 and 7 and divide by three, giving an 'average' rank of 6 for each one. The next value in the sequence will be allocated rank 8
- calculate the difference in rank (d) for each set of paired data
- square each difference
- add the squared differences together and multiply by 6 (A)
- calculate the value of $n^3 - n$ (B)
- divide A by B, and take the result away from 1

The answer should be a value between +1.0 (perfect positive correlation) and −1.0 (perfect negative correlation).

Some words of warning

- You should have at least 10 sets of paired data, as the test is unreliable if *n* is less than 10.
- You should have no more than 30 sets of paired data or the calculations become complex and prone to error.
- Too many tied ranks can interfere with the statistical validity of the exercise, although it is appreciated that there is little you can do about the 'real' data collected.
- Be careful about choosing the variables to compare — do not choose dubious or spurious sets of data.

Interpreting the results

When interpreting the results of the Spearman rank test consider the following:

What is the direction of the relationship?

If the calculation produces a positive value, the relationship is positive, or direct. In other words, as one variable increases, so does the other. If the calculation produces a negative value, the relationship is negative, or inverse.

How statistically significant is the result?

When comparing two sets of data, there is always a possibility that the relationship shown between them has occurred by chance. The figures in the data sets may just happen to have been the right ones to bring about a correlation. It is therefore necessary to assess the statistical significance of the result. In the case of the Spearman rank test the critical values for R_s must be consulted. These can be obtained from statistical tables, but Table 14.4 shows some examples.

According to statisticians, if there is a >5% possibility of the relationship occurring by chance, the relationship is not significant. This is called the rejection level. The relationship could have occurred by chance more than five times in 100, and this is an unacceptable level of chance. If there is a <5% possibility, the relationship is significant and therefore meaningful.

If there is a <1% possibility of the relationship occurring by chance, the relationship is very significant. In this case, the result could only have occurred by chance one in 100 times, and this is very unlikely.

How does this work? Having calculated a correlation coefficient, examine the critical values given above (ignore the positive or negative sign). If your coefficient is greater than these values, the correlation is significant at that level. If your coefficient is smaller, then the relationship is not significant at that level.

For example, suppose you had calculated an R_s value of 0.50 from 18 sets of paired data. 0.50 is

Table 14.4
Critical values for R_s

n	0.05 (5%) significance level	0.01 (1%) significance level
10	± 0.564	± 0.746
12	0.506	0.712
14	0.456	0.645
16	0.425	0.601
18	0.399	0.564
20	0.377	0.534
22	0.359	0.508
24	0.343	0.485
26	0.329	0.465
28	0.317	0.448
30	0.306	0.432

greater than the critical value at the 0.05 (5%) level, but not that at the 0.01 (1%) level. In this case, therefore, the relationship is significant at the 0.05 (5%) level, but not at the 0.01 (1%) level.

Location quotient

A location quotient is a measure of the degree to which a geographical activity is concentrated in an area. As in the case of the Spearman rank correlation coefficient, the end product is a number, which again gives an objective value to use for comparison. This value compares the concentration of an activity in a sub-region with the concentration in the whole region.

For example, when studying the concentration of employment in an industry in a particular region of a country, the calculation of the location quotient is:

$$LQ = \frac{X'/Y'}{X/Y}$$

where:
X' is the number employed in the given industry in the region
Y' is the number employed in all industries in that region
X is the number employed in the given industry in the country as a whole
Y is the number employed in all industries in that country

In this case the location quotient compares the proportion of employed people in a particular region in a given industry with the proportion in that industry nationally.

Sometimes data are given in percentage form, in which case the calculation is more straightforward:

$$LQ = \frac{(\% \text{ of workers in the given activity in the region})}{(\% \text{ of workers in the activity nationally})}$$

The key indicator in the use of location quotients is LQ = 1.0. A result of this nature indicates that the region in question has a fair share of that geographical activity compared with the rest of the country. The relative proportions, region vs country, are equal.

If LQ >1.0, this indicates a greater share of that activity in the region compared with the rest of the country. In other words, there is a concentration of that activity in that region.

Similarly, if LQ <1.0, then the region has less than its proportionate share — the activity is under-represented.

Nearest neighbour statistic

Nearest neighbour is used to analyse the distribution of individual points in a pattern. It can be applied to the distribution of any data that can be plotted as point locations. Consequently, it is often used to analyse the distribution of shop types in a town centre, the distribution of various sizes of settlement in an area and the distribution of some public services, for example doctors' surgeries, in an urban area.

The basis of the statistic is the measurement of the distance between each point in a pattern and its nearest neighbour. This must be done for each point identified within the area studied. Once all measurements have been completed, the mean distance $(\bar{d})$ between each point is calculated.

The nearest neighbour statistic can then be calculated using:

$$R_n = 2\bar{d}\sqrt{\frac{N}{A}}$$

where:

N is the number of point locations in the area

A is the area of study

The statistic can be any value between 0 and 2.15.

- 0 represents a pattern that is perfectly clustered; in other words, there is no distance between nearest neighbours — all the points are at the same location. If this is the case, there is no pattern to analyse.
- 2.15 represents a pattern displaying perfect regularity — all points lie at the vertices of equilateral triangles. All distances between nearest neighbours are identical. Again this is highly unlikely in the real world.
- 1.0 is said to represent a random pattern, although this is difficult to prove.

In practice, the outcome of such a calculation will be on the continuum between 0 and 2.15. Proximity to one of these will indicate the degree of either clustering or regularity.

Requirements of the method

When carrying out a nearest neighbour statistic calculation, it is advisable to map the activity(ies) on a transparent overlay first. This removes any potential distraction in the measurement and calculation process. It is important that the units in which distance and area are measured correspond — in most cases they will be either metres and square metres, or kilometres and square kilometres.

A further complication is the delineation of the area to be studied. In the case of settlement-based work, you have to decide on the boundary of the area you are studying. You can then establish a 'buffer zone' around this study area. When measuring the distance between each point and its nearest neighbour, if the nearest neighbour is within this buffer zone then you should measure *to* this point. However, this point in the buffer zone should not be counted in the overall study as a point *from* which to measure to its nearest neighbour. In short, you should only measure *to* points in the buffer zone, not *from* them.

Chi-squared test

This technique is used to assess the degree to which there are differences between a set of collected (or observed) data and a theoretical (or expected) set of data, and the statistical significance of the differences.

The observed data are those that have been collected either in the field or from secondary sources. The expected data are those that would be expected according to the theoretical hypothesis being tested.

Normally, before the test is applied it is necessary to formulate a null hypothesis. In the example given here, the null hypothesis would be that there is no significant difference between the observed and expected data distribution. The alternative to this would be that there *is* a difference between the observed and expected data, and that there is some factor responsible for this.

The method of calculating chi-squared is shown below. The letters A–D in Table 14.5 refer to map areas A–D in Figure 14.8. In the column headed *O* are listed the numbers of points in each of the areas A–D on Figure 14.8 (the *observed* frequencies). The total number of points in this case is 40. Column *E* contains the list of *expected* frequencies in each of the areas A–D, assuming that the points are evenly spaced. In the column *O – E*, each of the expected frequencies is subtracted from the observed frequencies, and in the last column the result is squared. The relevant values are then inserted into the expression for chi-squared, and the result is 4.0.

Table 14.5

Map	Observed (O)	Expected (E)	($O - E$)	($O - E)^2$
A	8	10	−2	4
B	14	10	4	16
C	6	10	−4	16
D	12	10	2	4
Sum	40	40	0	40

Figure 14.8

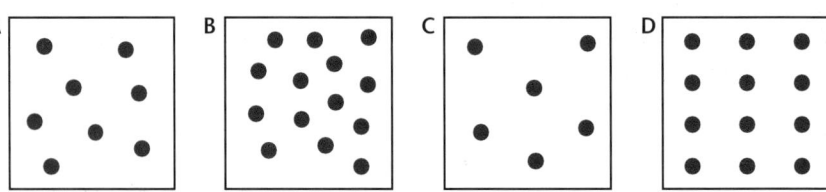

$$\chi^2 = \frac{\Sigma(O - E)^2}{E}$$

$$= \frac{40}{10}$$

$$= 4.0$$

The aim of a chi-squared test, therefore, is to find out whether the observed pattern agrees with or differs from the theoretical (expected) pattern. This can be measured by comparing the calculated result of the test with its level of significance.

To do this the number of degrees of freedom must be determined using the formula ($n - 1$), where n is the number of observations, in this case the number of cells which contain observed data (4). So, 4 – 1 = 3. Statistical tables give the distribution of chi-squared values for these degrees of freedom.

Then there are the levels of significance. There are two levels of significance: 95% and 99%. At 95% there is a 1 in 20 probability that the pattern being considered occurred by chance, and at 99% there is only a 1 in 100 probability that the pattern is chance. The levels of significance can be found in a book of statistical tables. They are also known as confidence levels.

If the calculated value is the same or greater than the values given in the table, then the null hypothesis can be rejected and the alternative hypothesis accepted.

In the case of our example, however, the value of chi-squared is very low (4.0), showing that there is little difference between the observed and the expected pattern. The null hypothesis cannot therefore be rejected.

Some further points on this technique

- The numbers of both observed and expected values must be large enough to ensure that the test is valid. Most experts state that there should be a minimum of five.
- The number produced by the calculation is itself meaningless. It is only of value for use in consulting statistical tables.
- As in Spearman's rank, only significance (or confidence) levels of 95% and 99% should be considered when rejecting the null hypothesis. Levels of confidence greater than these simply allow the null hypothesis to be rejected with even greater confidence.
- It is strongly recommended that you do not apply the test to more than one set of observed data because the mathematics become too complex.

Some words of warning

This test is difficult to understand and should not be attempted unless you are fully confident in your ability both to apply it and to understand the outcome.

Using analytical techniques

It is important to note that all the analytical techniques described here are only to be used as a *support* for your own ideas about the geographical significance of your study. All results, and the statistical analysis of them, should be related to the original hypothesis or established theory about that area of the subject.

Your results may support your hypothesis or established theory, or they may not. If they do not, there may be some reason or factor which is responsible. This could lead to further enquiry.

Above all, your project or coursework should make geographical sense, and this is far more important than demonstrating your ability to use mathematics or statistics.

Writing up the report

The final report of your enquiry should be well structured, logically organised, and clearly and concisely written. There are three aspects of this process that you should consider: structure, language and presentation.

Structure

The structure of the report should help the reader to understand it, and should also assist you in organising it logically. The following checklist provides a generalised structure for your report:

- report approval sheet/cover sheet
- title page and contents page
- executive summary
- aims and objectives
- scene setting
- research questions/hypotheses/issues being examined
- sources of information used
- methods of data collection and commentary on their limitations
- data presentation, analysis and interpretation
- evaluation and conclusion
- bibliography and appendices

You need not write these in the order given. Indeed, it may be easier if you do not. For example, the executive summary is perhaps best written at the end of the whole process, as it is only at this stage that the whole picture can be described. The following is a suggested order of completion.

Data presentation, analysis and interpretation

This is the section where you present and analyse your findings. At this stage you will have collected the data, sorted them and selected the most useful pieces. You will know what you have found out and what it all means. Your results will be complete and they will be most fresh in your mind at this time. You should be able to interpret each separate section of your results and formulate conclusions for each one. The whole picture may begin to appear in your head.

Sources of information and methods of data collection

Now you can write about what information you collected and the methods you used. Do not forget to discuss any limitations of the methods of collection you used and of the data sources themselves.

Conclusion

This should include a summary and an evaluation of all the major findings of your enquiry. Do not present anything new to the reader at this stage. Towards the end of this section, try to draw together all the sub-conclusions from each section of the data analysis into one overall conclusion — the whole picture.

Aims, objectives and scene setting (introduction)

Having written up the bulk of the enquiry, you can now write the introduction, making sure it ties in with what follows. This section is intended to acquaint the reader with the purpose of the enquiry and the background to it.

Appendices and bibliography

The appendices contain additional pieces of evidence that may be of interest to the reader but are not essential to the main findings. The bibliography provides details of the secondary sources used in your research, either as guides or as sources of information. You should give the author, date of publication, title of the publication, publisher and page number. When quotes are used in the body of your report, you should provide the name of the author and the date of publication and cross-refer it back to the bibliography.

Contents page
All the sections of the report should be listed in sequence with accurate page references.

Title page
This states the title of your report. Include also your name, candidate number, centre number and date of completion.

Executive summary
An executive summary should be a brief statement (no more than 250 words) covering all the main aspects of the enquiry. A good executive summary introduces the subject of the full report, refers to its aims and objectives, and provides a brief synopsis of the findings. A very good executive summary will tempt the reader into reading more by being comprehensible, interesting and stimulating. It should also make sense as a separate document from the full report.

Language

The quality of the language you use in writing up your enquiry is important. You are entirely in control of this aspect of the process, and your style of writing must be appropriate for this exercise. You should avoid poor or inaccurate use of English. In particular:
- your sentences should be grammatically correct and well punctuated
- your writing should be well-structured, with good use of paragraphs
- your spelling must be accurate (use a dictionary and your PC spellcheck)
- you must be clear in your use of specialist terminology and in the expression of your ideas
- you should be aware that the assessment of your work will take into account the above aspects of your writing

Proofreading is an important part of this process. Prior to submission, make sure you read through the draft from start to finish and mark any places where there are errors or inconsistencies. If possible, get someone to do this for you — parents may help. You need someone who is going to be highly critical of what you have written. A report littered with spelling mistakes or grammatical errors does not impress, and your PC spellcheck will miss many of these.

It is also essential to make sure that maps and diagrams are inserted in the correct place in the report — it is irritating to have to flick backwards and forwards when trying to read the document. Make sure all the references in the text are included in the bibliography.

Length

The examination board states the length of report it requires and will impose a penalty of some sort for overlong submissions. In such cases, it is the final parts of your report — dealing with the conclusions and evaluation — which are most under threat. These are important areas of your coursework and it is foolish to make the marks available for this section inaccessible.

Word limits are not designed to make life difficult — they should have the opposite effect. Reports written within the stated word limit tend to be better planned, structured and executed. No penalty is imposed for reports that are too short, but they are self-penalising because they contain inadequate material.

Presentation

It is a fact of life that most people are influenced by presentation, and that includes coursework examiners. Bear in mind the following:

- A neatly presented handwritten or word-processed report is going to create a favourable first impression, before its contents are read.
- Adequate heading and numbering of pages, with carefully produced illustrations, will make it easier for the reader to understand what is contained within the report.
- Layout is also important. Do not crowd the pages with dense text, which looks unattractive. Provide adequate margins, use either double or 1.5 line-spacing in word-processing, and make use of clear heading levels with short paragraphs.
- Make sure you allow enough time to add the finishing touches which give your work the 'final polish'. It goes without saying that this time will be available if you have not left completion too near to the final deadline.

You should now be in a position to submit your finished product confident in the belief that it is the best you could have done.

Example of an enquiry

Here is an outline of a possible enquiry to illustrate the principles identified above.

Title
Is the climate in Central Birmingham different from that of the rural–urban fringe?

Aim
The aim is to test the theory of urban heat islands by investigating the variations in temperature, humidity and wind speed between central Birmingham and the outer suburbs of the city.

Objectives
- To design a number of routes from the centre of the city to the urban fringe, identifying sample sites on each route.

- To measure the climatic variables of temperature, humidity and wind speed at each of these sites.
- To prove the existence of the urban heat island and offer explanations for it.

Hypotheses
- Central Birmingham is warmer than the rural–urban fringe.
- Central Birmingham has a lower average wind speed than the rural–urban fringe.
- Central Birmingham has a lower relative humidity than the rural–urban fringe.

Skills that could be demonstrated
- Drawing location maps.
- Sampling techniques.
- Primary data collection: temperature, humidity levels, wind speed.
- Construction of a variety of types of graph: scattergraphs, regression lines.
- Calculation of Spearman rank correlation coefficient and its significance.
- Calculations of mean and standard deviation.

To complete the enquiry
- Good presentation of all results.
- Analysis of results.
- Commentary on findings of results.
- Commentary on limitations of data collection.
- Overall conclusion.
- Bibliography.
- Appendix.

How will your investigation be marked?

The grid on pages 384–385 is used by your teacher when marking your report. Note that this mark will be subject to moderation by a moderator appointed by AQA.

Your investigation will be assessed according to five criteria:
- knowledge (10 marks)
- critical understanding (30 marks)
- knowledge and critical understanding (20 marks)
- skills and techniques (40 marks)
- quality of language (included within each of the above)
  Total = 100 marks

Examine the grid to see how each of these criteria is used in assessment.

Response Level 1	Response Level 2	Response Level 3
Knowledge Demonstrates basic descriptive geographical knowledge Uses basic geographical terminology Gives a general location of the area of study States briefly some of the geographical ideas, concepts, principles or theories on which this study is based	Demonstrates relevant geographical knowledge at a higher level Uses throughout the appropriate geographical terminology for this study Accurately locates the area of study, and provides some relevant geographical characteristics Provides details of the geographical ideas, concepts, principles or theories on which this study is based	
(0–6 marks)	(7–10 marks)	
Critical understanding An understanding of basic descriptive geographical knowledge A basic awareness that the sources of evidence, concepts and theories available for this enquiry have limitations and/or explanation of their potential Where appropriate, briefly states some attitudes, values, approaches, decision-making processes and/or some involvement of human activity	Shows, by explanation, understanding of geographical knowledge at a higher level Some explanation that the sources of evidence, concepts and principles or theories available for this enquiry have limitations and/or an explanation of their potential Where appropriate, shows by explanation an understanding of the effects of attitudes, values, approaches, decision-making processes and/or the effect of some greater involvement of human activity	Shows, by explanation, clear understanding of relevant high-level geographical knowledge Shows, by detailed explanation, a clear understanding that the sources of evidence, concepts and principles or theories available for this enquiry have limitations and/or provides an explanation of their potential Where appropriate, shows by explanation a clear and detailed understanding of the effects of attitudes, values, approaches, decision-making processes and/or detailed explanation of the effect of human involvement
(0–8 marks)	(9–20 marks)	(21–30 marks)
Knowledge and critical understanding Basic interpretation and analysis of data Basic synthesis from relevant geographical facts and ideas Simple descriptive conclusions drawn from the data with an awareness that they have some limitations A basic attempt at evaluation of the success or otherwise of the investigation	More detailed interpretation and analysis of data with some correct use, where appropriate, of basic statistical techniques More detailed synthesis from relevant geographical facts and ideas Developed conclusions drawn from the data with some development of their limitations A clear and detailed evaluation of the success or otherwise of the investigation	
(0–8 marks)	(9–20 marks)	

Response Level 1	Response Level 2	Response Level 3
Skills and techniques		
A limited range of skills and methods used	A wider range of skills and methods used	Uses and demonstrates most skills and methods appropriate to the enquiry. Most of the potential sources are used
Few of the potential sources used	A range of the potential sources used	
Only basic collection and sampling skills correctly used	Additionally, a single high-level collection and sampling skill correctly used	Additionally, some high-level collection and sampling skills correctly used
Only basic organisation and presentation skills used	Additionally, a single high-level organisational or presentational skill correctly used	Additionally, some high-level organisation or presentation skills correctly used
Basic awareness that the skills used and the evidence collected have limitations	Gives more evidence of the appropriateness and limitations of the skill used and of the evidence gathered, including some basic awareness of the relevance of scale (temporal and/or spatial) in the selection of location and the collection of data	A detailed evaluation of the appropriateness and limitations of the skills used and the evidence gathered, including a clear awareness of the relevance of scale (temporal and/or spatial) in the selection of location and the collection of data
Basic reasons given for the selection of the skills used		
(0–7 marks)	(8–15 marks)	(16–40 marks)
To obtain full marks at any level the appropriate **quality of language** descriptor below must be achieved. Use the same quality of language level as is used in the geographical element of the marking criteria under consideration		
Style of writing is suitable for only simple subject matter	Manner of dealing with complex subject matter is acceptable, but could be improved	Style of writing is appropriate to complex subject matter
Clear expression of only simple ideas, using a limited range of specialist terms	Reasonable clarity and fluency of expression of ideas using a good range of specialist terms	Organises relevant information and ideas clearly and coherently using a wide range of specialist vocabulary when appropriate
Reasonable accuracy in the use of English	Considerable accuracy in the use of English	Accurate in the use of English

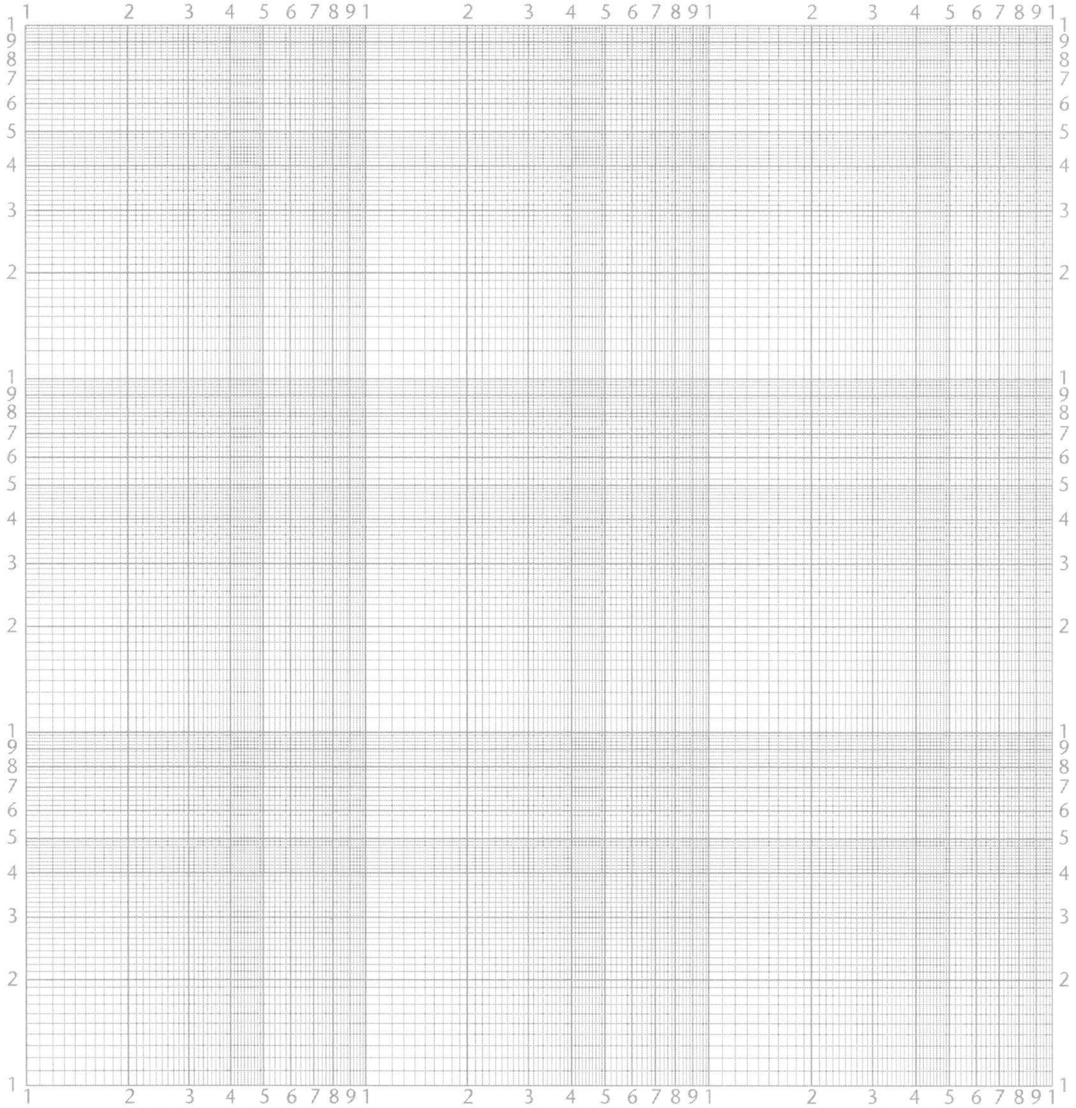

Figure 14.9 Logarithmic graph paper

Figure 14.10 Semi-logarithmic graph paper

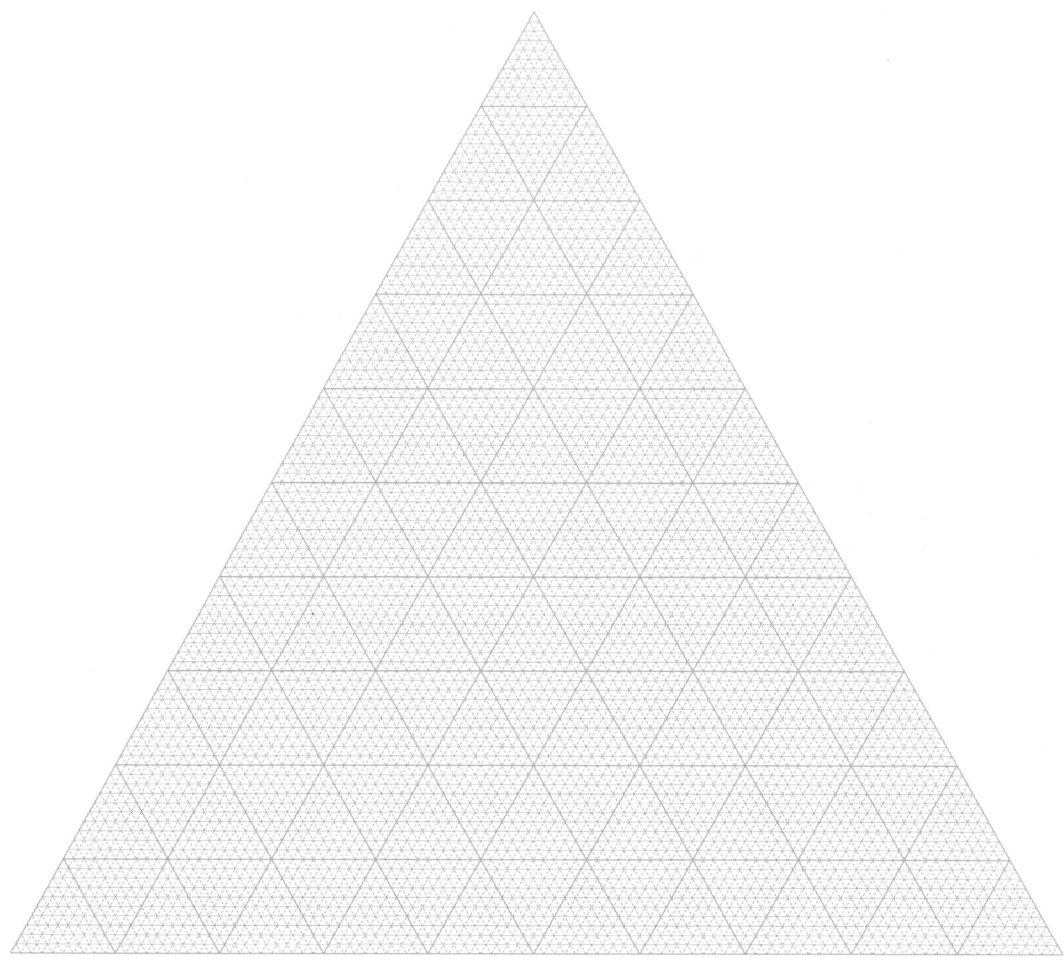

*Figure 14.11 **Triangular graph** paper*

Bibliography

Anderson, D. (2004) *Glacial and Periglacial Environments*, Hodder and Stoughton.

Bishop, V. (1998) *Hazards and Responses*, Collins.

Frampton, S. et al. (2000) *Natural Hazards*, Hodder and Stoughton.

Gilbert, O. (1992) *The Flowering of Cities: The Natural Flora of Urban 'Commons'*, English Nature.

Goudie, A. and Viles, H. (1997) *The Earth Transformed*, Blackwell.

Guinness, P. (2002) *Migration*, Hodder and Stoughton.

Guinness, P. (2003) *Globalisation*, Hodder and Stoughton.

Hill, M. (2004) *Coasts and Coastal Management*, Hodder and Stoughton.

Middleton, N. (1995) *The Global Casino*, Edward Arnold.

Nagle, G. (2002) *Climate and Society*, Hodder and Stoughton.

Nagle, G. (2003) *Rivers and Water Management*, Hodder and Stoughton.

Redfern, D. (2002) *Human Geography: Change in the United Kingdom in the Last 30 Years*, Hodder and Stoughton.

Redfern, D. and Skinner, M. (2000) *AQA Specification B Module 1: The Dynamics of Change Unit Guide*, Philip Allan Updates.

Redfern, D. and Skinner, M. (2002) *AS/A-Level Geography Coursework and Practical Techniques*, Philip Allan Updates.

Redfern, D., Skinner, M. and Farmer, G. (2002) *AQA Specification B Module 4: Global Change Unit Guide*, Philip Allan Updates.

Skinner, M. (2003) *Hazards*, Hodder and Stoughton.

Skinner, M. et al. (2001) *The Complete A–Z Geography Handbook*, Hodder and Stoughton.

Skinner, M. et al. (2001) *A–Z Geography Coursework Handbook* (including Investigative Skills), Hodder and Stoughton.

Index